Solid Sample Analysis

Springer

Berlin
Heidelberg
New York
Barcelona
Budapest
HongKong
London
Milan
Paris
Santa Clara
Singapore
Tokyo

Ulrich Kurfürst (Ed.)

Solid Sample Analysis

Direct and Slurry Sampling
using GF-AAS and ETV-ICP

With 95 Figures and 29 Tables

 Springer

Prof. Dr. Ulrich Kurfürst
Fachhochschule Fulda
University of Applied Sciences
Department of Nutrition
Marquardtstraße 35
D - 36039 Fulda, Germany

ISBN 3-540-62470-8 Springer-Verlag Berlin Heidelberg New York

Library of Congress Cataloging-in-Publication Data
Solid sample analysis : direct and slurry sampling using GF-AAS
and ETV-ICP / Ulrich Kurfürst (ed.).
p.   cm. Includes bibliographical references (p.   -   ) and index.
ISBN 3-540-62470-8 (acid-free paper)
1. Furnace atomic absorption spectroscopy. 2. Sample introduction
(Chemistry) 3. Solids--Analysis. I. Kurfürst, Ulrich, 1945-
QD96.A8S62   1998   543'.0873--dc21 97-37133

Cover Design: de'blik, Konzept & Design, Berlin
Production: ProduServ GmbH Verlagsservice, Belin
Typesetting: MEDIO INNOVATIVE MEDIEN SERVICE GMBH, BERLIN
SPIN: 10061121   52/3020-5 4 3 2 1 0 - Printed on acid-free paper

For my father

Preface

The analysis of solid materials by introducing solid test samples directly into the graphite furnace of an atomic absorption spectrometer must be regarded as a powerful analytical approach. Even if it is – of course – not the "ultimate method". After three decades of development, the instrumentation and the methodology are available to apply solid sampling successfully for the analysis of almost every material. Moreover, several tasks cannot be solved using other analytical methods as neatly as they can using direct solid sampling.

The conventional methods work more or less satisfactorily, so why do we suggest applying solid sampling much more extensively than it is today? To begin with, the features pointed out time and again should be named: *Rapidity of the analytical procedure, low susceptibility to analyte loss or contamination, very small quantities can be analyzed, and expenditure on instrumentation and personell is also low.* These properties are examined and the necessary conditions are discussed (Chapter 1) as are the analytical tasks (Chapter 6) for which use of this method is advantageous. Other features that are often overlooked are just as important: The *simplicity of the analytical procedures* allows the analyst to maintain an intimate relationship with the original scientific task that has to be solved with the analysis. Furthermore, the considerable *reduction of working place hazards and pollution* by avoiding the use of chemical reagents must nowadays be assessed as a feature as important as the others.

The essential aim of this book is to focus on the characteristic methodological features of solid sampling rather than a general discussion of the problems in trace element analysis. In Chapter 2, the reader will find an extensive general description of the solid sampling method and an in-depth discussion of all the serious objections that are raised to its application. Although most of these objections are based on earnest consideration, it has been shown that serious limitations only arise if adequate instrumental and methodological tools are not considered. This refers e.g. to the characteristics of sample handling, to the influence of sample heterogeneity, and to the use of certified reference materials for calibration. The great importance that is attached to the estimation of uncertainty in the analytical result is not only for proving that solid sampling results are generally as "certain" as these from the classical approaches but also for giving a practical application of the advanced "ISO-Guide for expression of uncertainy in measurement".

Most of the experience of solid sampling trace element determination was obtained using *graphite furnace atomic absorption spectrometry (GF-AAS)*. In Chapter 3, the preconditions in instrumentation and methodology are described that are specific for GF-AAS and this should be considered when direct solid sampling is applied.

Many features of *slurry analysis* are dependent on the properties of the solid sample (*SS*) as is the case with direct solid sample analysis. Chapter 5 gives a general overview of this method, whereas "cross connections" to attributes of direct solid sampling are given in other chapters. It is certain that the application of slurry analysis will become an essential routine method for laboratories in the future because, instrumentally, it is more closely related to the sampling of solutions making it easier to apply this technique using common AAS instruments. Consequently, the slurry method will develop into an independent analytical branch which would require a separate treatment.

Experience obtained with sample introduction into the *inductively coupled plasma by prior electrothermal vaporization (ETV-ICP)* is given in Chapter 4. The successfully performed analysis strongly points to the applicability of ETV-ICP for direct solid sampling, however, this experience is not as extensive as with direct or slurry sampling using GF-AAS. The authors consider that, in addition to the specific experience with solid samples, a general description of the ETV-ICP techniques should be within the scope of this book in order to facilitate further "goal-directed" research. The multielement capability of *mass sprectrometry (MS)* makes research into direct solid sampling in this field an extremely important topic.

My own involvement with direct solid sampling experiments since 1978 has its basis on the invention of "Direct Zeeman-AAS" by *Tetsuo Hadeishi*. With this advanced technique, the main barrier for direct solid sampling – extreme background absorption – was able to be overcome. The portrait of "Ted" shows him with the first prototype of his instrument at the Lawrence Berkeley Laboratory, University of California. He made his comprehensive knowledge available to the Grün Optik company and thus enabled the development of an advanced "solid sampling ZAAS instrument" (SM1), despite the limitations due to the smallness of this company. Over a decade of close cooperation a deep friendship grew. His death in 1991 – as a late consequence of the nuclear bomb on Hiroshima – was a very personal bereavement.

I hold *Wilhelm Grün* in high esteem for his courage in starting and keeping up the "SM-projekt". Because he relied on the small R&D-group he set up in his company, the instrument and method development was accomplished in a relatively short period. I remember with great pleasure the time spent working intensely with *Harald Schubring, Alfred Trost, Bernd Rues* and *Ralf Bernhard*. Although we had to learn that such a project can only be finally successful with the "stamina" of a much larger company, the inventions that were worked out at the Grün Optik company have obviously left deep marks on the road to more perfect solid sampling trace element determination.

When *Karl-Heinz Grobecker* visited the laboratory of Grün Optik for the first time, inspecting the prototype of the AAS instrument with its large "Zeeman-magnet", following my early solid sampling experiments – not really convincing, I must admit – he grasped the potential of this technique and the method. Looking back, I assess this event as being an "historical meeting": there was on one hand a young physicist having developed an instrument the utility of which he had no realistic idea and on the other, a post-graduate biologist faced with an analytical task with that he could cope with only if the great expectations were to become reality. Whenever I felt insecure about the success of our work, he made it clear how our work would help progress to be made in trace analysis of real samples. His work in the following years in bio-monitoring research and today for the production and certification of reference materials has contributed decisively to the solid sampling GF-AAS method.

We received a "baptism of fire" when *Anton Rosopulo* visited our laboratory, bringing with him his lifelong experience in trace analysis and a treasure chest of reference materials that were not widely available at that time. Although he was initially sceptical, he worked with us in a two day session of practical analysis and discussions about direct solid sampling trace analysis. In this way, a long and fruitful cooperation began and direct solid sampling analysis became a major part of his own scientific and practical work in food control. He gave us a realistic view of the "state of the art" when he claimed: "In trace element analysis of real samples the first decimal place must be true, the second should be as reliable as possible, the third is useful only for statistical evaluation; but analysts, who give and believe the fourth have not understood the serious problems of trace element analysis!" (To be honest, basically, the situation has not changed yet.)

It is thanks to the early users of the ZAAS "SM1" that convincing applications in routine analysis could be demonstrated within a short period after the market introduction. Even at that time, I acknowledged these pioneers for their courageous decision to get involved with an analytical method about which, at that time, the leading experts in Germany were more or less reserved – to say the least. These scientists thought we were guilty of ignoring the serious problems connected with direct analysis of solid samples. In an otherwise supporting letter *Hans Massmann* stated, "When an analyst reads the text of the brochure, cold shivers run up and down his spine. He fears that even more incorrect analytical results will now be produced. It only remains in the interest of everyone to demonstrate the possibilities but also to see the limits. The clearer that is done the better." (Translated by the language editor.) I hope now, 17 years later, with the descriptions in this book, his demands have been met. However, without the "ignorance" that was prevalent in the advertisement during the early period of the development of the instrument, the hard slog in the marketing activities would not have been maintained. Our "overemphasis" on the simplicity and the advantages of direct solid sampling was a reflex to the widespread neglect or at least the underrating of the problems that were connected with classical analytical approaches. Years later, I received a "private lecture" from *Günther Tölg* about the hard route of trace analysis in the previous decades for achieving accurate results and for getting general acceptance in the science community and belatedly I understood the psychological background to the resistance of experienced scientists to a hasty acceptance of direct solid sampling.

So I feel obliged to name some of the very first users, whose close and enthusiastic cooperation in method application and development has led not only to a fast advance but also gave us the mental support for going on. *Peter Esser* has developed numerous applications of solid sampling in the cement industry. His involvement in the interlaboratory experiments led to a broad acceptance of this method in this industrial branch. *Wolfgang Großmann* and *Karl-Heinz Tobis* integrated SS-GF-AAS into a variety of different methods of an environmental laboratory. They showed, that direct solid sampling can close "gaps" that may arise for particular elements or materials in routine analysis. The suitability of direct solid sampling for the analysis of soils and geological materials was investigated by *Wolfgang Gerwinski*, and this led to some methodological findings. *Jürgen Fleckenstein* used the SM1 for the investigation of heavy metal transfer in ecosystems, proving the special suitability of direct solid sampling for this purpose. *Bernd Claus*, *Norbert Chistian* and *Norbert Tismer* established systems of quality control in companies in the food industry, taking exemplary advantage of the fast and simple analytical process that allows the continual monitoring of low-contaminated raw materials. A comparable system of process quality control was established by *Rolf Erb* in a chemical factory, where elements other than the "common elements" were of interest. *Wolfgang Lichtenberg* realized that direct solid sampling facilitates new solutions when he developed a new method for the determination of "shot ranges" by the analysis of gunshot residues.

Herbert Muntau, M. Pinta, A.M. De Kersabiec, Markus Stoeppler, Robert Herber and *Jean Pauwels* are established scientists who were open-minded and willing to test the instrument and the proposed method at an early stage. Their positive judgement of the potential, their published results and statements among the scientific community aroused growing interest in direct solid sampling. Being aware of the scientific advance in SS-GF-AAS, *Wilhelm Fresenius* encouraged me personally to publish the results on instrumentation and methodology.

Finn Langmyhr and *Wolfgang Frech*, leading solid sampling analysts of the first generation (since 1970) have accompanied our work over a long period with advice and help especially with the scientific concept of the "Solid Sampling Colloquium". *Bruno Hütsch* has played an important role in the development of the "solid sampling graphite furnace" of the SM1: As have all scientists who have worked with the graphite furnace over the last decades, I have profited a great deal from his deep knowledge and the opportunity to deliver experimental graphite parts as well.

The above-mentioned merits in development and the experiments have been acquired from the second generation of solid sampling scientists and users since 1978. However, to avoid a "Germany-centered" interpretation, it must be pointed out that other groups have also worked on this field very successfully. *Ikuo Atsuya*, especially, has contributed instrumentation and methodology parallel and complementary to direct solid sampling over the same period; cooperation with him was inspiring and sustaining. Since 1985, the third generation of solid sampling experiments has elevated the method to a scientifically well-founded branch of trace element determination using the graphite furnace. *Douglas Baxter, James Holcombe, Nancy Miller-Ihli, Kenneth Jackson, Michael IIinds, Deborah Bradshaw* and *David Styris* gave me a lot of direct and indirect support for the preparation of this book. *Klaus Dittrich* contributed to the solid sampling method in his own scientific work and promoted it with the organisation of the CANAS-conference; his sudden death was a bitter blow for the scientific community and had prevented the further collaboration we had agreed upon. A breakthrough for applying the experiences of SS-GF-AAS to the ETV-ICP-technique has been achieved during the last few years by *Richard Dams, Peter Verrept, Luc Moens* and *Sylvie Boonen*; my stay as a guest scientist at Ghent University and the collaboration with these scientists was an unforgettable experience.

Last but not least I want to thank my close collaborators of the last few years at Fulda. The solid sampling investigations and the preparation of this book were also considerably supported by *Jutta Kress, Heike Hollenbach* and *Albert Rehnert*. In an atmosphere as friendly as it is in our laboratory not only the results count – we all enjoy working together. We regarded the cooperation with *Oswald Schuierer* who continuously worked on the perfection of instrumental part of the direct sampling process as a part of our working group.

I am aware, that a lot of solid sampling work has been published in languages other than English and German, e.g. in Japanese, Chinese, Russian and other East-European languages. I have to apologize to these workers for not discussing these papers, hoping, however, that their most important results are additionally published in international scientific journals that come to my attention.

Peter Enders from Springer Verlag encouraged me to tackle this project and he showed great patience in all delays of the production process. Without the cooperation of *Eddie Fulford* in the English language revision of my manuscripts, these would not be as "readable" as they are now.

I want to express my gratitude to all of the people I have named and the multitude of scientists and qualified users who have contributed to the development of "solid sampling" and I am happy that I have been able to find so many friends among them.

Hopefully, this comprehensive "account" of experience with and methodology of solid sampling trace element determination using the graphite furnace will promote a fourth generation of experiments that will achieve its considerable potential as a general method in scientific and routine analysis.

Fulda, November 1997   ULRICH KURFÜRST

Professor Tetsuo Hadeishi with the first laboratory assembly of the "Hyperfine Zeeman Effect Atomic Absorption Spectrometer" (1971)

Authors

DR. SYLVIE BOONEN
University of Ghent
Institute for Nuclear Sciences
Proftuinenstraat 86
B – 9000 Gent, Belgium

PROF. DR. RICHARD DAMS
University of Ghent
Institute for Nuclear Sciences
Proftuinenstraat 86
B – 9000 Gent, Belgium

DR. JÜRGEN FLECKENSTEIN
Federal Agriculture Research Center
Institute of Plants Nutrition and
Soil Science
Bundesallee 30
D – 38116 Braunschweig, Germany

PROF. DR. ULRICH KURFÜRST
Fachhochschule Fulda
University of Applied Sciences
Department of Nutrition
Marquardtstraße 35
D – 36039 Fulda, Germany

PROF. DR. LUC MOENS
University of Ghent
Institute for Nuclear Sciences
Proftuinenstraat 86
B – 9000 Gent, Belgium

DR. MARKUS STOEPPLER
Mariengartenstraße 1a
D – 52428 Jülich, Germany

DR. PETER VERREPT
Agfa-Gevaert
Medical Division, R&O
Septestraat 27
B – 2640 Mortsel, Belgium

Acknowledgement

The authors gratefully acknowledge the kind permissions to use figures from other sources in this publication.

Figures reprinted from *Spectrochimica Acta Part B*, from *Talanta*, and from *Science Total Environment* with kind permission from Elsevier Science - NL, Sara Burgerhartstraat 25, 1055 KV Amsterdam, The Netherlands

Figures reprinted from *Analytical Chemistry* with kind permission from American Chemical Society, Washington DC, USA

Figures reprinted from *Journal of Analytical Atomic Spectrometry* and from *The Analyst* with kind permission from The Royal Society of Chemistry, Cambridge, England

Figures reprinted from *Colloquium Atomspektrometrische Spurenanalytik*, and from *CANAS* with kind permission from Bodenseewerk Perkin-Elmer GmbH, Überlingen, Germany

Figures reprinted from *Microchemical Journal*, with kind permission from Academic Press, Orlando, Florida, USA

Figures reprinted from *Atomic Spectroscopy*, with kind permission from Atomic Spectroscopy, Florham Park, NJ, USA

Figures reprinted from *ICP Newsletter*, with kind permission from ICP Newsletter Inc., Hadley, MA, USA

Figures reprinted from *Pure & Applied Chemistry*, with kind permission from IUPAC, Research Triangle Park, NC, USA

The exact references are given in the figure captions.

Table of Contents

CHAPTER 3
Direct Solid Sampling with Graphite Furnace Atomic Absorption Spectrometry (GF-AAS)
ULRICH KURFÜRST 129

CHAPTER 4

**Solid Sampling by Electrothermal Vaporization-Inductively
Coupled Plasma-Atomic Emission and -Mass Spectrometry
(ETV-ICP-AES/-MS)**

PETER VERREPT, SYLVIE BOONEN, LUC MOENS AND RICHARD DAMS 191

CHAPTER 5
Introduction of Slurry Samples into the Graphite Furnace
MARKUS STOEPPLER AND ULRICH KURFÜRST 247

CHAPTER 6
Advantageous Fields of Application for Solid Sampling Analysis
JÜRGEN FLECKENSTEIN, MARKUS STOEPPLER AND ULRICH KURFÜRST 319

CHAPTER 1

Characteristics of Solid Sampling Using the Graphite Furnace

ULRICH KURFÜRST

The historical survey of direct solid sample analysis (SS) by graphite furnace atomic absorption spectrometry outlined below shows that the development of this method did not take a typical course. While the introduction of many analytical methods was followed by an rapid dissemination, the pathway of development and acceptance of direct solid sampling using the graphite furnace has dragged on for three decades in a hard step-by-step process. A general impression of the notable benefits that can be gained from this method is given below to explain why again and again researchers have made a great effort to develop it further. Today all the methodological and instrumental requirements necessary are available, however, for the successful application of direct solid sample analysis some "mental preconditions" must be considered and these are outlined in this chapter. Before this question is answered we must ask ourselves why the graphite furnace offers the potential for the direct analysis of solid material.

1.1
Specific Features of the Graphite Furnace

Graphite furnace systems as atomization devices for in-situ measurement using atomic absorption spectrometry are characterized by some basic differences in comparison to flame spectrometry:
"...it takes a thousand times longer to atomize a substance in a cuvette than in a flame. Indeed, the time taken to heat and atomize a substance in a cuvette is so long that refractory compounds have ample time to decompose..." and "The atomization of a sample in a cuvette takes place in a highly reducing medium (from the surface of incandescent graphite). The breakdown of the crystal lattice of the sample and the decomposition of compounds are therefore accelerated by the process of reduction of compounds" [1]. This feature explain the potential for a direct (or slurry) analysis of solid materials using a graphite furnace in contrast to the introduction of solid materials into a flame or a plasma (e.g. by the nebulization of a slurry or direct introduction). The atomization (or excitation) efficiency with these techniques is difficult or not at all possible to control (e.g. due to variations in particle size and distribution, viscosity, residence time). Hence, quantification would require quite complete matching of these parame-

ters as well as the chemical compositon of the unknown sample and the reference material – achievable only in very special applications.

Furthermore, the amount of sample that is introduced into the furnace with direct solid sampling can be determined exactly or can be achieved at least as accurately as it can in slurry sampling. Supposing complete vaporization/atomization (or more realistically: a constant rate) the amount of the analyte can, in principle, be deduced much more independently from the matrix composition.

The latter is the main advantage over the "classical" approaches to solid sampling by arc and spark emission spectrometry, where the ablated sample material cannot be determined exactly and may change significantly with different materials. Thus, for a quantitative analysis of an unknown sample, a reference material which is identical or at least very similar in regard to the parameters which determine the excitation rate is essential[1]. Similar problems still hamper the application of laser ablation as a sampling technique. The fundamental properties of the graphite furnace technique described here suggest that it has a high potential for matrix independent calibration. It has indeed been proven that solid sample analysis using graphite furnace AAS requires no complete matching of the sample and calibrant material, contrary to what has often been anticipated.

1.2
Historical Milestones in the Development of the Method

The potential of the graphite furnace technique for the direct analysis of solid samples was immediately recognized when this atomization technique for element determination by atomic spectrometry was invented. L'vov, in his pioneering paper [2] about the use of the graphite furnace, suggested that such an atomizer would be useful for the direct analysis of solids with AAS. In the following decade L´vov, Nikolaev, Katskov and other workers from the St. Petersburg University [1, 3] performed a large number of experiments with regard to the development of the graphite furnace system and these proved its suitablility for solid sampling.

Subsequent experiments from other groups of scientists also produced successful analytical results in different fields of reseach. Systematic investigations which outlined the basic methodological features of direct solid sample analysis at a very early stage were performed by Langmyhr and numerious coworkers at the University of Oslo over a period of ten years, mainly for geological samples. Lundberg and Frech at Umeå University studied the atomisation conditions for metallurgical samples, and Headridge and coworkers at Sheffield University performed a lot of investigations for metal reference materials. These investigations and a variety of additional papers from other groups gave the substance for the first reviews of the method (1975–1979) [1, 4, 5].

---

1) For metallurgical samples, these conditions can be achieved comparatively easily, leading to the analytical success of these techniques in this field.

At that time, spectroscopic limitations of the background correction techniques restricted the application of direct solid sampling with the graphite furnace to materials which do not produce extremely large background absorption, as generally occurs with the introduction of biological materials. With the invention of the background correction technique using the Zeeman effect, these materials also became directly analyzable. Since 1971, Hadeishi, McLaughlin and coworkers at the Lawrence Berkeley Laboratory have proven the enhanced correction capability of their invention, Zeeman-AAS, with the direct determination of biological materials [6]. During this period, this group has given a number of contributions to direct analysis of solid materials. However, this advance was not commonly used initially because commercial Zeeman-AAS instruments were not available at that time.

Hadeishi´s investigations led to the development of an atomic absorption spectrometer utilizing the Zeeman-effect for background correction and this was manufactured by the Grün-Optik company (for whom the editor of this book worked as head of R&D for nearly ten years). This instrument, introduced in 1979, was specially designed for the direct analysis of solid samples, it was optimized for the handling and introduction of solid test samples by using the platform boat technique, the software for data evaluation was tailored to the characteristics of direct solid sampling, and the different system components were combined to give a complete system. In the following years this instrumentation was used by early customers for the analysis of a large variety of materials, e.g. from biological and environmental origin, plastics, industrial raw materials. Based on these broad applications and their own systematic investigations Grobecker et al. published the first synopsis of the methodology for direct solid sample analysis using the graphite furnace and a platform sample carrier system [7, 8].

At the same time and independently Atsuya and coworkers from the Kitami Institute modified the graphite furnace of the commercial AAS instrument. This instrument (Hitachi) uses the Zeeman effect in an inverse mode, i.e. the magnetic field was applied to the atomizer [9]. With a graphite cup as the sample holder an adequate sampling procedure was achieved. With a series of investigations of the analysis of biological materials over several years this group made significant contributions to the development of the method.

The growing acceptance of solid sample analysis in the early 1980s is shown by the large number of papers published during this period. Systematic investigations concerning the direct analysis of biological materials were carried out by Chakrabarti and coworkers. In comprehensive papers [10, 11], platform and wall deposition of solid test samples were compared and in-situ matrix modification was applied for overcoming the limitations of the deuterium background correction technique.

The influence of analyte heterogeneity in samples in the microgram range was studied by Kurfürst et al. Their research which has been carried out for more than ten years now [12] shows that, although heterogeneity is a dominating factor for solid sample analysis generally, this effect is only in the order of random effects occurring in other analytical methods. These findings and the outline of a general sampling theory for this method [13] show that any serious doubts about its practical application are mere conjecture. This work has led to "homogeneity control" becoming a new field of application for the solid sampling techniques.

Advanced spectroscopic, instrumental, and methodological knowledge on the direct analysis of solid samples using graphite furnace atomic absorption spectrometry were presented by Baxter and Frech in 1990 [14]. These basic investigations of the method demonstrated that direct solid sampling with the graphite furnace at that time had reached a level that would allow its application as a general method of trace and ultratrace analysis.

The large instrument manufacturers also made some effort to apply the method to their own instruments with the development of "solid sampling attachments". However, some serious obstacles with the advanced instruments which were optimized for the analysis of solutions still hamper their extensive and rapid acceptance (the most important factor being the excessive reduction in size of the graphite tubes, that provides high senistivity but strongly hinders the introduction of solid test samples).

In this situation, as an alternative, the introduction of a slurry of the pulverized materials into the furnace aroused greater interest, because with this technique the common spectrometer and furnace design can be used. Brady et al. [15] have proposed the introduction of slurries into the graphite furnace since 1974. Basic investigations of slurry analysis with GF-AAS have been published by Hinds, Jackson and coworkers in a series of papers since 1983 mainly for soils and by Stephen, Littlejohn and Ottaway for foodstuffs. With ultrasonic agitation, the problem of particle segregation was reduced to such an extent that routine analysis using this approach has become feasible. Based on her invention in the mid-1980s, Miller-Ihli perfected the slurry method during the following years and gave comprehensive instructions on the methodology and its practical application [16]. As a result, a great number of results have been published, mostly using a commercial "slurry-attachment" for AAS instruments (Perkin Elmer).

In a number of recent investigations, Krivan and coworkers reported using direct and slurry sampling for the analysis of high purity materials, taking advantage of the blank-free procedure and the low levels which can be detected by this method. This work has perfected the methodology for this high-tech analytical task in what undoubtedly will become another preferred field of application for solid sampling.

1.3
The Method's Advantages

In the introductory section of most papers about trace analysis by solid sampling using the graphite furnace, the authors point out or emphasize the numerous advantages of this method, mainly in comparison to analysis after chemical sample preparation – exactly as it is in the foreword of this book. Figure 1.1 gives a visual comparison of both concepts. Numerous advantages can be traced back to the significantly fewer process steps required with the solid sampling methods.

The benefits of solid sampling analysis must be considered in detail to get a complete and satisfactory basis for deciding whether to apply the method in the laboratory. The statements in the following headings point to obvious, latent, and potential advantages of the method. These characteristics correspond to the advantageous practical application presented and discussed in Chapter 6 (where some further characteristics of solid sampling are discussed that may be of advantage in special applications, e.g. only extremely small quantities of sample material are necessary, the detection of heterogeneity effects).

Each of the positive features may have preconditions or may be limited to certain applications, which also are discussed in the respective section. Again it must be stressed, that one can only take advantage of this method if the instrumental and methodical particularities of the solid sampling are met.

1.3.1
Solid Sampling is Fast

Undoubtedly results by solid sample analysis are obtained much faster than those obtained by prior chemical sample preparation. If only a small amount of physical sample pretreatment is necessary for obtaining suitable test samples (e.g. only cutting off scrap) and the information on the approximate analyte content is available (i.e. specific instrument settings are known and the amount of calibration is significantly reduced) the time from the arrival of a sample at the laboratory to when a single result is known may be *less than 30 minutes.*

The time of the analysis is determined only by the furnace time/temperature program if the sampling operation is performed simultaneously with two sample carriers (one in the furnace, the other at the loading and weighing stage).

The relative analytical rate increases with the number of unknown samples analyzed with the once established calibration. "Assuming that the instrument is ready for measurement, and that the sample has been prepared for analysis, approximately ten solid samples can be analysed per hour" [17]!

These extremely favorable conditions are not maintained if more than a few elements are to be determined in the samples. The sampling procedure must be repeated for each analyte element and is more protracted than that for "liquefied" samples (weighing lasts longer than pipetting and more replicate measurements for one measuring sample are usually required). Consequently the time that is saved by the avoidance of chemical sample pretreatment is used up progressively with each additional element.

For a comparative quantitative assessment of the total time consumption of analytical methods a detailed "time analysis" was performed. This comprised all

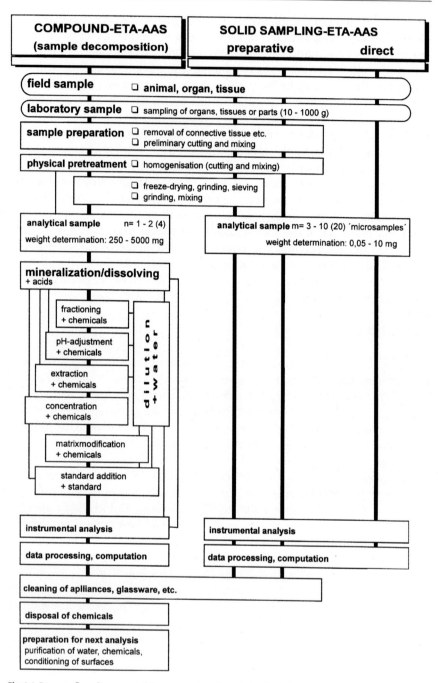

Fig. 1.1. Process flow diagrams of the conventional analysis of solid materials (here specified as animal organs and tissues) after prior sample decomposition ("compound-ETA-AAS") in comparison to that of direct solid sample analysis ("solid sampling-ETA-AAS"). The term "preparative" characterizes the normal physical sample preparation while the term "direct" here refers to the solid sampling analysis of unprepared test samples (from [35], with permission)

the individual steps in the entire procedures (e.g. also including the cleaning of the glassware) [18]. Figure 1.2 a shows the duration of direct solid sample analysis dependent on the number of laboratory samples and the number of elements to be determined in relation to that of the analysis after chemical transformation to solutions. The appraisal yields for the absolute time required for the analysis of one sample for one element (solid sampling vs liquefied sample) 70 min as against 440 min and for 7 elements 360 min as against 960 min. The analysis of 50 samples for one element lasts 810 min as against 2600 min and for 7 elements 4200 min as against 7200 min.

The considerations with regard to the time consumption are still valid if the analytical procedures are carried out manually or automatically. The situation will more favorable for the classical method of sample solution analysis if modern techniques are used in the samples pretreatment (e.g. microwave heating, continuous flow). Then the time of sample treatment may be shorter (in the above appraisal, the duration of sample digestion or decomposition is estimated to be 4 hours). However, if solid sample analysis is carried out with a modern multielement AAS instrument that allows simultaneous determination of 4–5 elements ("oligo-element") [19], solid sampling regains a significant lead in terms of the duration of analysis.

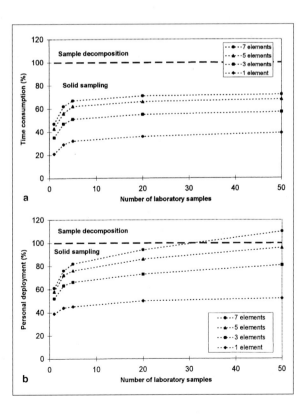

Fig. 1.2. Results of a study on expenditure in time and personnel of direct solid sample analysis relative to the methods based on prior sample decomposition (=100%) dependent on the number of laboratory samples and analyte elements.
a Total time consumption of the entire analysis.
b The duration of personnel activity during the analysis

1.3.2
Solid Sampling is Economical

In order to compare methods from an economical point of view, the expenditure
in terms of personnel deployment, investment required, and comsumables must
be considered. A complete economical analysis will be an important task if the
accomplishment of direct solid sampling is to be fully comparable with the classical approaches with respect to automation and widespread acceptance in legal
control analysis. Here, only some aspects of this issue will be considered.

In the comparative study mentioned above [18], the process steps that are connected with *personnel deployment* are summarized for both methods. In doing
so, the entire solid sampling procedure is assumed to be "handmade", whereas
automation with the instrumental analysis of solutions is standard today (however, the sample preparation procedure is still work-intensive). Figure 1.2 b
shows the relevant results. For the determination of only one element in up to 50
laboratory samples only 50% of the work which is necessary for the classical
analysis of solutions is required for direct solid sampling. This saving is progressively diminished by each additional element so that – with the assumptions
made – for 7 elements in 30 laboratory samples the analysis of digested or
decomposed samples will need less personnel expenditure.

If the advancement in the sample preparation process is taken into consideration, where the chemical treatment is also partly automated, currently the
"break even point" will be reached more or less earlier (approximately 3 to 4 elements in 20 laboratory samples). However, if the solid sampling process is partly or fully automated, the work involved would be extremely low (for more information on the automation of direct solid sampling see Sect. 2.1.4.3)!

Chemical sample preparation is not only work-intensive but also needs additional *equipment and instrumentation*, which takes up a lot of room in the laboratory. For the solid sampling procedure, significantly less investment and
space is required. For this method, only the spectrometer work station (including computer and autosampling system) and a microbalance are required.

However, comparing the different methods with regard to the expenditure of
time, equipment, personnel, and of course money will usually make no sense in
a qualified analytical laboratory because the decision will not be made *between
the methods* because the choice is whether to have complementary methods.
Only when the analytical tasks are well defined and can be solved advantageously
by direct solid sampling, is an installation of this method alone appropriate for
a laboratory (see Sect. 6.8).

Consumables for solid sampling are limited to the needs of the instrumental
analysis (spectral lamps, graphite parts, energy, cooling water). The avoidance of
(pure and ultrapure) chemical reagents and saving electricity needed for chemical sample pretreatment will reduce the costs significantly.

Moreover, today the costs per analysis are being increased more and more by
efforts made for analytical quality control. The relatively high price of Certified

Reference Materials (CRM) will be a considerable item. For direct solid sample analysis, one bottle (usually 50–100 g) of a CRM will last "forever", limited only by the durability of the product (with suitable storage conditions these will be stable for several years). Or, vice versa, for solid sample analysis one can justifiably purchase a large number of CRMs with different matrix composition and varying analyte content because of their durability– from the analytical point of view a favorable situation.

1.3.3
Solid Sampling Procedures are Simple

The entire analytical procedure of direct solid sampling can be outlined briefly: setting up the spectrometer, separation, weighing and introduction of solid test samples for calibration and the unknown laboratory sample to be analyzed, and finally data evaluation. If the laboratory sample is well known and all the effects that influence the accuracy of the results are controlled, the simple course described above may be a realistic description of the work involved.

However, with the enumeration of the process steps alone, an analytical method cannot be fully characterized. In addition, the qualifications required by staff in order to carry out all the analytical measures must be considered in any comparison.

Because turning solid materials into solutions is connected to some extent with a "standardization" of the matrix to be analyzed, the application of "cook book" procedures has been made possible [2]. Whereas with solid sampling, the differences in the original matrix composition may give rise to more distinct effects. *So, the simple procedure of direct solid sampling should not lead to the erroneous impression that the analysis can be carried out with less analytical knowledge. On the contrary, in each step of the solid sampling procedure, a good understanding of the inherent analytical problems is essential for the achievement of accurate results.* It is the aim of this book to give this helpful basic information in sufficient depth.

On the other hand, the appearance of effects in solid sample analysis is more specific to the respective analytical step because solid sampling results of single test samples are reached quasi-simultaneously. So the identification and the correction by appropriate measures may be simpler with solid sampling than in the situation where interfering effects can only be recognized after carrying out the complete analytical procedure. In this way, solid sampling is simpler and more transparent than other more complex methods. The experience of Belarra et al. with an investigation on the determination of lead in PVC samples may be regarded as typical; if the method is developed carefully for the specific sample type "… direct electrothermal atomization of sol-

---

2) Undoubtedly the carrying out of an analysis as a cook book procedure may be a reason for inaccurate results caused by differences in the matrix of the unknown sample.

id samples in a graphite furnace with calibration against standards in aqueous solutions is a simple and rapid method, which does not require any special training ..." [20].

1.3.4
Solid Sampling is a Powerful Detection Method

In analytical chemistry, continuous efforts are being made to obtain better limits of detection, i.e. to detect smaller and smaller amounts of analyte. However,
sample decomposition is generally connected with a dilution of the solid sample material. The respective factor is given by the relation of the solution volume
to the solid test sample mass, e.g. 100 ml used for 1 g solid material. The fact is
often overlooked that one or two orders of magnitude of the detection limit are
needed just to compensate for the reduced analyte content in the liquefied test
samples which are introduced into the instrument. (The dilution of analyte also
take place in the preparation of slurries, compare Eq. 5.5).

Consequently, one must distinguish clearly between the *limit of detection,*
that is a characteristic of the spectrometer (for a given element and a specific
analytical line), and which in the graphite furnace technique is usually
expressed in terms of analyte mass (e.g. 1 pg) and the *limit of analysis*[3] that is
characteristic of the sample to be analyzed with the given instrumentation, and
which is expressed in terms of analyte content in the original solid material (e.g.
1 pg g^{-1}).

While limits of detection for furnace systems that are optimized for direct solid sampling may be somewhat poorer, e.g. as a result of larger graphite tubes or
the influence of the sample carrier system, limits of analysis for direct solid sampling are generally significantly better than those of an analysis of "liquefied"
samples.

"There are few techniques that can offer such a low limit of detection as the
direct atomization of solid samples" [22]. This statement of 1979 from Bäckman
and Karlsson is still valid, even when the excellent limits of detection of the
advanced mass spectrometry are considered. When direct solid sample introduction with a graphite furnace is used with this technique (ETV-ICP-MS), the
limits of analysis are the lowest that can be currently achieved.

---

3) The term "limit of detection" and the methods of its assessment is defined exactly (e.g. [21]).
   Whereas a term for the smallest detectable analyte content in the original sample that can be
   determined with a given method is not standardized. For this measurant the term "limit of analysis" is used in this book. Both values can be converted to each other considering the largest test
   sample mass possible (compare Sect. 2.5.2).

1.3.5
Solid Sampling Prevents Analytical Errors

In round robin experiments, recurrently outlying values even emerge with the established methods, i.e. gross errors cannot be excluded completely in the complex analytical processes. Looking at the characteristics of solid sampling, it is safe to assume that with this method there is no danger of producing extreme outliers. The correctness of this statement again follows from the simple fact that the number of steps in the analytical process are very few and they are easier to control.

In this connection the *low danger of contamination and blank values* are repeatedly emphasized in the literature on solid sampling. This is the consequence of no reagents being added to the laboratory sample, the "contact surfaces" being small and easy to clean (actually only the tip of the spatula) and that the entire procedure from dosing to furnace introduction being extremely short.

Over and above these factors and for the same reason, the chances of "*human gross errors*" (e.g. miscalculations, use of incorrect units, misidentification, erroneous transcription) occurring are significantly lower. Undoubtedly such mistakes are a potential hazard to analytical results in routine analysis, all the more because the analytical procedure is highly complex. With direct solid sampling it is simple to keep an overview of the evolution of the final result and it is easy to reconstruct. For the analyst in charge, who, in any case, cannot supervise the actual analytical procedure, this may be regarded as an extremely important advantage.

1.3.6
Solid Sampling Reduces Working Place Hazards and Environmental Pollution

The production, use and disposal of ("pure" and "ultrapure") chemical reagents for sample decomposition are energy consuming, and are a burden on the environment (even in the case of correct waste disposal). Accordingly, the fact, that analytical procedures – and these are utilized extensively for environmental protection – become part of the problem with their frequent application is of increasing importance. Even considering the tendency to reduce the weight for the decomposed sample (down to 10 mg!) in order to reduce the pollutants, the question arises as to whether the analytical procedures should be assessed to see if they are environmentally safe ("Environment Impact Assessment").

Moreover, the working places in the laboratories are contaminated by the harmful substances, exposing the laboratory staff to potential health hazards. For this reason extensive protection standards must be met. However, even when the utmost caution is taken, accidents cannot be ruled out completely with the complex and often changing working procedures in an analytical laboratory.

So, last but not least, it should be borne in mind that the establishment of the solid sampling method in the laboratory can make a significant contribution to the reduction of these problems because, with this method, the use of dangerous and harmful chemical reagents is avoided (direct solid sampling) or less harmful substitutes can be used (slurry sampling).

1.4
The "Philosophy" of the Direct Analysis of the Solid Sample

This section considers some general viewpoints which the analyst should keep in mind for the assesssment of the method. For an objective examination of the suitability of solid sampling for a given analytical problem, in principle, a positive attitude towards the method is essential.

In the first place it must be pointed out, that this method follows the *basic rule of analytical science* completely, that is to perform the minimum number of steps possible consistent with the analytical process to be carried out. As discussed above, the avoidance of extensive sample pretreatment reduces the number of things which can go wrong and is also the reason for various advantageous features of the method. Despite this, a widespread "reluctance" to use direct solid sampling can be observed that can only be understood from the historical development: The analytical question was directed in the beginning to a specific material and was simply read as "How much .... is there?". It seems obvious to answer this question directly, e.g. to perform the analysis of a solid material in the solid state. The development of the instruments and analytical techniques for trace analysis has made this possible (arc and spark for AES, XRF, INAA). In many scientific fields there was and still is a strong preference for "solid sampling" methods – mainly fields which have a "physical orientation" (e.g. metallurgy). With the progress toward an independent analytical science, the problem was extended to a manifold of materials, the question became "How much .... is there in ...?". For this task, it was found that it was much easier to answer the question when the sample was in a liquid form. Because of this "great convenience" and the experience of the chemists who became involved in this new field of science, namely, trace analysis of solutions there was a rapid expansion of the methods. Therefore every time somebody needed to analyse a solid material it became the "habit" to transform it to the more "useful" liquid state.

Concerning trace element analysis, a statement by Welz [23] may be regarded as the consensus during that period: "If one considers ... the above described difficulties ..., then (direct solid sample analysis) seems not at all as attractive as it is on the first view. ... If one consider in addition, that today a number of fast and safe analytical methods exist, ... in many cases the dosing of solutions must be preferred" (2nd edn. 1975, p. 131, translated from the German by the author).

In view of the enormous problems connected with the analysis of unknown materials with a complex matrix composition, this was undoubtedly the correct approach for that time. Although successful direct analysis of real solid materials was frequently reported in the early years of SS-GF-AAS, at that time, the problems caused by different kinds of interference limited a successful general application.

However, it seems that the analytical community presumed that direct solid sampling is intrinsically connected with such serious problems and that further efforts would be unrewarding. For these reasons, instrumental and methodological advances were not investigated at once or extensively to see if they demonstrated any potential for a direct analysis of solid materials. These instrumentation advances took place in different fields, e.g. background correction techniques, furnace techniques, microweighing, electronic signal processing and these made possible the elimination or suppression of interferences that had previously hampered the application of direct solid sampling.

Moreover, with the growing experiences in separating and studying interfering effects in all steps of the analytical process carried out with the transformation of solid materials to a analysable liquid state, a basically better understanding of trace element chemistry was involved. So, solid sampling was able to – and will – benefit from the enormous progress based on the classical methods, i.e. above all, the experience in solution analysis!

A change of views about the applicability of direct solid sampling in the second half of the 1980s was characterized by Rettberg and Holcombe [24]: "An erroneous conclusion might be drawn that furnace atomic absorption techniques are fundamentally ill-suited for the direct microanalysis of solids. However, there is no inherent reason why the sample must be a liquid, and GFAA has other characteristics that actually suggest the strong potential for the direct analysis of solids."

With regard to the appearance of interferences and the methodological principles of overcoming them, the two approaches can no longer be regarded as being totally different. Actually, after the drying step in the furnace, the test sample from a solution is retransformed to the solid state, so that the difference to direct introduction of the solid test sample is not given by the absence or presence of concomitant substances but only by the character and the quantity of the concomitants always present.

However, there are, of course, *genuine differences* between the two approaches. Looking back, it can be recognized that attempts failed or erroneous opinions were formed concerning direct solid sampling because its "peculiarities" were not considered comprehensively. As an example, the confusion concerning the connection between analytical precision and analytical uncertainty of the analytical result may be mentioned: Direct solid sampling was often labeled a "semi-quantitative" method, because of the obvious worse precision due to heterogeneity effects. With such an judgement, it is overlooked that the generally large number of replicates which can be easily carried out with this method correspondingly improves the reliability of the analyte mean content. Even though this different feature was mentioned early at the development of direct solid sampling e.g. in [25, 26], it was generally not considered (compare Sect. 2.4.2.3).

This example illuminates the general statement, that *for a successful application of solid sampling, its characteristic methodical features must be considered.* This concerns all the steps of the analytical process, the handling and introduction of the test samples into the furnace, the problems connected with sample size and required modifications for calibration, and with the data evaluation process. Consequently the aim of this book is to clarify these differences and to describe the appropriate methodological "tools".

During the presentation of the advances in the instrumentation and method for solid sampling, the author found repeatedly that the target of acceptable accuracy was set to very close limits. The published results of element trace analysis of *known* samples with the classical methods often show impressive values in terms of precision and trueness (e.g. agreement with certified values).

In contrast, up to now, results of round robin or interlaboratory measurements for trace elements in *unknown* samples often show an amazing large spread. This may be still true even if the laboratories taking part are proven to be qualified, as the results in certification campaigns for reference materials often show. It is a merit of the European Community Bureau of Reference (BCR) that, with the release of a new reference material, all values on which the certified values are based are documented in the certification document (e.g. compare Fig. 2.26). So the routine analyst can get a realistic view of the accuracy that is achievable today with respect to element, content, matrix, and methods.[4]

A realistic example of a interlaboratory comparison experiment is documented in [27], where an original soil sample was spiked to larger contents and then mailed to 160 accredited routine laboratories: There, these sets were analyzed (As, Cd, Mo, Se, Tl) by different standard methods (after prior sample decompositon mainly by ICP-AES, FAAS, GF-AAS). The results can be roughly condensed as follows: the mean content of all laboratories was found generally at the expected value, however, the spread was mostly between 10–30% for 5000 mg/kg and between 50–200% for 5 mg/kg (standard deviation not range!) It is easy to imagine what the range would be if the natural contents of mostly around 1 mg/kg were to be tested. The reporting authors state: "What the results of this study show is how difficult it is to achieve those (proper) conditions under routine situations… Putting it another way, the results here can measure the robustness of the instrument" [27]. Or, putting it a third way, in trace element analysis for real samples one must still count on uncontrolled effects. The discussion on the XXVII-CSI Pre-Symposium (1992) about "Modelling of Graphite Furnace Processes: What do we Know?" [28] has thrown a spotlight on the discrepancy between what is desired and the reality of "absolute" analysis in trace and ultratrace analysis of real samples.

Therefore, judging the quality of analytical results from a particular method one must have (gain) a realistic idea about the "state of the art".

Figure 1.3 show results of interlaboratory experiments for the assurance of analytical quality in a large monitoring program of trace elements in food, performed by the German Federal Health Department (BGA) over several years [29].

---

4) In the first course of a campaign or a round robin experiment, the spread of the means from different laboratories and the number of outliers are still mostly significantly larger.

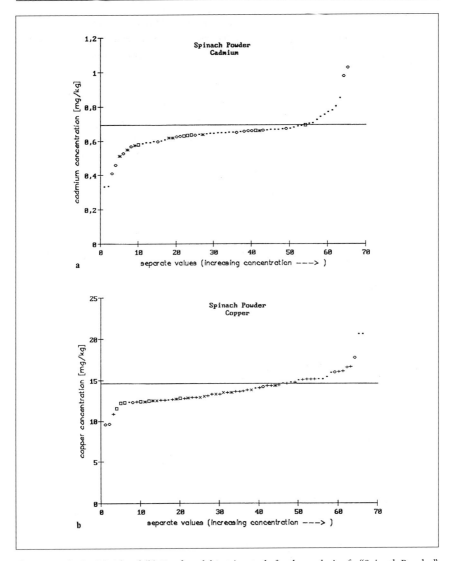

Fig. 1.3. Results for (**a**) Cd and (**b**) Cu of a colaborative study for the analysis of a "Spinach Powder" sample. 37 laboratories supplied approx. 60 results based on different analytical methods after sample decomposition: Flame-AAS (+), Polarometry or Voltametry (∗), GF-AAS (•), Zeeman-GF-AAS (◊), ICP-AES (x) and additional results from direct solid sampling GF-AAS (□). The values are arranged in increasing order; the horizontal line indicates the reference value (from [29], with permission)

The participating laboratories had been trained in previous "round robins" (see Chap. 6.3.2). From the plots of results, it can be recognized, that the solid sampling results are located generally in the range of the acceptable values, mostly in the "middle region" (Fig. 1.3 a) and sometimes on a "wing" (Fig. 1.3 b). (The reference values for this material were not known from the participants; as can be expected the results for "known" certified reference materials were even better). This situation can be regarded as typical, confirmed by other interlaboratory experiments (e.g. [30,31]). In the light of the realistic situation discussed above, the accuracy of the solid sampling results must judged as completely satisfactory!

Of course, solid sampling with GF-AAS shows no particular improvement in the accuracy that is generally achieved with other techniques or methods. However, if the preconditions for solid sampling are given, the final results from this method can be generally expected to be reliable.

Finally, it should considered, that the analysis of "unknown" materials is not an end in itself, but it is generally only a link in a chain of measurements which are required for clearing up a "primary problem" in other specific scientific fields e.g. in geology, metallurgy, biology, medicine.

For laboratories and institutes working in these fields a variety of the positive features of solid sampling may cumulate so that solid sampling will be conclusively the method of choice, because *solid sampling is undoubtedly more closely involved with the real problem.* From this attribute of the direct "characterization of powders" L´vov foresaw at a very early stage (1976) that "it will apparently find widest application in the research groups who are not directly connected with specialized analytical laboratories and which are therefore interested in solving their particular analytical problems without outside help" [1]. (See also the discussion in Sect. 6.8.2).

This prediction was confirmed by the experience gained with the market introduction of the "Solid Sampling GF-ZAAS" (SM1-SM30) from Grün Optik. This instrument concept aroused great interest, above all from those scientists responsible in industrial and research laboratories where the analytical task is only one part of a problem of overriding importance (e.g. quality control in a production process). A multitude of application examples for solid sampling given in Chapter 6 have been developed with this in mind.

1.5
Prospects

There can be no doubt, that solid sampling for distinct applications offers such clear advantages that, in these cases, will become the method of choice. Above all the ability for the assessment of sample homogeneity (overriding sampling error) and for the analysis of high purity materials (blank-free, low limits of analysis) should be emphasized.

Solid sampling with GF-AAS is, in principle, widely accepted today as being accurate, sensitive, and fast. However, under the existing circumstances this method cannot become standard in routine and monitoring laboratories. The reasons for this consist of certain ongoing problems that unfortunately are given today in the surrounding conditions:

First, instrumental developments during the last decade have been focused on the percularities of the analysis of liquid (decomposed) samples. Today the remaining AAS manufacturers shrink from the expense of designing an instrument that is optimized for direct solid sampling (furnace, sampler, software).

Second, with the legal regulations concerning environmental and health protection in many countries, the analytical methods to be used are often specified or at least strongly recommended, so that, at present, there is no market for such instruments large enough to justify an industrial enterprise investing the large amounts of money involved for research and development.

Furthermore, the infrastructure and the organization of routine laboratories has grown with sample decomposition as the standard procedure together with the analysis of solutions and will probably remain so for some time.

As a result, in the foreseeable future, direct solid sample analysis will remain a subsidiary method – or to express it more positively, it will be complementary, applied for specific analytical tasks if one of its features gives it a significant advantage over the standard methods.

According to the prevailing trends, a further development to a standard method (it would be justified from the analytical results which have been achieved) seems improbable. However, a situation is thinkable, where manufacturers and users are forced to make basic changes to the methods and techniques. A stimulus for such a response to alter the "paradigm" may be the deterioration of the environment and climate due to energy consumption and the production of waste. How fast a technology can be changed if necessary can be learned from the replacement of CFCs by substances less harmful to the ozone layer during the last 3 years in Germany and many other countries.

Slurry introduction may gain a greater acceptance and become more widely used because of its resemblance to solution analysis and because a commercial slurry sampling device is available. It remains to be seen whether further results will be so convincing that the method will be incorporated into the legal regulations for control analysis. This would be the precondition for widespread application in routine laboratories.

The techniques of ETV-ICP are still at the level where the laboratory has to build its own devices. Some problems connected with the coupling must still be overcome before this techniques are on a level of general applicability as it is with the GF-AAS techniques. However, the prospective of multielement capability of the detection techniques available when using solid sampling will make the method very attractive as the experiments on solid sampling using multielement AAS have shown [e.g. 19, 32]. The vision of a solid sampling process which is automated, multielement with the high detection power of modern mass spectrometry (ETV-ICP-MS) seems to be eminently attainable.

1.6
What can the Method do?

The criteria for characterizing an analytical method as being attractive were given by van Loon [33]: "Ideally a method for direct trace element solid sample analysis should have the following attributes:

(1) It should be applicable to a wide range of sample composition.
(2) The method should be relatively fast.
(3) Standardization should be simple.
(4) To avoid inhomogeneity problems, the method should be capable of handling fairly large samples.
(5) Simultaneous multielement analysis is desirable.
(6) Cost per analysis should be as low as possible.
(7) Repeatability and accuracy should be suitable for the particular application."

This book is intended to show that these targets can be achieved by solid sampling graphite furnace atomic spectrometric methods and describe how. Table 1.1 summarize the capabilities of the present method in ten points. This "mini review" was basically given by Langmyhr and Wibetoe [34] in their general review about solid sampling with atomic spectrometry. Here additions have been made to update the information and make it more specific for the graphite furnace techniques.

Table 1.1 Solid sampling graphite furnace AAS – What can the method do? (From: [34], with supplements by the author.) In the second column sections of this book are stated where information about that topic can be find

1. The technique is applicable to the analysis of inorganic and organic materials in the form of powders, drillings, cuttings from fibers, foils or sheets, samples of soft or hard tissues, as well as suspensions of solids in solid, liquid or gaseous dispersing agents. Liquid samples, such as biological fluids, may be analyzed after being transformed into the solid state by drying, dry or plasma ashing or lyophilization; these operations also serve the purpose of preconcentrating the analytes.	6.2-6.7 3.4 3.3.1 4.7 5.6
2. The method has been used for nearly all elements present as trace or minor constituents that are determinable with the technique used. In some instances, major components have also been quantified. The analytes are usually present in the µg/g and ng/g range, extreme values are analyte contents up to 50% and down to an analyte fraction of 10^{-10}.	2.5.2 3.3-3.4 4.7 5.6
3. The amounts taken for one test sample are normally in the range 0.1 to 10 mg with the extreme values down to 1 µg and up to 500 mg. Repeated measurements of 6–10 test samples are mostly carried out, for screening only 3, and for homogeneity testing >100 can be carried out easily.	2.4.2 2.5.1 3.2 4.5.1.3
4. Samples are weighed accurately and fast on semi-micro or microbalances. Modern weighing instruments are very simple to handle, showing stabilizing periods below 20 s even if placed on a normal lab bench.	2.1.2 5.2.1.2
5. The accuracy of the analytical results depends on the care taken by the analyst, on the quality, state and treatment of the apparatus and instrumentation, on the standards employed, and on the methods applied for analytical quality control. In general, the present method compares favourably with other methods for the quantification of trace and minor elements. The uncertainty of results is comparable or even smaller then those achieved with other methods.	2.2 2.4 3.1 4.5 5.2.2.3
6. The precision of the analytical data depends on a number of factors, the most important of which are the errors of sampling, weighing and measurement of absorbance. In general, a relative standard deviation of 5 to 10% for elements present at the 1 µg/g level, and 10 to 30% for analytes at the 1 ng/g level, have be considered as normal. The present technique is not recommended in those instances where samples to be analysed show extreme heterogeneity on the mg-level for the particular element.	2.3 2.4.1.2 5.2.2.2
7. The devices for the handling and the furnace introduction of test samples are either commercially available or can easily be made in a workshop. The use of a graphite sample carrier and a larger graphite tube that is easy to access is beneficial.	2.1.3 5.3.3
8. Compared to other instrumental techniques, the instruments for direct atomic absorption spectrophotometric analysis of solids are relatively inexpensive, the selection ranging from graphite furnace instruments priced from US $50,000 up to advanced computerized units including a microbalance and special data processing at a total of US $100,000.	1.3.2
9. The instruments may be placed on a lab. bench or a sturdy table, and should be arranged ergonomically to the microbalance, the only extra installation that may be required is a hood over the atomizer.	2.1.4
10. The number of samples that can be measured per hour will normally be 6 to 10, not including the time required for sample preparation that is normally required also for other methods (e.g. drying, milling).	1.3.1

CHAPTER 2

General Aspects of the Graphite Furnace Solid Sampling Method

ULRICH KURFÜRST

2.1
Sample Handling and Instrumental Requirements for Direct Sampling

The dosing of small amounts of solid test samples is totally different to that of liquid or slurry test samples in every step of the handling procedure which are separation, quantification and transportation. Using a manual or automatic pipette, all three steps are integrated with the dosing of liquid samples, while for solids these have to be carried out separately: For *separation* mostly a micro-spatula is used, for *quantification* a (semi- or full-) microbalance is required, and for *transportation* a sample carrier is highly advantageous.

All instruments and procedures involved must fulfil some analytical, technical, and/or ergonomical demands in order to operate them satisfactorily and obtain good results. These demands and approved techniques are discussed in the following sections.

2.1.1
Mechanical Sample Pretreatment

Usually the laboratory sample is finely ground to a more or less fine powder. If the analyte is not uniformly distributed, a grinding procedure is necessary to secure the representativity of the analytical result (see Sect. 2.3.1.2). Generally, dosing of pulverized samples in the micro- and milligram range is also simpler than that of sample pieces.

However, excessive grinding is generally not recommended, because the danger of (secondary) contamination grows with an extreme mechanical treatment. From a certain degree of particle size the grinding time (or force) must be increased drastically to get a significant further reduction of the particle size.

Usually, for the direct solid sample analysis, the *standard procedures of mechanical treatment* which are used for decomposition analysis are applied as well. For that reason, no further descriptions are given here.

Only one recommendation can be made for an approved technique for dried plant samples (e.g. leaves, grass). Good results have been achieved using a modified coffee mill [7]. The reduction of the upper free space of the milling room

Fig. 2.1. Vibration ball mill
MM 2 (Retsch, D-Düsseldorf)
delivers good milling and
homogenizing results for
fibrous, hard and brittle materials. Two milling containers
(2x10 ml) are operated simultaneously, so a larger number
of samples can be processed in
time. An attachment for cryro-
genic milling is available.

by a plate avoids the centrifugation of the particles away from the rotating
blades. No contamination was found for Zn, Cu, Cd and Pb on the environmental content level. No tests, however, were performed for the alloy components of the blades like Fe, Ni, Mn or Cr. The mill was found to be a significant
source of Cr contamination when grinding polymers [36]. A custom-made
grinding mill which was used for grinding filter discs is describe in [37]; all
parts which may be in contact with the sample material are made of titanium
or glass. In [38] it was reported that the shaft, stator,and rotor of a Turrax
homogenizer were replaced by ones made of titanium to avoid contamination
by Cr and Ni.

From the large number of types of mill on the market which give mostly good
results for special kind of materials, only one type should be mentioned. A vibration ball mill like the one shown in Fig. 2.1 provides good milling and homogenizing results for a large number of different materials. From the author's own
experience, this type of a mill can be recommended if the laboratory is con-
fronted with different types of sample material.

*Because of the great importance of a sufficient degree of homogeneity in solid
sampling, the analyst should pay special attention and care to the selection of the
grinding and homogenizing technique.*

Examples of the analytical effect of grinding/milling (sampling error) are giv-
en in Sect. 2.3.3. Because the particle size and its distribution is of even greater
importance with the slurry sampling technique, further experience and investigations are reported in Sect. 5.2.1.1.

2.1.2
Dosing and Weighing of Microsamples

The results of trace element determinations mostly refer to the mass of the sample material. So the quantification of test samples is commonly performed by
weighing. Only in special cases, are other physical terms used for reference, e.g.

the length or the volume of a wire [39], the area or the volume of a filtered medium [40, 41, 42, 43, 44].

The mass range which can be used for analysis is determined by different factors which are discussed in different sections of this book (see Sects. 2.3, 2.5, 4.5.1.3, 5.2.1.2). For direct sampling techniques, the upper limit for test samples is approximately 30 mg for dense materials (e.g. geological samples) or 10 mg for light materials (e.g. biological samples). The lower mass limit is approximately 0.02 mg and is found mainly in preparation and dosing problems of extremely low sample amounts (with regard to the dilution of powdered samples see Sect. 2.5.4.2).

These considerations show that for direct solid analysis by the graphite tube technique a microbalance must be available. Advanced *microbalances* show a stabilization period of typically less than 20 s – without rigid setting and positioning conditions. Preferably, a top loading microbalance should be used, because the area where the sample has to be deposited is easily accessible, the loading of a sample carrier is feasible directly onto the balance and stabilization is even faster (see Fig. 2.2).

Using a modern microbalance the duration of the weighing procedures (weighing of test samples and taring the sample carrier) including the time for transport and dosing of test samples is mostly shorter than the time for the measuring process in the furnace. If the operator uses two sample carriers (one car-

Fig. 2.2. A modern toploading microbalance (here: Sartorius MP5000, D-Göttingen). It is easily accessible and has a fast stabilization time.

rier being processed in the furnace, the other at the loading position), the time for the analysis is determined solely by the furnace time/temperature program.

Usually, separation and dosing of test samples from powdered materials are performed by a micro-spatula or by tweezers if the test samples are "chips". As the "main tool" of the practical operation a spatula with a heavy grip (to obtain an optimum center of gravity) slightly bent at its scoop end has proved itself as being universally applicable (see Fig. 2.5a).

With regard to the risk of secondary contamination using this tool, the author has heard about one conscientious analyst who plated the spatula with a layer of gold. However, such effects have never been reported with the dosing of dry powders. Tests have shown that no significant amounts of Cr from the spatula are transferred to biological samples on the pg-level.

If a specific powdered sample must be repeatedly dosed, we can use a dosing unit developed for an "automatic solid sampling system" [45]. It consists of a mililiter pipette tip as a sample container with a closeable cover on the top. A plunger formed of stainless steel (tantalum, PTFE) reaches through the cover down to the opening hole of the tip. The test sample often in the form of a powder pellet can be pressed out of the opening at the tip by a pull/push movement directly into a sample carrier for subsequent weighing. Because the sample container is totally closed the working place can be kept absolutely free of sample particles. Filling the container with 1–2 g e.g. of a CRM for calibration or control measurements lasts at least several weeks of routine analyses.

2.1.3
Sample Introduction Systems

2.1.3.1
Direct Loading of Test Samples into the Furnace

In early experiments of SS-GF-AAS, a micro-spatula was used for loading the sample into the furnace. The special *"solid sampling spoon"* is made of a tantalum boat that is loaded with the test sample, weighed on the balance, gripped with a spoon holder, introduced into the furnace from the side and then withdrawn, depositing the sample at the center of the furnace[25, 46].

These techniques are unsatisfactory for several reasons [47]: The position of the powdered samples cannot be reproduced exactly, thus the precision of repeat measurements suffers; sample particles may "stick" to the spatula surface, thus it must be reweighed to ensure that the test sample mass introduced is exact [48, 49]; the residues of the samples remain in the furnace after atomization, thus subsequent measurements can be adversely affected; and finally the weight and the dimensions of a spatula are very large in relation to the test sample, thus problems may arise with positioning, taring and weighing.

For the introduction of solid test samples a *"powder-pipette"* (Fig. 2.3) was developed [50]. Using this tool, the introduction of the sample aliquote through

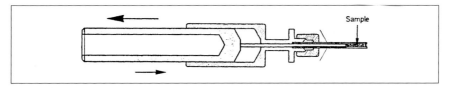

Fig. 2.3. The "powder pipette" works rather similary to a syringe pipette but has a PTFE plunger. The end of the capillary is pushed into the powdered or ground sample material several times, until the tip of the capillary is filled. After weighing the complete tool, the capillary tip is introduced into the sample port of the graphite tube and the test sample is dispensed.

the small center hole of the graphite tube of the furnace is feasible. However, in addition to the problems described above, during transport using this instrument, the quantification of the loaded test sample can also be adversely affected due to the loss of particles with movements of the powder pipette. Ali et al. [51] reported that if the solid sample is mixed with graphite powder (1+99), no losses occur during the transfer step. Moreover, filling by tapping a capillary glass pipette tip several times into the sample powder yields a fairly precise test sample mass (5–7% at ~0.4 mg coal/graphite mixture).

The so-called "*tape sandwich*" *technique* has been developed for measurement of longitudinal trace element concentration gradients in hair [52] (see Sect. 6.7.1) and has been extended for direct introduction of powdered materials into a graphite-cup furnace. A small amount (ca. 20 mg) of powdered material was spread uniformly over the middle region of a 18 cm long strip of Scotch tape (Scotch Magic Tape 810) and sealed with another strip of tape to make a sandwich (see Fig. 6.19). Segments were punched out using a specially designed tool and introduced individually into the graphite cup for atomization. The analyte blank values for Cr, Mn, Pb and Cd were found to be negligible. The average weight of blank segments were calculated and substracted from the weight of the individual test samples. The relative weighing error introduced depends on the standard deviation of the blanks (typically ±7µg) and the absolute mass of sample (0,1–1,5 mg). Generally (for sample weights >0,3 mg) the weighing error was <2,5%.

Despite the drawbacks described, such introduction techniques may be used for test measurements. However, if the results are unsatisfactory or the solid sampling procedure seems to be unacceptably difficult, the analyst must bear in mind that these techniques cannot generally be recommended for accurate and convenient analysis.

Generally no problems caused by *electrostatic charges* arise, because the solid sample analysis is performed mostly with sample materials containing moisture in equilibrium with the surrounding humidity (compare Sect. 2.4.1.1). Only if very dry samples (e.g. coarse inorganic materials that show no capillary effects) are to be dosed and transferred, may problems appear. Conducting plates for the deposition of sample vessels and dosing tools, "grounding" of all instruments including the hand of the operator can help to overcome such difficulties. The required items (e.g. grounding bracelet) are available in electronic stores.

Test samples in form of single pieces (lumps, chunks, scaps, nuggets), may be introduced with the help of a transfer instrument (e.g. tweezers) directly through the enlarged side-hole, possibly with the help of a funnel. This technique has been used satisfactorily for metals [48, 53, 54], gallium arsenide [55], PVC [20] and animal feed particles [56].

2.1.3.2
Utilization of a Sample Carrier

Most of the problems mentioned above can be avoided, if the test sample is loaded into a carrier which is transported into the furnace and remains there during the atomization or vaporization [47].

If the material of the carrier can only be used once (e.g. when it is made of paper), only the residues remaining can cause interferences. If the sample carrier is of refractory material (e.g. graphite, ceramic, metal) it can be removed from the furnace after the measurement process, the sample residues can be removed and it can be used again for subsequent test samples.

With the advancement of the SS-GF-AAS method during the last 15 years, several *sample carrier systems* have been developed – some of them in the laboratory stage, some have been developed by instrument manufacturers for use with their specific AAS instruments. Figure 2.4 shows a variety of commercial and laboratory-built sample carrier systems.

For ease of handling and interference-free operation, the system should meet some demands, namely:

- It must be easy to place the test sample on the sample carrier prior to the weighing step. It is strenuous to place powder-microsamples on a very small area, apart from that, fallen particles are a potential source of imprecision. It is a great advantage if the carrier can be loaded directly onto the balance after taring.
- The volume of the carrier should be large enough to accept test samples up to the maximum mass for which a spectroscopic limitation is given (e.g. background) in order to give flexible analytical conditions (a carrier volume of 10 mm^3 corresponds approximately to a loading of 5 mg of biological samples).
- The sample carrier must be able to be gripped easily by a tool (such as pincers), to reduce the danger of losing particles. If the carrier is lifted from the balance it should be transported directly into the furnace without being gripped again.
- The introduction into the furnace must be smooth, avoiding touching of contaminated parts of the furnace (e.g. cooled ends of the graphite tube). Because

Fig. 2.4. Commercially available and laboratory made graphite furnace carrier systems for solid test samples (Note: The dimensions of tubes and carriers correspond approximately to the real sizes relatively to each other).

a microboat, version 1 (from: [199]);
b microboat , version 2 (from: [200]);
c platform boat (from: [201]);
d miniature cup (from: [201]);
e cup-in-tube (from: [155]);
f sampling platform (from: [202]);

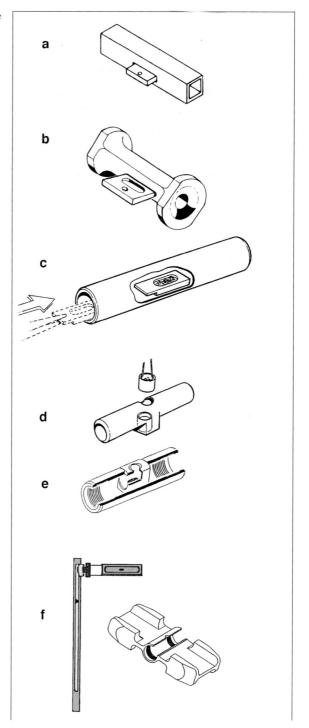

Fig. 2.4. Continued

g solid sampling probe (from: [203]);
h solid sampling tube and platform (from: [204]);
i solid samping cup (from: [24]);
j ring chamber tube (from: [205]);
k mini-massmann graphite cup (from: [137]);
l sampling rod (from: [192]).

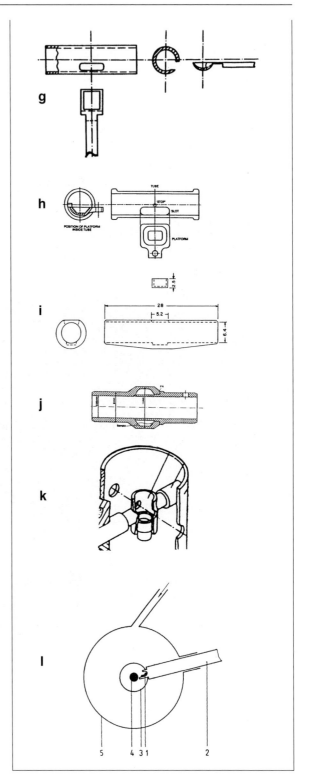

Fig. 2.5. Graphite furnace and introduction system for a sample carrier. Special pincers that keeps the graphite carrier in the "hold positon" are exactly and smoothly moved in the axial direction into the furnace by a carriage running guide.
a The Platform boat system (see. Fig. 2.4 c) of the SM20 (Grün-Analysentechnik, D-Ehringshausen).
b The solid sampling platform system (see Fig. 2.4 f) of the AAS 5, (Analytik Jena, D–Jena)

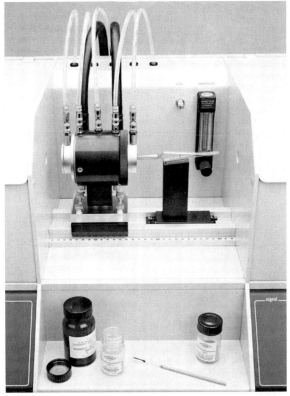

a

b

of the limited dimensions, a mechanical guide for the introduction into the tube is recommended (see Fig. 2.5).
– The sample residues which remain in the carrier after atomization (vaporization) must be easily and completely removable.

- In AAS-analysis, the sample carrier should also create or support isothermal conditions during the atomization step (e.g. acting as a "L'vov platform") in order to reduce gas phase interference. This property is discussed for the different carrier systems in Sect. 3.1.2.2.

The survey of the commercially available and laboratory made systems given in Table 2.1 describes the properties of the carrier and introduction systems for solid samples which are shown in Fig. 2.5 referring to the above discussed criteria.

Table 2.1 Survey of sample carrier and graphite furnace introduction systems for direct solid sampling

---

Microboat (see Fig. 2.4 a, b) [199, 200]

Instrument Type	Instrumentation Laboratory
Loading	simple, large deposition area (4x6mm)
Sample amount	ca. 10 mg of biological material
Transportation	special instrument, graphite grip on boat body
Introduction	radial through an enlarged introduction hole, small spatial allowances, not free of jerks
Cleaning	simple (large and open deposition area)
Atomization	reduced platform effect by square cross section of the tube
Similar systems	Solid Sampling Platform (see Fig. 2.4 h, Philips) [204]

Platform Boat (see Fig. 2.4 c) [206, 128]

Instrument Type	Grün Analysentechnik
Loading	simple, large deposition area (4 x 7 mm)
Sample amount	ca. 10 mg of biological material
Transportation	simple using special tweezers, graphite grip on the boat body
Introduction	axial introduction, precise and free of jerks with the use of a mechanical guide
Cleaning	simple (large and open deposition area)
Atomization	acting as a L'vov platform
Similar systems	Graphite Cup [47]

Miniature Cup (see Fig. 2.4 d) [207, 9, 208]

Instrument Type	Hitachi
Loading	small cup aperture (i.d. 2.5--4 mm)
Sample amount	2--5 mg of biological material
Transportation	special tweesers, gripped from the top, no direct take off from balance possible
Introduction	radial through an enlarged introduction hole, small spartial allowances
Cleaning	special tool required (small aperture)
Atomization	delay of vaporization/atomization
Similar systems	modification by a "graphite lid" [210], Solid Sampling Cup (see Fig. 2.4 i, Varian) [24]

Cup-in-Tube (see Fig. 2.4 e) [151, 150]

Instrument Type	Perkin Elmer
Loading	laborious, because of covered deposition area, small side hole (3.4 x 4.2 mm)
Sample amount	2-4 mg of biological material
Transportation	special instrument, grip from the top, no direct taking off from the balance possible
Introduction	radial through an greatly enlarged introduction hole, small spartial allowances
Cleaning	deposition area not accessible, special tool required
Atomization	reduced isothermal conditions because of shielded absorption volume

Solid Sampling Probe (see Fig. 2.4 g) [203, 128]

Instrument Type	Philips
Loading	simple, large deposition area
Sample amount	ca. 10 mg of biological material
Transportation	graphite stam on the boat body
Introduction	radial, enlarged introduction hole, precise and free of jerks with the use of a el.-mech. guide
Cleaning	simple (large and open depositon area)
Atomization	acting as a L´vov platform, enhenced effect if introduction of the probe is delayed
Similar systems	Autoprobe [211], Graphite Rod (see Fig. 2.4 l) [192]

Ring chamber tube (see Fig. 2.4 j) [205]

Instrument Type	Zeiss
Loading	large aperture (diameter of the tube)
Samples amount	15 mg of biological material
Transportation	closed sample deposition volume
Introduction	laborious, re-setting of the entire tube
Cleaning	simple
Atomization	delayed sample vaporization

Mini-Massmann Graphite Cup (see Fig. 2.4 k) [137]

Instrument Type	Varian
Loading	small cup aperture (i.d. 4.5 mm)
Sample amount	5 mg of biological material
Transportation	special tweesers, gripped from the top, no direct take off from balance possible
Introduction	into the "outer" cup of the "mini-massmann", small spartial allowances
Cleaning	special tool required
Atomization	delay of vaporization (depending on type of form cup)

Sampling platform (see Fig. 2.4 f) [202]

Instrument Type	Analytik Jena
Loading	narrow, but open deposition area (3 x 12 mm)
Sample amount	ca. 6 mg of biological material
Transportation	simple by special tweezers, graphite grip on the platform body
Introduction	axial introduction, precise and free of jerks with the use of a mechanical guide
Cleaning	simple (open depositon area)
Atomization	acting as a L´vov platform, spartially isothermal (side heated tube)

2.1.4
Working Place Conditions

2.1.4.1
Data Support

It is a particular advantage of direct solid sampling that data evaluation can occur virtually simultaneously with the performed analysis and consequences can be drawn for the subsequent analysis.

A precondition is the interconnection of spectrometer, microbalance and data station for automatic data transfer of sample weight and instrument response. Such a system relieves the operator of data transfer which diverts his attention from the analysis (see below, Sect. 2.1.4.2).

The operator should be informed at any moment during the analytical session about the quality of the current analysis by appropriate software. In addition to the notification of test sample number, sample weight and instrument response, essentially the sample content should be calculated continuously. If the operator is informed of other data such as the mean and standard deviation from the test samples analyzed up to that point, the drawing of relevant conclusions for the running analysis is facilitated, e.g. to decide whether to reduce or increase the number of replicate measurements or to change the sample mass for consecutive runs. It is conceivable that advanced software could perform "quasi-online" statistical tests for distribution (e.g. outlier recognition), signal diagnosis for the recognition of matrix effects and deliver "help" functions for method development. However, because of the complex interrelations between the instrumental, analytical and statistical parameters and because the amount of experience the operator has must be considered, such a software tool cannot be developed by conventional program techniques – it needs an "expert system".

2.1.4.2
Ergonomics

As long as the analytical procedure of direct solid sampling is not carried out automatically, the operator is continuously working with the spectrometer, weighing and transporting test samples. Moreover, if he uses two sample carrier at the same time to increase the efficiency of analysis, this implies a great deal of manipulation.

In order to reduce the workload of the operator, the work stations (sample vessels, balance, furnace, keyboard) should be reachable in a sitting position, without uncomfortable stretching. In particular, the height of the furnace should be identical to the height of the balance, consequently the spectrometer must be installed at a lower position than on a normal laboratory bench. The optimum ergonomical configuration of the balance and spectrometer (and PC) has proven to be perpendicular to each other as shown in Fig. 2.6.

Fig. 2.6. Sketch of an ergonomical set-up of the required analyical instruments and tools
for direct solid sampling.
The spectrometer, the balance,
keyboard and the furnace are
in an easily accessible position.

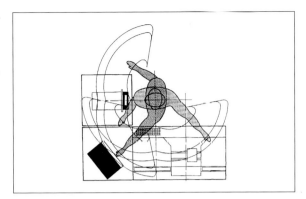

2.1.4.3
Automation

Sample-handling techniques for liquids using automated pipettes cannot be
adapted to powders for obvious reasons. For a long time, automation of sample
handling for powdered solids in the milli- and microgram range seemed to be
impossible.

Four basic steps must be performed separately in an automated process for
solids: *separation* of the sample aliquot, *weighing* with a microbalance, and the
transport of the sample to the balance, to the furnace, and the *introduction* into
the tube. Additionally the electronic control of the automatic solid sampling system must be linked to the software of the analyzer controller and data processor.

Some attempts at automation of test sample transport and introduction can
be found in the literature. In [53] a modification of an automatic sample dispenser for liquids is described by Lundberg and Frech. Pieces of metal are held
by the dispenser arm – whose capillary end is covered by a platinum gauze – by
means of suction. Using this instrument, the samples can be dropped automatically even into the preheated furnace. In [57], there is a description of a sampler
which can automatically introduce up to ten platform boats carrying predosed
and weighed solid test samples. Karanassios and Horlick [58, 59] described an
automatic apparatus for "direct sample insertion technique (DSI)" where the test
sample is introduced in a graphite probe directly into an ICP-torch. A carousel
capable of holding up to 24 probe assemblies is part of the autosampling system.
Although with this technique no graphite furnace is used and the introduction
conditions are more restricted, the construction principles may be instructive
also for the construction of an automated system using a graphite furnace, e.g.
the use of pneumatic devices. However, with all of these systems, weighing (and
taring) was still performed manually.

In an engineering masterpiece, weighing, transportation and introduction
were combined into a "*solid sampling workstation*" (Ing.-Büro O. Schuierer, D-
Ismaning) which frees the operator from the main and most tedious parts of the

sample handling procedures (Fig. 2.7)[1]. A precondition was the development of a very stable top loading microbalance M2P (Sartorius, D-Göttingen) whose working head is integrated with the automaton (see Fig. 2.2)

Attaching this instrument to the furnace, the ergonomical conditions described above become more or less unimportant because the only duty for the operator is to load the sample carrier which is placed in a convenient position. The sample throughput is increased considerably compared to manual operation, an advantage for analytical applications such as quality control, homogeneity studies, screening etc.

The solid sampling workstation has been commercialized as an attachment to the AAS-Spectrometers of the SM-Series (Grün Analysengeräte, D-Wetzlar) and the "AA5solid" (Analytik Jena, D-Jena). Of course the automation can also be used in combination with a side loaded graphite furnace used as a sample introduction attachment to an ICP-source.

The operational flexibility of the workstation is enhanced by the implemented "firmware" which offers a variety of operating modes from a simple "step by step" procedure to a fully automated batch-process with up to 10 platforms including the burn-out treatment ("cleaning"), transportation and registration of tare and sample masses. The workstation communicates with the software of the system regarding transfer of data, synchronism with the spectrometer and sampling sequence.

Fig. 2.7. Solid Sampling Workstation (System Schuierer)
Front: Displays and keyboards of balance and workstation for manual control;
Middle left: Case for the mechanical parts of the balance; location of platform boats on top (batch operation possible) and windshield of balance pan;
Right: Transport drive with rail for movement between deposit, balance pan, and furnace; case for driving motor and mechanisms of the "automated" tweezers;
Rear: Graphite furnace of the spectrometer (here: from SM20) for axial introduction of a platform boat by the tweezers on the transport drive.

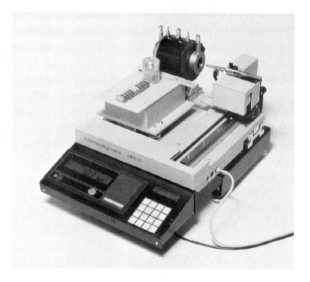

1) A precondition was the development of a very stable top loading microbalance M2P (Sartorius, D-Göttingen) whose working head is integrated with the automaton (see Fig. 2.2)

The whole solid sampling process can be fully automated by combining the workstation with a "solid dispenser" as described in [45].

Up to 25 dosing units for different samples are arranged on a turntable and the whole dosing process is automated. It fulfils all analytical demands, namely: contamination free operation; no carry over by the separating device; no demixing of particle fractions; positive separation features for different sample types; adjustable quantity control for a large mass range; sufficient reproducibility of the desired mass; sufficient sample supply for multiple sampling; and easy handling (e.g. filling). An improved "solid dispenser" has recently been developed as a commercial attachment to the "solid sampling workstation". It is considerably smaller and therefore handier in its usage yet offering the same features as the one described above. The principle of test sample separation is somewhat different: A rotary piston has a cavity which which when at the top is filled from the reservoir for powdered materials and when at the bottom is emptied onto the graphite platform.

With the developement of a sampler for liquids more than 15 years ago [582] the precision was improved considerabely, mainly because coincidental quantification errors caused by manual pipetting are avoided. Although the quantification of a solid test sample is as accurate using hand dosing as by using automatic dosing, it can be assumed that errors caused by the physical and mental efforts of the operator during the sample handling procedure – which are not possible to quantify – are reduced as well.

Indeed, when a large series of trace element determinations was carried out using the "solid sampling workstation" the data obtained are more reliable than manually obtained data [61, 13, 62].

2.2
Calibration in Solid Sampling Analysis

Calibration is the key issue for the quality of the analysis, as it is with all comparative analytical techniques. The first concern of the analyst must therefore be to find a suitable calibration material and an adequate calibration method. This applies to all atomic spectroscopic techniques, but due to the strong influence of the matrix, this task is of special importance with the solid sampling technique.

Whether a calibration method is adequate is determined by the demands of the analysis. The analyst should consider carefully whether the efforts which may be necessary for a certain calibration procedure are worth the required degree of accuracy ("fitness for purpose").

2.2.1
Reference Materials

The trueness of the final result is strongly determined by the correspondence of calibrant and unknown sample material with respect to the physical and chemical behavior of matrix and analyte during the analytical process.

Ideally the calibrant material should have a composition identical to the sample material, preferably both having a similar analyte content. These conditions may be so if the sample compounds are more or less pure substances as it is often the case with industrial products and raw materials (e.g. pharmaceuticals, cement, polymers). The composition of "real samples" e.g. from the ecosystem, the environment or from human and animal bodies is often unknown and/or shows large variations, so that no completely matching calibrant material or no similar material is available at all.

On the other hand it is a characteristic of the graphite furnace technique that the influence of the matrix on the analytical signals is not as strong as it is with other techniques. *The advances made in the recent years in instrumentation and methodology has given generally a greater independence of the choice of calibration material and method.*

2.2.1.1
Calibration Using Reference Solutions

The advantages of the use of an aqueous solution as the calibrant material are obvious: The materials for preparing a reference solution are readily available and inexpensive, the calibration procedure is fast and easy to perform and heterogeneity (i.e. sampling error) need not be taken into account.

Styris et al. [63] have formulated the criteria that are necessary for a reference solution to be successfully applied to the analysis of solid materials: "One of these is the complete release of analyte from the solid and from the aqueous samples during a single atomization cycle. ... The second and most obvious criterion is that the numbers of free atoms that enter the analytical volume during the atomization cycle should be equally proportional to the quantities of analyte in each sample. Of course, the same analyte atomization mechanisms could be involved for the aqueous and solid samples; this would indeed be fortuitous. It is more likely that the different types of samples exhibit dissimilar atomization mechanisms."

Although the matrix of solid materials may have a strong influence on the vaporization/atomization process, satisfactory results are achieved frequently by using calibration curves established with aqueous reference solutions. Even in the early days of this technique (GF-AAS), successful experiments were reported apart from the failures or limitations that have been described (see Table 3.5).

Because *peak area* is a better measure of the total amount of analyte released, its use in the evaluation should help to overcome the problems caused by differences in the vaporization/atomization mechanism. Indeed, a lot of published data show that with peak area evaluation calibration with a reference solution leads to true results even with strong matrix effects being present.

Remaing problems may have their origin in differences in the analyte fraction released, different atomization efficiency (AAS), or different transport efficiency (ETV-ICP). Often these effects can be overcome by the influence of matrix modifiers (see Sects. 2.5.4, 4.6, 5.4).

The establishment of the advanced techniques for background measurement, vaporization/atomization, matrix modification and fast electronic signal acquisition (in GF-AAS usually summed up as the "STPF-concept") has led to analytical progress which has extended the field for the application of reference solutions as the calibrants in solid sampling as well.

However, in addition to the potentially interfering matrix effects there are some other disadvantages associated with the use of prepared reference solutions:

It is generally known that there are problems of *analyte instability* (mostly overcome by acidifying the reference solution to 1% HNO_3), there can be contamination by container surfaces and blank values (which can be overcome in ultratrace analysis only by high expenditure, e.g. double destilling of dilutants, high purity reagents, clean bench and clean room conditions).

It is seldom mentioned – and never published at all – that in practice, gross errors may be caused by *mistakes during the preparation*. In particular, if only one reference solution is prepared for calibration such an error is not apparent from the measurements (these may be "precise" but not "true").

It is often overlooked that when using a calibration established with a reference solution, the *moisture content* of the solid sample leads to a bias if the dry weight of the material is used as a reference (whereas the influence of the moisture may be compensated for if the calibration is based on a similar solid material, see Sect. 2.4.1.1).

The influence of different matrix compositions of calibrant and sample material to the vaporization/atomization mechanism in the graphite tube have been documented and discussed by many investigations over the years and are discussed and documented for the different solid sampling techniques in the respective chapters of this book.

2.2.1.2
Calibration Using Certified Reference Materials

The idea of using (natural) solid reference materials for calibration is to compensate for potential matrix effects which occur with the vaporization/atomization if both calibrant and sample material are affected in the same way.

If a certified reference material (CRM) is available whose matrix composition and analyte content match that of the sample, ideal analytical conditions are fulfilled.

A huge number of investigations have documented successful analyses based on a calibration using CRMs, leading to excellent or at least acceptable trueness. So it should be applied in routine analysis whenever possible.

In Tables 2.2 a, b, c are shown the available CRMs certified by NIST (National Institute of Standards and Technologie, SRM Program; Gaithersburg MD20899 USA), BCR (Community Bureau of Reference, IRMM; Steenweg op Retie, B-2400 Geel, Belgium), NRC (National Research Council Canada, Institute for Environ-

Table 2.2 List of certified reference materials (CRM) from NIST, BCR, NRC and NIES for different kind of materials and the certified contents of some selected elements (uncertified but indicated values in brackets); units in mg/kg

Material Name Code No.	As	Cd	Cr	Cu	Hg	Mn	Ni	Pb	Se	Tl	V	Zn	No. elements certif./uncertif.
a. Soils, Sediments, Sludges													
Calcareous Loam Soil BCR CRM 141 R		14.6	195	46.4	0.025	683	103	57.2				283	9 / 9
Light Sandy Soil BCR CRM 142 R		0.34	(113)	69.7		970	29.2	40.2				(101)	6 / 2
San Joaquim Soil NIST SRM 2709	17.7	0.38	130	34.6	1.40	538	88	18.9	1.57	0.74	112	106	27 / 22
Montana Soil No 1 NIST SRM 2710	626	21.8		2950	32.6	1.01	14.3	5532			76.6	6952	22 / 27
Montana Soil No 2 NIST SRM 2711	105	41.7		114	6.23	638	20.6	1162	1.52	2.47	81.6	350.4	24 / 28
Sewage Sludge amended Soil BCR CRM 143R		71.8	(577)	130.6	1.10	904	299		(0.6)			1055	7 / 2
Estuarine Sediment BCR CRM 277	47.3	11.9	192	101.7	1.77		43.4	146	2.04			547	10 / 24
Estuarine Sediment NIST SRM 1646A	6.23	0.148	40.9	10.01		234.5		11.7	0.193		44.84	48.9	19 / 19
Buffalo River Sediment NIST SRM 2704	23.4	3.45	135	98.6		555	44.1	161	1.12	1.06	95.0	438	28 / 21

Marine Sediments													
NRC BCSS-1	20.7	0.24	106	39.3	0.092	365	49.3	21.9	0.72	(0.98)	252	172	19 / 1
NRC MESS-2	11.1	0.25	123	18.5	4.57	229	55.3	22.7	0.43	(0.6)	93.4	119	14 / 3
NRC PACS-1	211	2.38	17.5	452		470	44.1	404	1.09		127	824	16 / -
Pond Sediment													
NIES 2	12	0.82	75	210			40	105				343	13 / 13
Sewage Sludge													
BCR CRM 145R		3.50	(313)	696	2.01	156	247	286			(101)	2.12	8 / 1
b. Biological Materials, Agriculture, Food Ingredients, Food Products													
Aquatic Plant													
BCR CRM 060	(8)	2.2		51.2	0.34	1759	(40)	63.8	(0.7)			313	7 / 7
BCR CRM 061	(7)	1.07		720	0.23	3771	(420)	64.4	(1)			566	7 / 8
Plankton													
BCR CRM 414	6.82	0.383	23.8	29.5	0.276	299	18.8	3.97	1.75		8.10	112	11 / 6
Sea Lettuce													
BCR CRM 279	3.09	0.274	(10.7)	13.14	(0.05)	(2090)		13.48	0.593			51.3	6 / 11
Chlorella													
NIES 3				3.5		69						20.5	9 / 12
Apple Leaves													
NIST SRM 1515	0.038	0.013		5.64	0.044	54	0.91	0.470	0.050		0.26	12.5	24 / 22
Olive Leaves													
BCR CRM 062	(0.2)	0.10		46.6	0.28	57	(8)	25	0.1)			16.0	7 / 8

Table 2.2 Continued

Material Name Code No.	As	Cd	Cr	Cu	Hg	Mn	Ni	Pb	Se	Tl	V	Zn	No. of certif. elements/indic. values.
Peach Leaves NIST SRM 1547	0.06			3.7	0.031	98	0.69	0.87	0.12		0.37	17.9	20 / 20
Tomato Leaves NIST SRM 1573A													
Spinach Leaves NIST SRM 1570A	0.068	2.89		12.2	0.030	75.9	2.14		0.117		0.57	82	19 / 8
Rye Grass BCR CRM 281	0.057	0.12	(2.1)	9.65	0.200	81.6	3.0	2.38	0.028			31.5	12 / 5
White Clover BCR CRM 402	0.093		(5.19)				(8.25)		6.70				4 / 4
Pine Needles NIST SRM 1575	0.21		2.6	3.0	0.15	675		10.8					15 / 11
Beech Leaves BCR CRM 100		(0.34)		(12)		(1300)		(16.3)				(69)	8 / 7
Spruce Needles BCR CRM 101		(350)	(2700)	(5000)		915		(2600)				35	9 / 6
Hay Powder BCR CRM 129				(10)		70			(0.025)			32.1.	9 / 5
Cod Muscle BCR CRM 422	21.1	0.017		1.05	0.559	0.543		0.08	1.63			19.6	10 / 9

Material												
Dogfisch Muscle NRC DORM-2	18.0	0.043	34.7	2.34	4.64	3.66	19.4	0.065	1.4	(0.004)	25.6	14 / 2
Tuna BCR CRM 463					2.85							1 / -
Tuna BCR CRM 464					5.24							1 / -
Mussel Tissue NIST SRM 1974a	1.2	0.17	0.322	1.14	0.024	1.26	0.124	1.2	0.247	0.191	11.3	23 / 11
Oyster Tissue NIST SRM 1566A	14.0	4.15	1.43	66.3	0.064	12.3	2.25	0.371	2.21	4.68	830	26 / 5
Bovine Liver NIST SRM 1577B		4.15		160		1.05		0.129	0.73		127	18 / 7
Bone Ash NIST SRM 1400								9.07			181	8 / 7
Bone Meal NIST SRM 1486								1.33			147	8 / 9
Human Hair BCR CRM 397	(0.31)	0.521		(110)	12.3		(46.0)	33.0	2.0		199	5 / 3
Wheat Flour NIST SRM 1567A		0.026		2.1		9.4			1.1		11.6	15 / 10
Wholemeal Flour BCR CRM 189	71.3			6.4	(1)	63.3	(380)	379	132		56.5	7 / 10
Brown Bread BCR CRM 191	(23)	28.4		2.61	(2)	5.90	(440)		(25)		19.5	6 / 11
Rice Flour NIST SRM 1568A	0.29	0.022		2.4	0.0058	20.0			0.38		19.4	17 / 11

Table 2.2 Continued

Material Name Code No.	As	Cd	Cr	Cu	Hg	Mn	Ni	Pb	Se	Tl	V	Zn	No. of certif. elements/indic. values.
Rice Flours NIES 10		0.023 0.32 1.82											13/9
Bovine Liver BCR CRM 185	0.24	0.298		189	0.044	9.3	(1.4)	0.501	0.446			142	9/9
Bovine Muscle BCR CRM 184		0.013		2.36	0.026	334	(0.27)	0.239	0.183			166	8/109
Pig Kidney BCR CRM 186	63	2.71		31.9	1.97	8.5	(0.42)	0.306	10.3			128	9/9
Dogfisch Liver NRC DOLT-2	16.6	0.043	0.37	25.8	1.99	6.88	0.2	0.22	6.06			85.8	14/1
Single Cell Protein BCR CRM 274	0.132	0.030		13.1		51.9		0.044	1.030			42.7	8/3
Skim Milk Powder BCR CRM 063 R				0.602				0.0185	(0.129)			49	12/3
Skim Milk Powder (spiked) BCR CRM 150		0.218		2.23	0.0094	(236)	(0.615)	1.0	(0.127)			(49)	6/5
BCR CRM 151		0.101		5.23	0.101	(223)	(0.056)	2.002	(0.125)			(50)	6/5
Milk Powder NIST SRM 1549		0.0005	0.0026	0.7	0.0003	0.26		0.019	0.11			46.1	17/10
Total Diet NIST SRM 1548		0.028		2.6		5.2			0.245			3.08	14/10

c. Fuel, Ashes, Hazardous Materials

												other elements in preparation
Coal (bituminous) NIST SRM 1632B	3.72	0.0573	6.28		12.4	6.10	3.67	1.29			11.89	24 / 17
Coal (subbituminous) NIST SRM 1635	0.42	0.03	2.5	3.6	21.4	1.74	1.9	0.9	5.2	4.7		14 / 10
Blend Coal BCR CRM 040	13.2	0.11	31.3	0.35	139	25.4	24.2		(0.79)	30.2		10 / 2
Gas Coal BCR CRM 180	4.23	0.212	(13.5)	(9.1)	0.123	34.3	(16)	17.5	1.32	19.3	27.4	12 / 17
Coking Coal BCR CRM 181	27.7	0.051	(5)	(12.3)	0.138		(8.6)	2.59	1.15	12.0	8.4	11 / 14
Steam Coal BCR CRM 182	(1.47)	0.057	(20)	(12.3)	0.040	195	(39)	(15.3)	0.68	24.3	33.3	9 / 18
Fly Ash BCR CRM 038	48.0	4.6	(178)	176	2.10	479	(194)	262		(17.3)	(334)	581	10 / 4
Coal Fly Ash NIST SRM 1633B	136	0.784	197	113	0.14		121	68.2	10.3		295	9 / 3
City Waste Incineration Ash BCR CRM 176	(93.3)	470	836	1302	31.4	(1500)	123.5	10.87	41.2	2.85	(43)	25.77	12 / 4
Urban Particulate Matter NIST SRM 1648	115	75	403	609			82	6.55	27		140	4.76	14 / 22
Vehicle Exhaust Particulates NIES 8	2.6	1.1	25.5	67			18.5	219			17	1.04	16 / 15
House Dust NIST SRM 2583								170				

mental Research and Technology; Ottawa, Ontario, Canada K1A 0R6) and NIES (National Institute for Environment Studies, Japan Environment Agency, Yatabemachi, Tsukuba, Ibaraki 305 Japan) showing the certified (mean) contents. [2] Of particular interest for scientific research may be that NIES now only supply their CRMs, free of charge, directly to the user, however, requests must include a detailed description of the intended use, and results must be shared with NIES scientists.

Many other national and international organizations have also released certified or uncertified reference materials. These organizations also offer special reference materials e.g filter, catalyst, paint, plastic, cellulose, lyophilized blood and urine. Voluminous lists of biological and environmental reference materials currently available are given by Zeisler et al. (IAEA) in [65] and by Roelandts in [66, 67, 68, 69]. For reference materials in geological/mineralogical analysis see [70, 71] and for the analysis of metallurgical samples see the lists for reference materials of metals and alloys in [72, 73, 74, 75, 76, 77, 78]. The COMAR database provides a comprehensive overview of more than 7000 reference materials now available.

The analyses of *biological materials* using CRMs for calibration have been highly successful. Even though the matrix composition from plant and animal materials can vary, the influence of matrix differences to the measured peak area can be usually neglected. Thus other viewpoints may taken into consideration for the choice of a suitable CRM, e.g. the homogeneity or the analyte content.

For *environmental materials*, the composition of the matrix can vary substantially (e.g. soils, sediments, sludges, ashes and waste materials). Fortunately many materials with different compositions are available today. However, if the influence of the matrix on the analytical signal is significant the validity of a calibration should be verified even if a CRM of the same material type is used for calibration.

Metallurgical and geological/mineralogical materials using appropriate CRMs have also been analyzed successfully. The influence of these matrices on the peak shape is generally strong. Because the matrix composition of specific materials is often well known or does not vary over a large range, adequate matrix matching may be achieved with the limited number of CRMs available.

Unfortunately the practical application of calibration using CRMs can be hampered by some *restrictions*:

For unusual types of sample materials, CRMs having an equivalent matrix composition may not be available.

---

2) The data is taken, with permission, from the Promochem catalogue "Reference Materials for Micro, Macro and Trace Element Analysis" [64]. The PROCHEM group of companies, from their offices in Europe, India and North America, distribute reference materials, including those from BCR, NIST, LGC(UK), NRC, CANMET, NWRA (Canada), IAEA (Austria) and other CRM producers. Their catalogue gives valuable information about matrix, contents, use and handling of these materials.

With the use of solid reference materials, an uncertainty due to the sampling error may be introduced into the analytical process. Because of the low test sample mass, a heterogeneous distribution of the analyte may lead to increased imprecision. So a larger number of test sample measurements may be necessary to obtain a sufficiently small uncertainty band in the established calibration curve (see Sect. 2.4.2.4).

This property of a reference material must be checked with its use in solid sampling. Unfortunately the certified values of reference materials and the assumed sampling error are mostly based on a subsample size of at least 200 mg. A reliable characterization of the micro-heterogeneity of these materials in the milligram range is not documented in the certification reports. A homogeneous material may be more advantageous than using a material with a corresponding matrix, e.g. for this reason, milk powder (natural contents or spiked) has been chosen for calibration in the solid sample analyses of a variety of biological materials [79]. Materials which show a "nugget-effect" are unsuitable for calibration purposes (see Sect. 2.3.2).

Not only the matrix composition, but the matrix amount which is introduced into the furnace also influences the analytical signal. So it may be necessary to ensure that the analyte content of sample and calibrant material are in the same range (at least in the same order of magnitude). Fortunately the producers of CRMs are releasing more and more materials with a similar matrix but with a different analyte content.

One (or more) of these problems can make it difficult to use the advantages of calibration using CRMs.

The use of two or more CRMs for establishing a calibration curve can help us to recognize matrix effects and to find suitable materials. If the measurements show the same slope with the regression (considering the uncertainty in the measurement and the certified value), it can be concluded that the matrix (and/or the content) has no dominant influence [22] (see Fig. 2.8 a).

In Fig. 2.8 b a slight but clearly recognizable difference in the slopes from two CRMs can be recognized. However, without further investigations it cannot be decided if this systematic effect is due to a matrix effect or due to a bias in the certified contents (see Sect. 2.4.2.4). However, if such effects occur in routine analysis, the simultanious use of both materials for one calibration curve may give a higher certainty of an acceptable result from the analysis of unknown samples. In Sect. 3.3.1 examples are discussed for the problem of the uncertainty in certified contents when reference materials are used for the detection of "matrix effects".

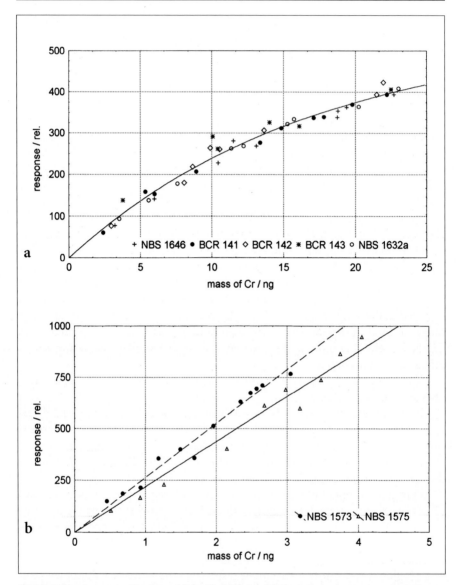

Fig. 2.8. Calibration curves for Cr established by different CRMs. As instrument response peak area is used, the analyte mass is calculated from the test sample mass and the reference content (from:[193], with permission). The certified or indicative contents for Cr from BCR and NIST (former NBS) are given in paranthesis:
a NIST SRM 1646 Estuarine Sediment (76±3 mg/kg); NIST SRM 1632a Coal (34.4±1.5 mg/kg); BCR CRM 141 Calcareous Loam Soil (75 mg/kg); BCR CRM 142 Light Sandy Soil (74.9 mg/kg); BCR CRM 143 Overfertilised Soil (228 mg/kg);
b NIST SRM 1573 Tomato Leaves (4.5±0.5 mg/kg); NIST SRM 1575 Pine Needles (2.6±0.2 mg/kg).

2.2.1.3
Preparation of Synthetic Solid Reference Materials

Preferably only physical properties and quantities should be used for the determination of the analyte in a reference material (as it is with the preparation of aqueous reference solutions). For these measurands, the accuracy which can be achieved is at least one order of magnitude better than analytical results. This is one reason why the use of synthetic reference materials can be advantageous.

The preparation of a synthetic reference, however, is only suitable, if the composition of the sample is simple or well known (at least for the components which determine the spectroanalytical behavior). For example, for the determination of Au and Pd in *silver*, Hinds [80] added dissolved Ag (400 µg) to the reference solution in order to match the matrix and to overcome signal suppression. However, the preparation procedure for a solid reference material generally needs a lot more effort than the preparation of reference solutions. Some special approaches have been reported for the preparation and use of synthetic reference materials:

For the analysis of *sulfide ores*, Langmyhr et al. [81] prepared a solid synthetic reference material by intensive grinding of solid matrix components (FeS, ZnS, PbS, CuS, S) and adding the analyte as cadmium sulfide in two dilution steps. The analytical result using this reference material was in good agreement with results using other methods.

Gries and Norval [39] used synthetic reference materials for the determination of Tl in an *aluminium* matrix, produced by ion implantation. In [82] the authors reported that no difference was detected between implantation and "dip standards". They prepared the dip reference materials for the analysis of refined cadmium metal by simply depositing aliquots of 10 ml reference solutions of Tl on pieces of foils from spectrally pure cadmium which were left to dry. Additionally they used a mixture of $TlNO_3$ in pure urea as "secondary standards" for a check on the stability of the experimental conditions throughout the work.

Marks et al. [48] prepared standards by doping a *nickel-base alloy*. Trace metal additions were mixed with 25 g of a nickel-4% aluminum powder and pressed into a pellet. Then the alloy was vacuum-induction melted and the pellet was added. The melt was held at 1482 °C for 2 min for optimum mixing without severe loss of volatile trace elements. The standards are characterized by other methods e.g. XRF and flame and nonflame AAS and wet chemical techniques. No information was given, however, for the analytical results achieved in comparison to the added and original analyte amounts.

Hiltenkamp and Jackwerth [83] used electrolytically *spiked solid gallium* samples for the analysis of high-purity gallium. For the preparation of the synthetic reference material they introduced 2-3 g exactly weighted high purity gallium, 500 µl 37% HCl and a reference solution into an electrolytic cell for 45 min. Possibly, analyte traces which are not registered are below the limit of detection by GF-AAS.

For the determination of Ag in copper, Pauwels et al. [84] prepared a set of synthetic metallurgical reference materials by alloying *high purity copper* material with pure silver by high frequency levitation melting. The blank value of Ag in the copper material was determined by an iteration procedure using the doped content of the synthetic reference materials and the content of the copper material determined by using these reference materials. Because the calibration was based on physical properties traceable to SI-quantities, the analytical results of the Ag determination for two copper candidate reference materials was accepted as part of the certification by the BCR. A similar procedure for adding metal impurities to (molten) silver was reported by Hinds [80].

For the analysis of *polyethylene samples*, Brückner et al. [85] prepared synthetic reference materials by adding metal-containing stabilizers and pigments to polyethylene granulate. For homogenization these materials were extruded and cut several times. The contents of Ni, Cu, Pb, and Cr were calculated considering the portions, the stoichiometry and the dilution of the components. Good agreement for calculated contents of analyzed materials was achieved with these materials as the calibrant.

Atsuya and Akatsuka [86. 87] prepared synthetic reference materials for the direct analysis of plant materials by *coprecipitation* of metal ions e.g. with magnesium(II) 8-quinolinate. Complexes were formed and approx. 100% recovery was found in the filtered and dried precipitate for 8 elements (Al, Cd, Co, Cu, Mn, Ni, Pb, and Zn). Using these precipitates as solid calibrants the results of the analyses of plant reference materials were in good agreement with the certified values. The role of the magnesium was characterized as a "matrix modifier" which matches the matrix of plant materials sufficiently [88].

In [89] the synthetic addition of analyte elements (Cu, Fe, In) to CdTe-*semiconducter material* was reported. The calibration curves obtained were found to be in very good agreement with those obtained by reference solutions. However, no description of the preparation procedure for the synthetic reference materials was given.

Záray et al. [90, 91] used "purified" *silicon carbide powder* in ETV-ICP-AES in order to create vapor phase matrix effects comparable with the unknown sample. "5 mg SiC-powder (ESK-933) is initially purified at 2200 °C for 40 s using a graphite boat with two holes. After cooling, an aliquot of standard solution is injected into the free hole of the graphite boat and dried." [91]. The author concluded that the matrix effect is thus eliminated and the calibration becomes reliable for quantitative determination (Ca, Ti, Fe).

Also for the analysis of *ceramic powders* Nickel and Zadgorska [92, 93] prepared synthetic matrices for the determination of Al, Fe, Ti and V. A simulation of silicon carbide is performed based on the ability of the modifier CoF_2 + BaO (1+1) to decompose silicon carbide completely. The simulation is performed by in situ chemical processing a mixture of high purity silicon dioxide, graphite powder, the modifier and oxides of the analyte elements in a definite stoichiometric proportion. The calculations of the isothermal-isobaric potential for the real and the simulated matrix systems supported the assumption that isoformation could be expected.

2.2.2
Calibration Methods

Calibration using one of the materials described above, can be performed using several approaches. The following presentation of the different calibration methods follows the amount of effort required for their application. This corresponds not necessarily with the degree of accuracy that can be reached with the respective methods.

2.2.2.1
Direct Comparison of Response Data

The analytical measurements can be directly evaluated by a direct data comparison method using *mass-specific response R '*(related to *unity mass*, usually 1 mg) which is defined for the i-th solid test sample (of calibrant or sample material) of *mass m_i* as

$$R'_i = \frac{R_i}{m_i}$$

Eq. 2.1

The *sample content* can be estimated by the averages of specific response data from the unknown test samples (subscripts) and the calibrant material with the known *(reference) content c_r*

$$c_s = \frac{\overline{R'}_s}{\overline{R'}_c} c_r$$

Eq. 2.2

This procedure is possible with the use of solid or liquid reference materials. With the use of a reference solution for calibration, the dimension of c_c is a concentration (unit e.g. mg/ml), accordingly the calibration response data must refer to the volume of the liquid test samples (volume-specific response).

Example (Evaluation of test data in Table 2.11)

The ratio of the average form the mass specific response values of the sample Maize Leaves (0.831 s/mg) and calibration measurements from the BCR CRM 281 Rye Grass (0.13 s/mg) is 6.39, multiplied with the reference content (2.38 mg/kg) yields 15.2 mg/kg as the estimate of the unknown sample content (Eq. 2.2).

A precondition for the application of this method is a linear correlation between response and analyte amount. However, if all response values lie within a small range, linearity is not essential.

Alternatively the sample response R_s can be "bracketed" by calibrant response signals which are slightly lower and higher *(bracketing method)*. The procedure is described in [94, 43] for the use of a solid reference material. Using a CRM with the certified content c_r, the sample content can be estimated by

$$c_s = \frac{c_r}{m_s}\left[\left(\frac{R_s - R_1}{R_u - R_1}\right)(m_u - m_1) + m_1\right]$$ Eq. 2.3

where
m_s=mass af the unknown solid test sample
m_l, m_u=lower (l) and upper (u) test sample mass of the reference material
R_l, R_u=respective response data.

With the use of a reference solution the unknown sample content can be estimated by

$$c_s = \frac{V}{m_s}\left[\left(\frac{R_s - R_1}{R_u - R_1}\right)(c_u - c_1) + c_1\right]$$ Eq. 2.4

where
V=volume of the calibrant test samples
c_l, c_u=concentration of the lower (l) and upper (u) reference solution

Equations 2.3 and 2.4 can also be applied with the mean values from larger sets of measurements for the respective terms.

These methods can be regarded as a linear appoximation for the correlation of response and analyte amount (i.e. calibration curve). So, if linearity is not ensured, the difference between the low and high response values should be sufficiently narrow. With an appropriate choice of the "brackets", nonlinearity may have only a small – negligible – influence on the final calculated result.

2.2.2.2
Calibration Curve Method

A complete calibration curve can be constructed by introducing various amounts of the analyte into the furnace. If the range of analyte amount covers a sufficiently wide range, a regression fit using an appropriate mathematical function can give a reliable calibration function which can then be used for the analysis of samples of widely varying contents.

Usually in solid sampling with the graphite furnace, a regression curve for signal response R_c on analyte mass a_c is calculated.[3] The *test sample mass* $m_{o,i}$ is

---

3) The regression of response R_c on the *subsample mass* m_s is recommended only if the content of the unknown sample is calculated by comparison of the slopes of the regression lines for the calibrant *and* the sample material [95]. With this technique the uncertainty for the result may be enlarged by the error of two regressions procedures

transformed into the *analyte mass* $a_{c,i}$ using the *known analyte content* c_c of the solid calibrant material by

$$a_{c,i} = m_{c,i}\, c_r \qquad\qquad \text{Eq. 2.4a}$$

and correspondingly for liquid samples by

$$a_{c,i} = V_{c,i}\, c_r \qquad\qquad \text{Eq. 2.4b}$$

With direct solid sampling it can often be assumed that the calibration curve passes through the origin (R=0 if a=0) (see Fig. 2.8 a, b). An intercept caused by blank values of reagents can be excluded (except in the case of matrix modification). Particularly with the Zeeman-AAS, no shift of the baseline due to background should appear, so it is obvious that if no analyte is introduced into the furnace (no sample loading), no instrument response may occur [51, 96, 97, 98].

Examples

In Fig. 2.8 b the test sample measurements from two CRMs are fitted with the function R=b·a where the regression yields for SRM 1573 b=263 s/ng and for SRM 1575 b=219 s/ng respectively. In Fig. 2.8 a the test sample measurements from different CRMs are fitted in this case of unlinearity with the function $R=b_1(1-\exp(-b_2 a))$ which gives a good fit with the estimated parameters b_1=55 s and b_2=0.056 ng^{-1}.

The "turned" regression function is used as the analytical function for calculation of the analyte mass $a_{s,i}$ from the response values of the unknown test samples (compare Eq. 2.46). The *content* of the unknown sample from a *number* n_s of measurements can be estimated by the mean of the test results by

$$\bar{c}_s = \frac{1}{n_s}\sum \frac{a_{s,i}}{m_{s,i}} \qquad\qquad \text{Eq. 2.5}$$

(Examples are given in Sect. 2.4.2.4)

If samples with a similar analyte content are to be analyzed it can be advantageous, to establish only a section of a calibration curve by varying the analyte amount (i.e. the calibrant test sample mass) in a limited range. Then linear regression can lead to accurate results even if the true calibration function show slight curvature (the bracketing method is the border case of this approach). However, in this case the condition for a reliable linear approximation is a pronouncedly small error of the calibrant signals (e.g. by the use of reference solutions instead of a CRM) in order to get a sufficiently narrow uncertainty band for the regression fit and acceptable uncertainty limits for the sample measurements (see Sect. 2.4.2.4).

2.2.2.3
Analyte Addition Methods

With the analyte addition method (AAM) [4] known analyte amounts are combined with the unknown test sample. If an aliquot of a reference solution is added
to the solid test sample, it can be assumed that – under specific circumstances –
the vaporization/atomization of the added analyte atoms is equally affected by
the sample matrix. With this method, the calibration procedure is included in the
measurement of each sample.

The classical approach of *"analyte variation-AAM"* is performed by adding
reference solutions containing various amounts of the analyte to test samples of
constant amount (volume). By a subsequent linear regression and extrapolation
to "zero-response" the content of the sample can by calculated (after verifying
that the response is sufficiently linear).

Three approaches of the analyte addition method are feasible with the direct
analysis of solid samples:

Because dosing of constant test sample masses is rather difficult, it may be
better to perform it vice versa as *"sample variation -AAM"*, i.e. by *varying the test
sample mass* and keeping the added analyte amount constant [96].

The data evaluation can follow the usually applied AAM technique. The
regression of the response data on the sample mass and the extrapolation to R=0
yields the test *sample mass* m_a which corresponds to the *added analyte amount*
a_a. By these values the sample content c_s can be estimated by

$$c_s = \frac{a_a}{m_a} \qquad\qquad\qquad \text{Eq. 2.6}$$

Example

Figure 2.9 a shows the graphical evaluation of this AAM first carried out by
Eames and Matousek [96] for a Ag determination in the silicate rock reference
material G248 (data points reconstructed from the original figure). The linear
regression analysis yields to b=0.304 mg^{-1}, b$_0$=0.138. Using these values, the mass
m_a which corresponds to the added analyte amount of 1.5 ng is calculated to
0.455 mg, thus yielding a sample content of 3.3 mg/kg. The authors in [96] concluded that the AAM has improved the trueness compared to the result based on
a calibration from a reference solution (1.5 mg/kg), however "the analyte addition technique was not entirely successful, since a discrepancy remained
between ... the established values" (3.8 mg/kg). Although the difference may
seem to be unacceptable and may have been caused by inappropriate atomiza-

---

4) *Analyte addition method* (AAM) is be a better designation of this method [99] than the commonly
   used term *standard addition method* (SAM)

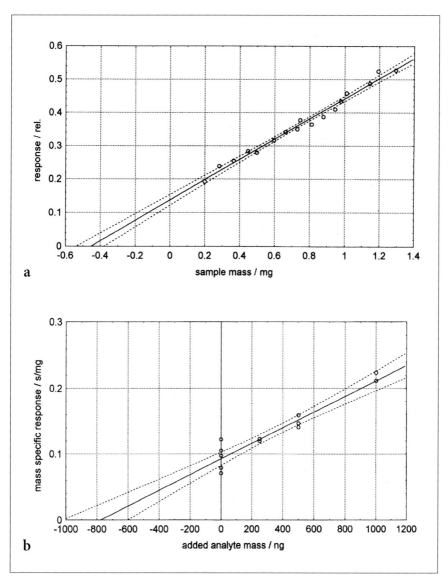

Fig. 2.9. a "Sample variation-AAM" experiment of a silicate rock sample (addition of a 1.5 ng Ag as reference solution). The test sample results of solid sampling GF-AAS are entered, the regression line and uncertainty band are extrapolated to the R=0 line.
b "Sample & analyte variation-AAM" experiment of the sample bovine liver sample for the determianton of Se. The test sample results of solid sampling GF-AAS are entered, the regression line and uncertainty band are extrapolated to the R´=0 line.

tion conditions (not isothermal conditions), such a statement only can be verified, if the uncertainties in the sample content *and* the reference content are considered (see Sect. 2.4.2.4, where the uncertainty interval for this analysis is estimated as being ±0.48 mg/kg). Already from the 95%-uncertainty lines in Fig. 2.9 a it can be seen that AAM as an extrapolation method may lead to an extended uncertainty.

Because the result with this approach of AAM depends only on a single reference solution a systematic error due to unstability, contamination or a gross error during its preparation might not be recognized.

If both, the *test sample mass and the added analyte amount* are varied the data evaluation for the "*sample & analyte variation-AAM*" must be modified. Three different procedures are described in the literature.

a. The following procedure is given by Baxter [100]:

The *responses $R_{0,i}$ for no analyte addition* are transformed to mass specific response (Eq. 2.1). The *responses from test samples with added analyte R_i* are transformed to mass-specific data by

$$R'_i = \frac{1}{m_0}\left[R_i - \overline{R}'_0\left(m_i - m_0\right)\right]$$ Eq. 2.7

where
$\overline{R}'_0$=mean of mass-specific response from the data with no analyte addition
m_0=unity mass

A linear regression yields the relationship between the transformed response data ($R'_{0,i}$ and R'_i) on added *analyte amount a_a*. The content of the sample c_s can be estimated by extrapolation to R=0, yielding the analyte amount a_0 which corresponds to a test sample of unity mass (e.g. 1 mg) according to

$$c = \frac{a_0}{m_0}$$ Eq. 2.8

Example

For the illustration of the described procedure Table 2.3 shows a data set from the determination of selenium in NIST SRM 1577a Bovine Liver (from: [100]). Linear regression yields an intercept of 0.0926 s/mg and a slope of 0.000119 s/ng ·mg, thus the content for the sample is estimated as 778 mg/kg (certified 710 mg/kg); Fig. 2.9 b shows the respective diagram.

A disadvantage of this approach of AAM is that by applying Eq. 2.7 the uncertainty in the average of the R_0 are propagated to all values of R'_i, which under unfavorable conditions may contribute a significant part to the total uncertainty [100], (see example in Sect. 2.4.2.4). A large number of test sample measurements without analyte addition would provide a small uncertainty in

Table 2.3 Data from a "sample & analyte variation AAM" experiment (direct solid sampling GF-AAS) for the determination of Se in Bovine Liver (NIST SRM 1577a) (analytical data from:[100])

test sample no.	added analyte a_a (ng)	sample mass m (mg)	instr. response R (s)	specific response R_0' (s/mg)	specific response R_0', R' (s/mg)
1	0	0.43	0.034	0.079	0.079
2	0	0.79	0.077	0.097	0.097
3	0	0.58	0.061	0.105	0.105
4	0	0.58	0.071	0.122	0.122
5	0	0.39	0.028	0.072	0.072
6	250	0.67	0.094		0.125
7	250	0.48	0.072		0.122
8	250	0.25	0.053		0.124
9	500	0.47	0.108		0.159
10	500	0.70	0.112		0.141
11	500	0.83	0.122		0.138
12	500	0.74	0.122		0.147
13	500	0.87	0.136		0.148
14	1000	0.62	0.188		0.224
15	1000	0.68	0.180		0.212
mean		0.60		0.095	0.134

the mean *and* a smaller uncertainty in the regression line estimate on the "extrapolation side".

b. The extensive data conversion via Eq. 2.7 was avoided by the "three-point estimation-AAM" described by Atsuya et al.[101, 102]. With this approach three (or more) series of constant analyte addition on various test sample masses were evaluated each separately by linear regression. From the regression curves (one for each added analyte amount) the mass specific response for a constant test sample mass of 1 mg is estimated; these values are then plotted against the amount of analyte added. A further linear regression yields the relationship between the mass-specific response data and the amount of analyte a_a added. The content of the sample can then be estimated by extrapolation to $R=0$, yielding the analyte amount a_0 which corresponds to a test sample of unity mass, according Eq. 2.8.

Example

Figure 2.10 a shows the plot of the experimental data of an AAM measurement for the determination of Cr in NIST SRM 1571 Orchard Leaves. From the regression lines (line I-IV) for each amount of analyte added, the mass specific responses at the sample mass 1 mg can be read, approx. 0.75 for no analyte addition, 1.13 for 1 ng, 1.35 for 2 ng, 1.7 for 3 ng added amount. The regression line between these pairs and the extrapolation gives $a_0=2.6$ ng (Fig. 2.10 b), thus the Cr content can be estimated as 2.6 µg/g. The Cr content separately estimated from each series of different analyte amounts added (see above method a.) yields 2.1, 2.9, 2.8 µg/g respectively (certified 2.6±0.3 µg/g).

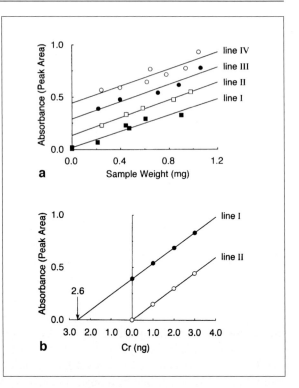

Fig. 2.10. a "Three-point estimation-AAM" experiment of a
orchard leaves sample for the
determianton of Cr. Line I: 5.0
µl portion of pure water added
to sample; Lines II, III, and IV:
1.0, 2.0, 3.0 ng of chromium as
a 5.0 µl portion of chromium
standard solution added to the
solid test sample.
b Line I: calibration curve for
Cr by three-point estimation
analyte addition method established with the measure-
ments shown in Fig. 2.10 a;
line II: extensive application of
the calibration curve proposed
of one unknown sample to
other unknowns samples
(from: [102], with permission).

The authors state that the uncertainty with the proposed method is less [102],
however this is achieved solely by the large number of test samples that is
required for this evaluation method (here: n=22). Statistically this evaluation and
the resulting uncertainty in the final result is aquivalent to AAM of method b.
described above. Obviously, an advantage of this method is the convincing visu-
alization of the measurement results, so that inconsistent results can be recognized easily, particularly whether each point exists on the regression curve.

Furthermore, the authors propose that the extensively prepared (final) *AAM-
regression line can be used as the calibration line* for the analysis of other samples with similar matrix composition without performing further analyte addition, because it can be regarded as a "matrix-matching method". For this purpose, the regression line must be shifted through the origin in order to compensate for the original analyte content (line II in Fig. 2.10 b).

c. With a third approach the *"generalized analyte addition method (GAAM)"*
the response data were not transformed, but instead a two dimensional linear
regression for the instrument response R on both independent variables added
amount aa and test sample mass m_s yields the relationship

$$\hat{R} = S\ a_a + D\ m_s + B \qquad\qquad\qquad Eq.\ 2.9$$

where
S=*response per unit of added analyte mass*
D=*response per unit of test sample mass*
B=*constant (blank) value*

The sample content simply can be estimated by the terms S and D, which are achieved as the regression coefficients, by

$$c_s = \frac{D}{S}$$

Eq. 2.10

Example

In Fig. 2.11, the geometric representation of GAAM is demonstrated for a determination of Cu in the sample Aquatic Plant (BCR CRM 60) using ETV-ICP/AES [from 103]. From the data points (n=17) a regression plane is calculated. The slope of the regression plane passing through the signal/sample mass face of the cuboid is D=113,780 counts/mg and through the signal/addition amount face is S=2,098 counts/ng. The ratio of these slopes gives the estimatimated content 54.2 mg/kg (certified 51.2 mg/kg).

As Baxter and Frech [100, 14] pointed out with the theoretical and practical adaptation of the GAAM to the solid sampling method, problems associated with the common approaches of AAM are overcome by the GAAM, particularly the

Fig. 2.11. "Generalized analyte addition (GAAM)" ETV-ICP-OES experiment of an aquatic plant sample for the determination of Cu. The *points* represent the test sample results, the *lines* sections indicate the distance from the regression plane (from: [103], with permission).

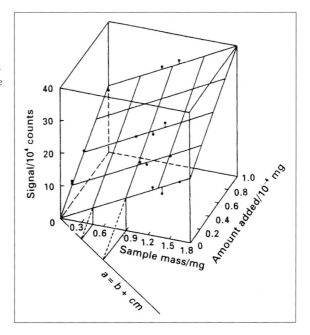

influence of unlinearity of the calibration curve is reduced and blank values are compensated (e.g. in the case of matrix modification).

Applying an analyte addition method the range and the absolute amount of test sample mass and added analyte must be chosen thoughtfully [95], because the experimental design influences strongly the uncertainty of the final result because these are achieved via an extrapolation procedure (see Sect. 2.4.2.4).

Successful analyses by analyte addition have been reported even if the direct comparison with reference solutions has yielded no satisfactory results (see lists of applications (Tables) in Chaps. 3, 4, 5). Beside the chemical matrix effects the physical effects caused by the expanding vapor of the solid material resulting in a shorter residence time of the analyte [224] may be compensated for by applying AAM. An impressive example for the latter effect is given by Holcombe and Wang in [49] for the determination of vanadium in coal fly ash samples (italics added): "The analyical results are severely depressed for V using calibration curves prepared from simple V solutions (-68%). ... The depression ... may be caused *by the expulsion of the matrix vapor products by the sample matrix.* ... When the standard addition method is employed, this expulsion can be accounted for if the added V and that in the coal fly ash behave similar. The agreement between the experimental results (using standard additions method) and the certified data suggest that the mechanism for V loss is similar for the two sources of V"[49].

However the capability of compensating or surpressing interfering effects should not be overrated. Welz has discussed the limitations of the analyte addition technique for the elemination of "multiplicative errors" [99].

2.2.2.4
Special Calibration Approaches

Pauwels et al. [104] have developed the method of *extrapolation to zero* matrix for direct solid sampling (based on the method of Thompson et al. [105]). It takes advantage of the "independent determination", because it is based on aqueous *reference solutions*, i.e. calibration only based on physical properties. In contrast to the direct calibration on the basis of standard solutions this technique has the potential for compensation the influence of the matrix.

From a set of measurements a linear regression curves for (test sample) mass-specific response on test sample mass is calculated and extrapolated to the R-axis. This gives the hypothetical *signal per sample mass* $R'_{s,0}$ at $m_s=0$ ("zero-matrix"). Correspondingly for a set of measurements from reference solutions a linear regression curve for (analyte) mass-specific response on test sample mass is calculated, where the extrapolation yields to the hypothetical signal per analyte mass $R'_{c,0}$. The content of the unknown sample is than estimated by

$$c_s = \frac{R'_{s,0}}{R'_{c,0}}$$

Eq. 2.11

Example

Figure 2.12 shows the test sample results and the regression line for the determination of Cd in the sample Pig Kidney (BCR CRM 186) (reconstructed from: [14]). Extrapolation to zero matrix yields $R'_{s,0}$=0.125 mg^{-1}. The declining slope of the regression line is due to nonlinearity response function. The respective experiment using a reference solution gave a yield of $R'_{c,0}$=0.044 ng^{-1} hence, applying Eq. 2.11 gives a Cd content of 2.86 mg/kg (certified 2.71±0.15 mg/kg).

The authors [14] conclude that this calibration method is independent of other analyses (like it is with the use of a CRM) and thus can contribute results to the certification of new reference materials. To some extent, the influence of nonlinearity of the AAS response (absorbance) is compensated for by this method.

Fig. 2.12. "Extrapolation to zero matrix" experiment of a pig kidney sample for the determination of Cd. The test sample results of solid sampling GF-AAS are entered, the regression line is extrapolated to the m_s=0 line. This yields a mass specific response of 0.125 mg^{-1}, the uncertainty of this estimation is marked on the response axis (from: [104], with permission).

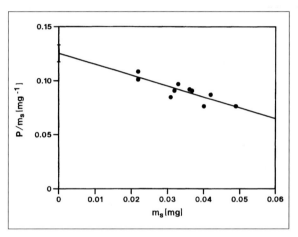

The *vaporization rate method*, developed by Rettberg and Holcombe [106], focuses on the different atomization process for analyte elements from solid materials, considering the often-observed impeding of vaporization (broadening and tailing of absorption signals) and stepwise vaporization (double- and multipeak) specially in metallurgical and geological materials.

To determine the total analyte amount in a test sample – even if a complete vaporization cannot be obtained within a reasonable (integration) period – "these studies evaluate a method to associate the *rate* of analyte evolution (vaporization) with the total amount of analyte initially present in the sample atomic absorption peak area measurements were used to define a curve describing the rate of release. This unique extrapolation approach required only a knowledge of the rate of vaporization at a given time; hence only *partial* vaporization of the analyte from the sample was necessary for quantification" [106]. Practically, they measured the integrated absorbance from up to five heating cycles of one solid test sample. The logarithem of the *integrated absorbance* ΔP_t of the *heating period* Δt was plotted over the entire time of the test sample analysis, devided into the period of the heating cycles. From a (first-order) kinetic model the following mathematical expression was derived

$$\ln(\Delta R_t) = -kt + \ln\left[R_0\left(e^{k\Delta t} - 1\right)\right]$$ Eq. 2.12

This relationship allows us to estimate the response R_0, which represents the total amount of analyte. A linear regression of the measured values yields the *vaporization rate constant* k (slope) and the second term on the right of Eq. 2.12 (intercept), which can be exploited to determine R_0. The conversion into the total analyte amount a_0 can be performed by assuming that the peak area is proportional to the analyte mass and correlating the absorbance data with those of a reference solution.

Example

Figure 2.13 a shows the results of a vaporization rate experiment for the determination of Pb in one test sample (2.6 mg) of tin with $\Delta t = 15s$. The extrapolation of the regression line gives an intercept of −1.8 and a slope of 0.085 s^{-1} (see

Fig. 2.13. "Vaporization rate method" experiment of a tin test sample for the determination of Pb by the direct solid sampling GF-AAS (from: [106], with permission); **a** peak area absorbances of four atomization cycles of 15 s each, **b** ln(absorbance) versus time of the same data.

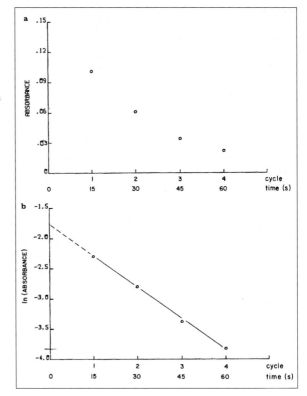

Fig. 2.13 b, reconstructed from: [106]). Using these values in the intercept term of Eq. 2.12 gives a total response of 0.58 s^{-1}. For the estimation of the sample content, this value must be transformed into the corresponding mass of Pb and divided by the test sample mass.

Although the result of the example above of 58 mg/kg, based on several test samples analyzed by this method (RSD 12%), is not in satisfactory agreement (bias -32%) with the determination based on the dissolved Sn material and analyte addition, the proposed approach may promise satisfactory results for materials out of which the analyte is not completely vaporized in one atomization cycle, frequently observed with the analysis of metal chip samples (analyte metal soluble in the matrix metal, see Sect. 3.3.3), or if extremely large contents must be determined (e.g. Mn in steel up to 0.05%) [107, 108].

2.3
Sample Heterogeneity and Sampling Error

In the early period of the present technique, a lack of sample homogeneity was supposed to be the most serious limitation for its application. However, this conjecture was not based on reliable data about the distribution of trace elements in "real samples" on the micro- and milligram level.

Using methods with prior sample decomposition, the contribution of sample heterogeneity to the analytical imprecision cannot be isolated, because contribution of other random effects in the entire analytical procedure are not known exactly (methods not under "statistical control", see Sect. 2.4.1.2). Whereas for other methods, the required test sample mass is too large for the identification of heterogeneity effects.

It has been studies using the solid sampling method which have delivered the first exact information about the micro-heterogeneity of laboratory samples. *These data – and the results of hundreds of applications in routine laboratories – show that the sampling error is at least one order of magnitude smaller than previously expected.*

So it can be stated even at this stage that the usually recommended test sample mass, e.g. for the use of CRMs 100-250 mg, is not required for sufficiently precise test results. Concerning other materials and methods, Langmyhr [4] has already early pointed out the fact that "in most analyses the amounts atomized correspond to the masses arced or sparked in atomic emission spectrographs, so the sampling errors of the two methods can be expected to be the same. It is surprising that the critics of the direct-atomization technique do not seem to have similar objections to the use of atomic emission".

Although sample heterogeneity is generally not the limiting property for the use of test samples in the milligram range, it can be the decisive factor regarding the validity of solid sampling results. So the knowledge of effects which are caused by the heterogeneous distribution of the analyte is required in order to apply appropriate methodical and statistical tools.

2.3.1
Sampling Characteristics of Powdered Samples

2.3.1.1
Influence of Physical Sample Properties

The source of uncertainty caused by the random differences in the amount of analyte in test samples due to sample heterogeneity is commonly called the sampling error.

Two different effects can be responsible for the appearance of an error when taking test samples from a pulverized material: first, various analyte contents of different particle fractions and second, segregation of a certain particle fraction. It should be pointed out that particles of different size and shape can have different analyte contents and this effect may cause bias if segregation cannot be excluded.

Examples
The occurrence of a systematic error due to segregation is discussed by Holcombe and Wang [49] for coal fly ash samples. With the sampling procedure used preference was given to particles with smaller diameters. Because the average content of Cd in smaller particles is higher than in larger particles (found by other investigators), this unsuitable sampling yielded an analytical result which was too high (4.6 µg/g for 0.1-mg test samples instead of 3.3 µg/g for 2-mg test samples). A similar problem is described in [109] for the determination of Pb in soil using the slurry method. The effect of different analyte content in particles of different sizes was suspected to be the origin of differences in the overall content determined for sieved (70×10^{-6} m) and unsieved Urban Dust (IAEA-396) [110]. While the Pb content of the unsieved material was found to be approx. 20% lower, for Cu the respective content was approx. 30% higher, pointing to a significantly higher Cu content in the larger (rejected) particles.

On the mass-level of laboratory samples, generally segregation effects can be avoided by a carefully performed mixing before separating test samples [5].

The following discussion considers only the effects caused by various analyte content of particle fractions in otherwise proper mixed samples, i.e. the particles of different analyte content are statistically distributed so that the sampling gives rise only to random effects.

The first approach of a (sub-) sampling theory for a mixture of two components was given by Baule and Benedetti-Pichler and later in a more developed form by Benedetti-Pichler [111]. Wilson gave similar expressions extended to more than two components [112]. For the *standard deviation due to the sampling*

---

5) On the mass-level of field or bulk samples, the segregation effect can be the dominant factor of the total sampling error

error s_H as a function of the *number of particles n* in one test sample, he derived the expression:

$$s_H = \sqrt{\frac{1}{2}\sum\sum\left[\left(c_i - \bar{c}\right)\rho_i - \left(c_j - \bar{c}\right)\rho_j\right]^2 \frac{P_i P_j}{\rho_i \rho_j}\frac{1}{n}}$$

Eq. 2.13

where the sample properties are characterized by
$P_{i,j}$=*portion of the sample components i and j,*
$c_{i,j}$=*analyte content of particle fraction i and j,*
$\bar{c}$ =*average content of the analyte in the sample,*
$r_{i,j}$=*density of particle fraction i and j*

For mineralogical samples, the sample properties are often available or can be achieved. Trace elements are present in rocks as minerals or at the boundary of the minerals as oxides, sulfides, etc., i.e. they may be strongly heterogeneously distributed. In this case it is desirable that the number of particles exceeds 10^5 to obtain an acceptably low sampling error (e.g. RSD<10%). A model calculation according to Eq. 2.13 shows that 1 mg of powder for spherical particles of 50 μm diameter contains 7 x 10^3 particles but for 10 μm as many as 8 x 10^5 particles. For the realistic situation of non-uniform particle size the total number of particles can be replaced by a weight reciprocal mean for groups of different sizes [112].Wilson gave some examples for the practical application of Eq. 2.13.

Example

Pieczonka [113] prepared the different fractions of a large number of wheat grains and had analyzed these parts separately in a carefully performed study using SS-GF-AAS. Table 2.4 shows the results and an estimate of the precision

Table 2.4 Cadmium content in different fractions of burdened wheat grains and estimation of imprecision due to heterogenious analyte distribution (analytical data from: [113])

grain fraction	mass fraction (%)	content of Cd (mg/kg)
Pericarp	3	1.90
Embryo (germ)	3	8.20
Endosperm (flourbody)	84	0.52
Rest	10	4.00
Whole grain:		
Particle number N (estimated)	2500 mg⁻¹	
S_H (rel., calc. from Eq. 2.13)	9.8%	
RSD (found)	7.8%	

due to the sampling error, calculated from Eq. 2.13 and they are in good agreement with the analytical precision. Similar findings are reported in [114] also by using solid sampling.

However, for real samples of biological or environmental origin, the sample properties are not easy or, in most cases, even impossible to assess quantitatively.

It has been considered that a real mixture which is composed of several sample components which are different in all physical properties "may be duplicated into a hypothetical mixture, of uniform grain size, which contains only two minerals, each of different X-content" [115].

So for a practical estimation of the standard deviation due to sample heterogeneity the simplified form of Eq. 2.13 for only two components (subscript 1, 2) can be useful:

$$s_H = (c_1 - c_2) \sqrt{\frac{\rho_1 \rho_2}{\rho^2} \frac{P_1 P_2}{n}} \qquad \qquad \text{Eq. 2.14}$$

where
ρ = *overall density of the sample.*

Example

This relation was applied by Solberg [4] to estimate the imprecision due to the sampling error of a synthetic mixture if particles of cadmium sulfide are distributed in a matrix of iron(II) sulfide (this simulates the worst possible situation where the total analyte amount is located in only one small sample fraction). Table 2.5 shows the calculated precision in dependence on the particle size and the Cd content. For the extreme cases "the error may be so large under a combination of unfavorable conditions that reliable analytical data are unobtainable. On the other hand, ...the sampling error may be reduced to small and acceptable values when the size of the particles is reduced to below 5 mm"[34].

Practical applications of Eq. 2.14 may be given if the content and the portion of the main fraction and the fraction with the largest analyte content of a sample is known or can be estimated or if pure substances have been mixed (e.g. for

Table 2.5 Predicted imprecision due to the sampling error for cadmium in a mixture of cadmium sulfide in a matrix of iron(II) sulfide (from: [34])

particle diameter (μm)	no. of particles (mg^{-1})	rel. sampling imprecision s_H (%) average content of Cd in the mixture		
		10μg/g	1μg/g	0.1μg/g
10	3.8×10^5	51	162	513
5	3.0×10^6	18	58	183
2	4.8×10^7	4.6	14	46
1	3.8×10^8	1.6	5.1	16.2
0.1	3.8×10^{11}	0.05	0.16	0.5

a pharmaceutical product). An example for the application of Eq. 2.14 is given in Sect. 2.3.2.

2.3.1.2
Concept of the Homogeneity Constant

To overcome the problems of a lack of information about the sample properties, Ingamells and Switzer [116] gave another approach for the estimation of the imprecision due to the sampling error by introducing the *sampling constant K_s*:

$$RSD = \sqrt{\frac{K_s}{m}} \qquad \text{Eq. 2.15}$$

In this equation the experimental *relative standard deviation (RSD)* of test samples can be used to calculate the sampling constant. K_s comprises all the internal sample properties determining the heterogeneity, which is constant for a given laboratory sample (c, P, ρ of the particle fractions). So the independent variable is only the test sample mass m, which can be influenced (to a certain degree) in the actual analysis.

On the condition that the analytical procedure is free of other sources of errors, K_s can be determined by a reliable analysis performing a sufficient large number of replicates. However, often other random effects inherent in the analytical procedure are dominating the overall standard deviation of the measurement. If the uncertainty components due to these effects are known (e.g. expressed as the respective RSD, i.e. method under statistical control, what is often *not* true for techniques with prior sample decomposition) the *relative sampling standard deviation S_H* can be isolated by

$$S_H^2 = RSD^2 - \sum \left(\text{rel. effect sdv.} \right)^2 \qquad \text{Eq. 2.16}$$

With the advance of the solid sampling technique the determination of the sampling constant has become easier and more reliable, because the contributions to the precision due to the effects inherent in the analytical procedure are determinable or are even negligible compared to that due to the sampling error.

For the use of the solid sampling technique Kurfürst et al. [12, 117, 13] have defined the *relative homogeneity constant H_E* (for the element E) that includes all properties of a given sample (compare Eq. 2.13)

$$S_H = \frac{H_E}{\sqrt{m}} \qquad \text{Eq. 2.17}$$

Basically H_E is identical to the square root of K_s. However, the use of the homogeneity constant is especially advantageous in direct solid sampling because the numerical value of H_E represents the imprecision due to the sampling error for test samples of unity mass (usually 1 mg), thus giving directly information about the degree of homogeneity, and the sampling error that can be expected with the sample, respectively. The numerical value of K_s represents the test sample mass necessary to achieve a RSD of 1%, an approach that may be advantageous if the method applied allows us to choose the test sample mass over a wide range in order to attain precision in a small series of test samples (e.g. only 3 decomposition solutions).

Example

Figures 2.14 a, b, c show three series of AAS determinations for Zn in a cod fish material. The pronounced increase in the (im-) precision with decreasing test sample mass can be recognized, confirming the relationship which is described by Eq. 2.17 (the qualitative evaluation of this experiment is given in Sect. 2.4.1.2, see Table 2.8).

A determined value for the homogeneity constant is valid in the mass range of the *representative sample mass M_r,* which can be calculated for direct solid sampling by the (average of the) test sample mass (for slurry sampling see Sect. 5.2.2.1, Eqs. 5.1, 5.2, 5.4)

$$M_r = n_s \overline{m}_s \qquad\qquad\qquad \text{Eq. 2.18}$$

where
n_s=*number of replicate test sample measurements*

It must be pointed out that on a significantly larger mass scale, additonal heterogeneity effects may occur (e.g. due to extremely rare analyte "nuggets", see Sect. 2.3.2). So for a reliable and representative determination of H_E a large number of replicate measurements are required (see Sect. 6.6.4).

The concept of the homogeneity constant can be advantageous for several purposes: For different samples, the degree of homogeneity (for an analyte element E) can be compared. Moreover, the influence of the test sample mass on the precision can be estimated; this can be helpful if an alteration of the test sample mass is necessary (e.g. when using an alternative analytical line). Comparing the precision achieved by different analytical techniques, the contribution of the sampling error can be recognized and statistically monitored. Finally, if the homogeneity contant were given by the producers of CRMs it would be possible to estimate whether a CRM is suitable for calibration purposes in solid sampling (i.e. acceptable uncertainty in a calibration curve) or what the minimum sample mass used for decompositon must be so that the result can be expected in a "tolerance range" [118] (see Sect. 6.6.2).

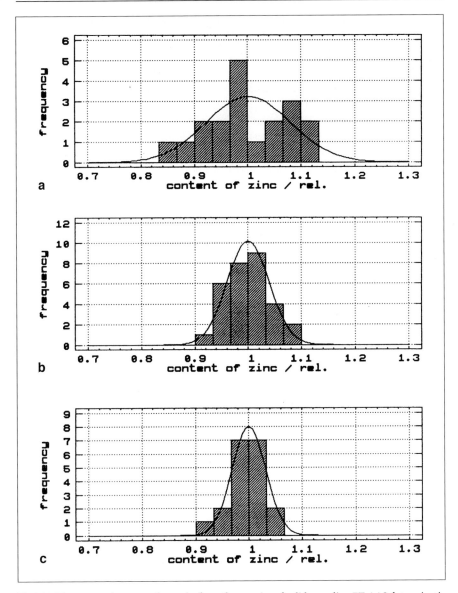

Fig. 2.14. Histogram of test sample results from three series of solid sampling ET-AAS determinations of Zn in a cod fish material using different test sample masses; **a** m=0.35 mg, **b** m=1.0 mg, **c** m=1.9 mg.

Figure 2.15 give the estimated imprecision due to the sampling error over a large range of test sample mass for different homogeneity constants according to Eq. 2.17.

Example

Assuming a sample powder that can be characterized by a homogeneity constant of about 10 mg$^{1/2}$ i.e. 10% S_H for test samples of 1 mg, as it was found to be typical. If test samples of 100 mg of such a material are chemically decomposed and analyzed, only a relative sampling error of about 1% can be expected between different sample solutions (compare Fig. 2.15). However, in decomposition analysis an RSD >3% typically arises. This points to the fact that the imprecison in "conventional" analysis is not due to analyte heterogeneity but due to unknown and/or not totally controlled effects (e.g. incomplete dissolution, cf. [119]).

The probability of obtaining certain particles when taking test samples follows a *multinominal distribution* – or binominal for only two fractions. For pulverized samples, typically a large number of particles (>>100) of each fraction are taken – even with microgram test samples. In this case, the multi- or binominal distribution can be approximated by a normal distribution.

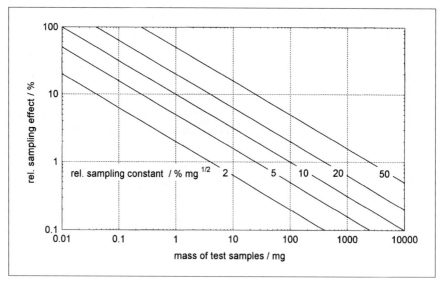

Fig. 2.15. Imprecision due to the sampling error depending from test sample mass on logarithmic scales for different relative homogeneity constants. Typical values of H_E found with direct solid sampling from 2 to 50 mg$^{1/2}$ are entered as the parameter. Even for laboratory samples with a poor grade of homogeneity the contribution of heterogeneity the overall precision is small or negliable if test samples of more than 100 mg are used (e.g. <2% for H_E<20 mg$^{1/2}$).

Although only few authors have given histograms of solid sampling data, it can be supposed that the data are mostly distributed normally. As long as the precision of analytical result is sufficiently good, a slight deviations from normality does not affect the statistical evaluation significantly (e.g. uncertainty interval).

Most important is the fact – empirically found and documented in the solid sampling literature nearly unanimously – that the uncertainty due to the sampling error applying the solid sampling method typically ranges between 5% to 15% RSD for different materials, e.g. for mussels [46], other animal tissues [120], for plant materials [9], for metallurgical materials [121, 84, 268], for silicate rocks [96, 122], and plastics [123].

Langmyhr and Wibitoe concluded in a review on the solid sampling method with GF-AAS : "However, from the applications of the method, it appears that most materials have been analyzed with acceptable precision, very few analyses have been reported to give such a poor precision that the data had to be rejected" [34].

2.3.2
Samples Containing "Analyte Nuggets"

In 1985 Mohl et al. [124] presented a series of solid test sample results from some CRMs whose respective distributions are obviously not of the gaussian type (normal), but are onesided and for a few test samples, high contents in single test samples were found which appeared as "outliers". The precision of these results are partly unacceptable. Figure 2.16 shows an example [6].

Initiated by these observations, a "worse case study" was started. The CRMs for which micro-heterogeneity effects had been observed were investigated with the aim of finding a general analytical and statistical description of micro-heterogeneity effects.

Because, at that time, the automated solid sampling system for pulverized materials was available as a prototype, it had become feasible to carry out a large series of replicate measurements of one sample with a highly reliable procedure. Figures 2.17 a, b show two histograms of the results found for different reference materials. Both distributions are skew, one is even multimodal (more than one maximum). With the slurry method, skewed distributions are found as well [125], Figure 2.17 c shows an example.

As Ingamells et. al [116] have shown, such an effect can appear if the sample contains a particle fraction which is very small *and* has an extreme analyte con-

---

6) For analysts who were against the solid sampling technique at that time, these results proved a presumed serious drawback for its application. Of course, the advocates of the solid sampling technique would have preferred that these results had been found to be irregular (e. g. due to a contamination source)

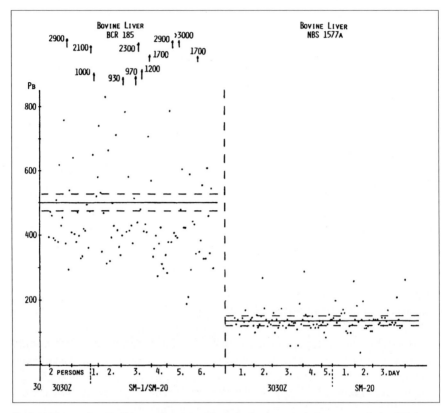

Fig. 2.16. Time series of solid sampling GF-AAS results for test sample from an analysis of Pb in two CRMs Bovine Liver (from: [124], with permission). The certified values and the confidence limits are indicated by the *horizontal lines*. While for NIST SRM 1577a the test sample results are symmetrically distributed, for BCR CRM 185 the distribution is significantly unsymmetrical with a number of "outliers" (*arrows*).

tent compared to the main fraction(s). With an illustrative analogy these particles are called "nuggets".

In the case of small *probability p for the event x* (here: getting nuggets in a test sample) and a large number of samples (here: particles), the binominal distribution can be approximated by a *Poisson distribution*. The characteristic distribution parameter z is given by the average number of events, where $\mu=z$ and $\sigma=z^{-1/2}$ [126]. The shape of the Poisson probability function is skewed on right-hand side if $z<9$ and even one-sided if $z<1$ (compare Figs. 2.17, 2.18, 2.19).

In this specific case z *represents the average number of nuggets* in test sample of mass m. An estimate of the *content c_s* of the laboratory sample and the *standard deviation due to sample heterogeneity s_H* can be expressed by the distribution parameter z and the values evaluated by the determination via the Poisson distribution of the analyte nuggets in the data set [116, 61, 13]:

Fig. 2.17. Histogram of test sample results from the analysis of nugget containig samples solid sampling GF-AAS (in a. and b. the mean contents are normalized to 1, the *light hatched pattern* indicates the designated fraction with one analyte nugget).
a determination of Pb in a spruce shoot sample with direct solid sampling;
b determination of Cd in a cod fish sample with direct solid sampling;
c determination of Mn in a river sediment sample with slurry GF-AAS (from: [125], with permission).

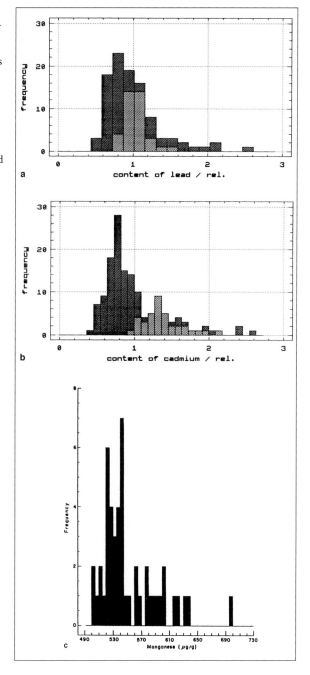

$$c_s = z c_n + c_B \qquad\qquad\qquad \text{Eq. 2.19}$$

$$s_H = \sqrt{z c_n} \qquad\qquad\qquad\qquad \text{Eq. 2.20}$$

where
c_n=*contribution of one nugget to the content of test samples of mass m,*
c_B=*content of the basic (matrix) material.*

Ideal nugget conditions are given with the simulated analytical data in Figures 2.18 a,b,c. It elucidates the connection between the terms that give the different content values c_s, c_B, c_n and the nugget distribution paramerter z for a given test sample mass and the change of the values of c_n and z if the test sample mass is chosen significantly larger. (Note: The value of z and c_n depend on the test sample mass used for its determination. These terms can be transformed to properies of the given laboratory sample by Z= z/m that represents the *average number of nuggets in test samples of unity mass* and C_N=c_n·m that represents the *contribution of one nugget to the content of test samples of unity mass* [13]).

As a "guinea pig" for the appearance of the nugget effect and the evaluation by the model drawn above, the reference material Bovine Muscle (BCR CRM 184) has been used [61]. Figure 2.19 a shows the histogram for test sample results of a Pb determination (m=0.55 mg), using SS-GF-AAS . The designated fraction with different numbers of nuggets (x) are not separate, but are overlapping, and a broadening effect due to the size distribution of the nugget particles is taken into consideration as an additional assumption. The histogram for the so-designated fractions (Fig. 2.19 b) and the fitting with a Poisson probability function yields z=0.508, i.e. Z=0.92 mg^{-1} (nuggets per mg).

Some values for the bovine muscle sample obtained from the nugget-evaluation are given in Table 2.6. The agreement of the sample content calculated by the overall mean and via the nugget evaluation (Eq. 2.19) shows the reliability of this model. The difference in the overall RSD and the calculated relative standard deviation using the nugget model (Eq. 2.20) is caused by the fact that, with the nugget evaluation, only the imprecision due to the sampling error is determined (s_H). The "broadening" by other random effects (e.g. instrument fluctuation) – included in the analytical RSD – have no influence on this value.

Pauwels et al. [127] have applied this model for a microheterogeneity study of a candidate reference material (cod fisch), giving further impressive examples for the distribution of test results that can be approximated with a Poisson probability function (see Sect. 6.6.1).

The evaluation of solid sampling data by the "nugget-model" allows the calculation of some useful values for reference materials, such as: the minimum representative sample mass (e.g. used for other analytical techniques, see Sect. 6.6.2), the test sample mass for which the distribution of results can expected to be nor-

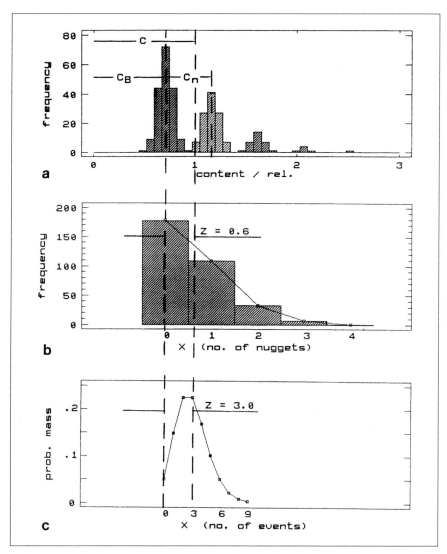

Fig. 2.18. Simulated analytical data and evaluation of an "ideal" nugget effect.
a Frequency histogram with a basic fraction of the content c_B and fractions with different number of nuggets (x=0, 1,2,3,4). The distribution of one nugget c_n is chosen so large, that the fractions are totally separated.
b The distribution of the number of nuggets x in the subsamples in a and the Poisson fit which gives an average number of nuggets of z=0.6 (The scales of the abscissas of a, b and c are chosen so that both types of distributions correspond: the values of c_B lies on x=0 and the value of c lies on x=z so that Eq. 2.19 can be directly read from these figures.)
c Poisson probability function with an average of z=3. This correspond to a five time larger test sample mass compared to a/b. Although higher number of nuggets are probable in the test samples, the decreasing contribution of one nugget to the total content is dominating, so that the scatter of results is smaller (Eq. 2.20).

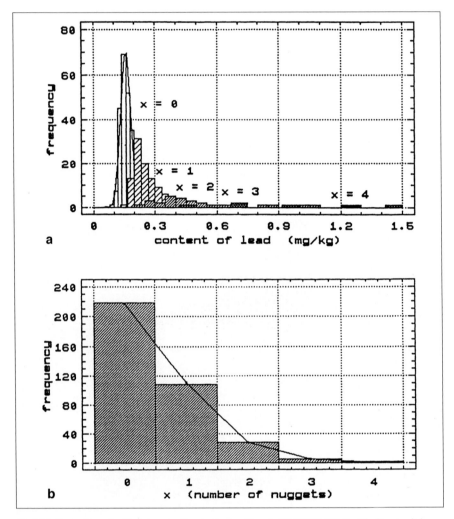

Fig. 2.19. Evaluation of a Pb determination in Bovine Muscle (BCR CRM 184) with direct solid sampling GF-AAS by the nugget-model;
a. histogram of n=360 test samples with the destination of subsample fractions probably containing different numbers of "lead nuggets" (miscellaneously hatched);
b. distribution of the subsample fractions and the Poisson probability fit (z=0.51).

mal, the uncertainty which can be expected if a recommended test sample mass is used or vice versa the test sample mass which is necessary for achieving a desired precision (compare Table 2.6).

The *origin of nugget particles* in pulvarized bovine muscle samples are shown in [128]: Calcificated capsules of dead larvae of the beef tapeworm accumulate Pb with a content up to 700 times higher than the surrounding muscle tissue (see Chapt. 6.6.4). So "ideal nugget conditions" are met: A very small sample fraction

Table 2.6 Application of the nugget model of the direct solid sampling data from the Pb determination in Bovine Liver (BCR CRM 184)

quantity	calculated from	value	unit
certified			
n_r	accepted results	11	
c_r	mean of means	0.2394	mg/kg
s_r	distr. of means	0.016.9	mg/kg
s_W	within lab. std. dev.	0.0205	mg/kg
CI (Confidence interval) 95%		0.0113	mg/kg
evaluated by the nugget model			
n_S	–	360	–
m_S	–	0.55	mg
c_B	Fig. 2.19 a	0.141	mg/kg
c_n	Fig. 2.19 a	0.182	mg/kg
z	Fig. 2.19 b	0.508	
H_{Pb}	Eq. 2.17	41	$mg^{1/2}$
c_S	Eq. 2.19	0.233	mg/kg
c_S	mean (n=360)	0.236	mg/kg
s_H	Eq. 2.20	0.130	mg/kg
s_S	std. dev. (n=360)	0.168	mg/kg
predicted values from nugget model evaluation			
normal distr. (z=9)	$m_2 = z_2/z_1 \times m_1$	9.7	mg
M (min. repr. mass)	Eq. 6.2	20	mg
m_S for S_H=1%	Eq. 2.17	1700	mg
S_H for m_s=200 mg	Eq. 2.17	2,9	%

Table 2.7 Model calculation for the sampling imprecision in bovine muscle material containing nuggets from calcified capsules from the beef tape worm

supposed values
typ. content total muscle (c)	0.15 mg/kg
typ. content muscle tissue (c_B)	0.10 mg/kg
max. content of capsules (C_N)	50 mg/kg
capsules mass fraction ($a \cdot P_N$)	0.001
no. of particles (N)	$1000\ mg^{-1}$

calculated values
homogeneity constant (H_{Pb})	$33\ mg^{1/2}$
S_H for test samples of 1 mg	33%
S_H for test samples of 1 g	1%

contains a large portion of analyte. In Table 2.7 a model calculation for such a pulverized material is documented, using Eq. 2.21.

The exogenous origin of nuggets in spruce needles (compare Fig. 2.17 a) was explained by Wittenbach et al. [129]. The inclusion of particles from anthropogenic aerosols creates the nugget effect when small test samples are used and leads to a significant increase of the overall analyte content in the material.

Theoretically the Poisson distribution evolves to the normal distribution if z>9, i.e. the nuggets then create no skewed distribution (but can still be responsible for an extended width) [126]. However, for practical reasons (a low number of replicate measurements, superimposed broadening effects) the histogram of test samples that includes an average number of nuggets with z>4 will not appear significantly different from normal, and the evaluation via the nugget-model will not be possible (cf. [116]).

If the physical properties of the nugget material are known or can be determined, the imprecision caused by the sampling error can be estimated by [13]:

$$s_H = c_N \sqrt{\frac{a}{N} \frac{P_N}{\sqrt{m}} \frac{1}{\sqrt{m}}} \qquad \qquad \text{Eq. 2.21}$$

where
c_N =content of the nugget material,
P_N =fraction of the nugget material,
a=ratio of densities (nugget to matrix material)

(Note: For the "nugget-conditions" $c_N >> c_B$, $P_N << 1$, $\rho_N \sim \rho$ Eg. 2.14 approaches to Eq. 2.21)

Example

In solid sampling it may be advantageous to prepare a mixture of the pulverized sample material with graphite powder for the purpose of "matrix modification" (compare Sect. 2.5.2.3). Assuming a sample portion of 0.1 (that is then the nugget fraction) in a mixture with graphite powder 1+9, $N=10^6$ mg^{-1}, $a=5$, an imprecision due to the sampling error of only 0.7% can be expected with a test sample size of 1 mg (calculated by Eq. 2.14 or 2.21). At $N=10^5$ mg^{-1} it increases to 2.2%, which is still small compared to the overall analytical imprecision (typically greater than 5%).

For mineralogical and geological samples the model of nuggets of identical size and content can be a good approximation of the real sample composition as well.

It must be pointed out again that the described nugget model may be rather artificial and does not reflect the real physical properties. However, despite its limitations, the application of the nugget model to solid sampling data can provide reliable values because the sampling error is dominated by the fraction of particles which contain a most significant portion of the analyte element: "The sampling characteristics of most mixtures, in which a single element X is of interest, may be duplicated by a hypothetical mixture of uniform grain size which contains only two minerals, each of different X-content" [115].

2.3.3
Influence of Grinding/Milling

The influence of particle size to the precision (due to the sampling error) is described theoretically by Eqs. 2.13 and 2.14 respectively. The number of particles n can be converted to the (average) test sample mass m by

$$m = \frac{n}{N} \qquad\qquad \text{Eq. 2.26}$$

where
N=the number of particles per unit mass

N is a property of the laboratory sample which can only be increased by grinding.

Because the characterization of the particle size (distribution, form and dimensions!) of real samples is extremely difficult, no exact correlation between experimental data and a theoretical description can be accomplished.

The decrease of the sampling error for the determination of Ag in silicate rock material with increasing grinding time and different mills respectively was studied from Eames and Matousek [96] and are visualized in Figs. 2.20 a,b.

From the data reported by Nakamura et al. [130] for calcium carbonate scale samples, homogeneity constants of 8 $mg^{1/2}$ for a mean particle size of 1 µm, 12 $mg^{1/2}$ for 10 µm, but 35 $mg^{1/2}$ for 100 µm can be calculated; consequently they recommend for materials that may include the analyte elements in minor crystalline inclusions a diameter for most of the particles smaller than 10 µm. In [131] the authors studied the influence of the grinding time (Ishikawa AGA ginder, agate mortar and pestle) on sampling error and concluded: "Based on data obtained so far, the following grinding conditions are considered to be applicable: a 3.0 mg rock sample is ground for 20 min and the particle size range of the powder is 0.3-25 µm, 12% of the particles being larger than 10 µm. According to Wilson ([112], see Sect. 2.3.1.1), the size distribution is suitable for obtaining sufficiently accurate results".

Fecher and Malcherek [132] documented the results of Cd determinations obtained routinely over several weeks for Wholemeal Flour (BCR CRM 189). The original reference material show a RSD between 12-29%. Additional grinding reduces the sampling error significantly so that a RSD of approx. 5-10% was achieved (n~10). The authors pointed out that, for the original material, despite the worse precision, the average content generally shows a good agreement with the certified content (0.071 mg/kg).

Pauwels et al. [133, 134] have investigated the homogenizing effect of a cod fish material for various trace elements during production as a reference material. Figure 2.21 show the accentuated decrease of the distribution width for Hg of test samples results after each separate production step.

Kramer et al. [135] documented experiments which give an impression of the correlation between particle size and sampling imprecision. The particle size

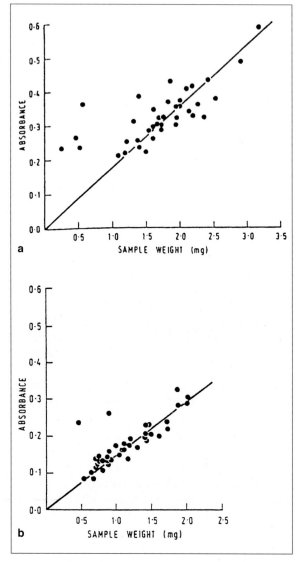

Fig. 2.20. Plots of absorbance vs test sample mass for a silicate rock sample after ginding a sample/graphite mixture in a Grindex mixer/mill; **a** grinding for 5 min ; **b** grinding for 10 min (from: [96], with permission).

of Rye Grass (BCR CRM 281) was reduced by about one order of magnitude by additional impact milling, e.g. the mode of the particle size distribution at 100 μm is shifted to 10 μm. Accordingly the standard deviation of direct solid sampling results (0.1–1.0 mg) was reduced by factors of about two (Cd, Zn) and of three (Pb).

In [136, 110] the effect of the advanced "air-jet" milling technique on Urban Dust (IAEA-396) is documented. The mode of particle size is reduced from 20 μm to 5 μm. The respective median values (10 μm to 3 μm) cannot be com-

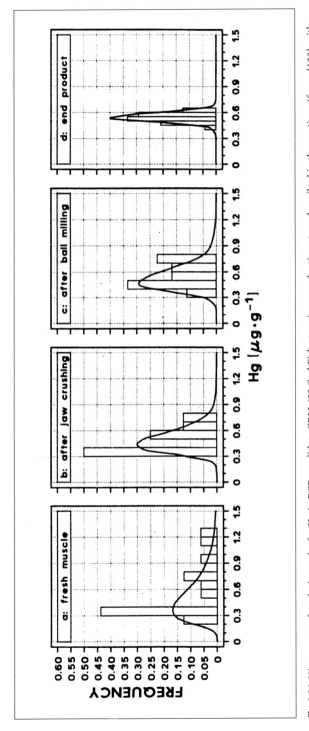

Fig. 2.21. Histograms of analysis results for Hg in BCR candidate CRM 422 Cod Fish at various production steps described in the captions (from: [133], with permission).

pared, because the material was sieved through a mesh size of 70×10^{-6} m, while for jet-milling the original material was used. Thus, the homogeneity constant was lowered from 9 to 3 mg$^{1/2}$.

Astonishing results are described in [137]: For a number of CRMs, the precision in Cd results of the final ground sample are found to be significantly larger than for the original materials. For Coal Fly Ash (NIST SRM 1633), RSDs of 9.6% before and of 25% after the final grinding of the sample were found. For Orchard Leaves (NIST SRM 1571), an RSD of 33% was found for test samples of 1-3 mg. However, this SRM was investigated in some other investigations with direct solid sampling. Approximately 0.15 mg has yield a RSD of 14%, which decreases to 9% for this low test sample size after further grinding [8]. For test sample of 1 mg a RSD of 7.7% was determined. Other results have confirmed that for Cd this material can by characterized by a homogeneity constant of about 5-7 mg$^{1/2}$ [138, 139, 140]. Obviously the increased analytical imprecision occurring with the experiments in [137] are not due to sample heterogeneity but must be due to other uncontrolled effects.

After homogenizing absorption coal using a mortar and pestle, the sampling error for Au determination decreases drastically [141]. The analysis of the original coal scraps gave a relative homogeneity constant of 16 mg$^{1/2}$, whereas the ground sample could be characterized as being 2.2 mg$^{1/2}$.

Because of the small effective sample mass pipetted from a *slurry* (compare Eq. 5.3) the precision may be influenced by sample heterogeneity as well. Several investigations showed the dependence of particle size on slurry results [142, 143, 144]. The experiments and the requirements are discussed in Sect. 5.2.1.1.

2.4
Accuracy Assessment of Solid Sampling Analysis

2.4.1
Errors and Accuracy

The flow chart diagram in Figs. 2.22 a, b display the steps in the measuring process of solid sampling using different calibration methods. Imperfections that give rise to an error appear in all stages of the analytical process, namely in the sample preparation, the dosing and quantification of test samples and in the instrumental analysis, all in calibration and in unknown sample measurement as well. Furthermore, in solid sampling analyis the error due to sample heterogeneity must be considered, as already discussed in Sect. 2.3.

Although it is the main task of the analyst to identify and to exclude all sources of error, an analytical procedure is never free of effects which may restrict the accuracy of analytical results.

In this section, the influence of effects on the accuray of solid sampling results is discussed that customaryly are designated as *errors*. The "error concept" can be used only for the *qualitative* characterization of imperfections in the mea-

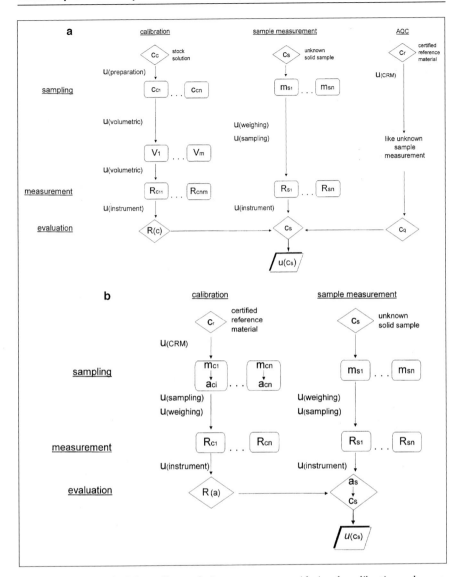

Fig. 2.22. Flow chart of solid sampling analysis process steps considering the calibration, unknown sample and AQC (analytical quality control) measurement, the uncertainty components (u "effect") in the measurement steps are indicated (compare Sect. 2.4.2.3);
a calibration based on reference solutions, AQC by the analysis of a CRM; **b** calibration based directly on a CRM (inherent AQC).

surement. For the *quantification* of a lack in accuracy that follows from various *effects* in the analytical process the *uncertainty* in the result of the measurement must be derived (see discussion in Sect. 2.4.2.3).

According to [145] the *accuracy* [7] of a measurement describes in general the closeness of agreement between the result of the performed measurement (test results) and the "accepted reference value" (the best estimate of the "true value"). Because the accuracy can be affected by two kinds of errors (random and systematic) that influence the test result in different ways (scatter and bias) it must be distinguished between the terms "trueness" and "precision" [145], where

- *trueness* is defined as "the closeness of agreement between the average value obtained from a large series of test results and an accepted reference value", and
- *precision* is defined as "the closeness of agreement between independent test results obtained under prescribed conditions".

The point of issue for analytical results generally is the *trueness* of the determined analyte content [147]. In the first place it is influenced by *systematic* effects inherent of the applied analytical procedure which measure can be expressed in the terms of bias.

The detection of systematic effects in the analytical method is not possible by the use of statistical tools but only by the analytical methodology. If an unacceptable bias to the accepted reference value appears, often a carefully performed investigation is necessary to identify the source of systematic effect (or effects) which is (are) responsible for the lack of trueness. If a systematic effect can not be excluded from the analytical method, a unbiased result may be achieved by a *correction factor* which can be estimated e.g. by the analysis of a suitable CRM (see Sect. 2.4.2.1).

The *precision* is a measure of the *scatter of results* within a set of test samples and is expressed in terms of imprecision, mostly computed as the standard deviation of test results. The precision is affected by the *random effects* which are inherent to the analytical process. If the number of test results is large, the trueness of the mean results is only made worse insignificantly by random effects, however from a limited set of test samples, the average randomly shows a more or less significant deviation from the accepted value ("run bias"). Consequently, the realization of the relative large number of replicates (typically 5-10 solid test samples) must be judged as an advantage of direct solid sampling compared to the decomposition methods (typically 2-5 decomposition solutions) (see discussion in Sect. 2.4.2.3).

---

7) The existing ISO 5725 (2nd edn., 1986) gives a different definition of accuracy which corresponds to the difinition of the term "trueness" given below. However, the proposed definition enables better differentiation between the results of random and systematic effects. Furthermore, the term trueness is used in other guides from ISO (e.g. [146]) that are relevant for this topic. For this reasons, the above defined terms are used here.

Up to now, mostly the influence of random effects has been subject of the statistical treatment of measurement data, e.g. if a "confidence interval" is estimated. Even recognized systematic effects are not considered in this concept, although these can contribute a larger deviation from the "true value" than random effects. It is the improvement in the new concept of the "combined uncertainty" from ISO [60] that both kinds of effects can be considered in an estimate of the final analytical result. The required procedures and statistical tools are described and applied in Sect. 2.4.2.3.

Attention must be paid to the fact that each effect (compare Figs. 2.22 a, b) may contribute a *systematic and a random component* (e.g. contamination of test samples by different analyte amounts may cause a bias and contribute to the scatter of test sample results). However, one component usually dominates, so in the sections below the errors due to systematic and those due to random effects are discussed separately.

2.4.1.1
Systematic Effects

Because atomic spectrometry techniques do not allow an absolute measurement, a permanent responsibility of the analyst is the control of the *calibration procedure* (i.e. quantification of the spectrometer response in terms of analyte mass or content). This is the main part of method development and requires a high analytical qualification. It is still the key point for the quality assurance of solid sampling results – i.e. in first place the assessment of trueness.

In addition to the calibration of the spectrometer, the exact calibration of all instruments which are used for the determination of "input values" prior to the analytical process must be ensured. For the solid sampling technique this concerns in particular the *microbalance* used for quantification of the test samples. Systematic effects can be attributed to the uncertainty of the internal calibration weight and in difference of the densities of internal weight and sample material. As Hinds [148, 149] has shown, these effects can practically neglected in trace element analysis. While the tolerances of calibration weights can give bias of max. 5 ppm of the nominal value, the influence of buoyancy can be estimated to be <<1% (e.g. 0.12% for water, 0.04% for soil). However, attention must be paid to the directions given in the manufacturer's manual with regard to the handling and the setting conditions (e.g. temperature range, the "levelling with the plumbline").

This is also valid for *pipettes* if liquid aliquotes are taken from reference solutions for calibration. Micropipettes are comparatively inaccurate instruments. If both – calibrant and sample – are pipetted with an identical instrument (solution analysis) a bias in the calibration of the pipetted volume is compensated. However, because the quantification of solid test samples is performed with a microbalance, the pipetting error influences the final analytical result directly. This systematic error can be avoided if the accuracy of the pipette is checked by weighing aliquots of the reference solution with the balance, whose accuracy is generally two orders of magnitude better [150].

Since the content of the analyte normally has to be related to the *dry sample mass c(d.m.)*, the *moisture content* c_w (related to the total sample mass, mostly stated relatively in %) must be determined and taken into consideration [14, 151], particularly so if the calibration is based on a reference solution. The determined content related to the equilibrium wet mass c(w.m.) must be corrected by the moisture content of the sample according to

$$c(d.m.) = \frac{c(w.m.)}{\left(1 - \frac{c_w(rel.)}{100\%}\right)}$$

Eq. 2.27

Alternatively, a correction factor ($f_w > 1$) can be determined to multiply c(w.m.) with, and this is given by the relation of the sample mass before drying and after drying to constant mass of the sample [159].

Systematic negative bias to reference values, frequently documented for analysis based on reference solutions (i.e. solid sample results are too low, e.g. [152, 9, 8]), may be caused by the moisture content, because no dry weight correction was performed.

Using a solid material as the calibrant, this effect will be partly compensated for, assuming that the moisture contents of calibrant and sample material are approximately identical. However, this can not be taken as guaranteed as Langmyhr and Aadalen [153] have shown with coal samples; while Coal (NIST SRM 1632a) contained only 0.43% water, an Australian coal contained 6.55%. In any case, reporting analytical results, it should be stated if a correction for the moisture content is performed [e.g. 14] or not [e.g. 154].

The influence of different *matrix compositions* of calibrant and sample material to the vaporization/atomization mechanism in the graphite tube that can result in a systematic error of the analytical result have been documented and discussed by many investigators over the years and are discussed and documented according to the different solid sampling techniques in the respective chapters of this book (see Sects. 3.1, 4.6, 5.4).

In trace and even more in ultratrace analysis, a bias can arise from secondary *contamination*, if test samples are contaminated by amounts of exogene analyte. While this can be a serious problem with decomposition procedures (e.g. by blank values of reagents, vessel surfaces), for direct solid sampling it can be almost disregarded as a source of bias. A constantly recurring contamination of test samples is extremly unprobable because of the large number of replicates, the short and limited sample handling and the very small sampling contact areas. Only if extreme low contents of ubiquitous elements present are to be determined (e.g. aluminium [155, 156]) even with the solid sampling technique, the analyst must be aware of contamination.

Example

In fifteen years of solid sampling experience, the author has observed only one case of a bias by contamination: for the determination of Cd the sample boat had been placed on a small block of red plastic for sample loading. At that time, the colour was originated by a large amount of cadmium oxide in the plastic raw material (up to several percent of Cd!). As the sample boats had not competely cooled down, they "soaked up" a large (trace) amount of Cd. It took two days to identify this source of contamination! Only later did it become generally known that the additives to plastic materials can lead to problems with trace element determination (e.g. with the use of yellow pipette tips).

Loss of analyte can cause a bias as well. In electrothermal techniques, analyte loss can occur if the temperature of the sample pretreatment step is chosen too high with respect to the volatilization of the analyte or its compounds (e.g. Cd, As). The method for recognizing such an effect is recovery experiments visualized by pretreatment curves (see Sect. 2.5.3).

Sudden *changes of instrument sensitivity* caused by instability of the lamp, the furnace, the plasma or the electronics may cause a bias. By repetitive measurements of a reference material, such effects can be recognized (see Sect. 2.4.2.2).

Additional sources for a bias must be reckoned with *slurry sampling* since there is a *sedimentation error*. Contamination by *analyte blanks* of the dilutent and the reagents must be considered, as it is with conventional decomposition techniques [157] (see Sect. 5.2.2).

2.4.1.2
Random Effects

With the direct solid sampling method, we can only identify three main random effects which contribute significantly to the imprecision of the results from the analysis of test samples. These are the *sample heterogeneity* (which cause the sampling error, discussed in Sect. 2.3.1), the *baseline scatter* ("instrument error") and the *readability of the balance*.

Each random effect provides a contribution to the overall experimental standard deviation which can be calculated from a set of test sample measurements. So for direct solid sampling the *relative standard deviation RSD* (measurement imprecision) can reasonably be expressed as

$$RSD^2 = S_H^2 + S_R^2 + S_W^2 \qquad \text{Eq. 2.28}$$

where
S_H=relative "sampling" standard deviation
S_R=relative "instrument response" standard deviation
S_W=relative "weighing" standard deviation

The imprecision of weighing and the instrumental response are of significant extent only under unfavorable conditions:

With the use of a modern "full" microbalance, the standard deviation for a number of readings is less than 1% (test sample mass larger than 0.1 mg). It may become of importance if the sample mass is extreme low or a "half" microbalance or an older unstable one is used.

The precision of response values (for a constant amount of analyte) of modern spectrometers is $\ll 3\%$ including the instability of furnace or plasma. Only if the determination is performed at the limit of detection, may the baseline scatter due to noise from the lamp, the photomultiplier or the electronics contribute a larger instrumental error.

Example

Figures 2.14 a, b ,c shows three series of Zn determinations in a cod muscle sample. The weighing imprecision can be ignored as its value was <1% in all three series. The instrumental error, determined by measurements without sample introduction, however contribute a significant value to the total error with low test sample mass (analysis performed with the unsensitive line 307.6 nm). Table 2.8 show the data evaluated from the measurements and and calculated applying Eq. 2.28. In Fig. 2.23, the calculated sampling standard deviation is plotted against the test sample mass, the fitting gives a homogeneity constant of 3.7 mg$^{1/2}$. This diagram may be regarded as an example for the validity of the concept of the homogeneity constant.

The considerations above show that the direct solid sampling technique is under "statistical control" [158] – in contrast to analytical techniques with prior sample decomposition, where the sources of random effects are difficult to determine or are even unknown. It is obvious that the precision achieved with the solid sampling technique is mostly dominated by the sampling error. This property allows us to determine the degree of heterogeneity of a sample.

With the *slurry* method additionally the *volumetric error* S_V and *slurry preparation effects* must be considered (see Sect. 5.2.2.2).

Table 2.8 Experimental and evaluated data from a homogeneity determination experiment for Zn in a cod fish sample (see Fig. 2.23)

	experimental				calculated	
m (mg)	n –	RSD (%)	S_R (%)	S_H (%)	H_{Zn} (mg$^{1/2}$)
1.90	19	3.10	0.60	3.00	4.10
1.00	30	3.90	1.20	3.70	3.70
0.35	19	7.20	3.90	6.00	3.60

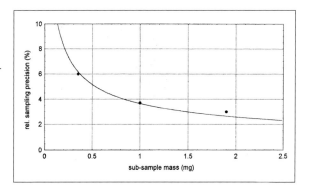

Fig. 2.23. Plot of the relative imprecision due to the sampling error for three test sample masses from the Zn determination in a cod muscle sample (see Fig. 2.14), and the fitting using the relation of Eq. 2.17.

2.4.2
Analytical Quality Control in Solid Sampling

The key issues of analytical quality control (AQC) is the assurance of getting true results [8]. This demand should be discussed for two different tasks as there is the assurance of trueness with the actually performed analysis and the repeatability over the whole analytical session.

2.4.2.1
Assessment of Trueness

Although a deliberate method development may have been performed, for most laboratories it is difficult to check the trueness of the analytical result by other – independent – methods or the continious participation in colaborative or interlaboratory experiments. However, since a large variety of CRMs are available, the situation has changed for the better. The use of CRMs has become an easy, reliable and accepted tool for quality assurance at the laboratory stage.

A certified reference material (CRM) can be used with th solid sampling method in two ways for the purpose of trueness assessment:

– The instrument resonse can be directly calibrated by the CRM. Thus the trueness of the analysis of the unknown samples is based directly on the quality of the certified content (if the CRM is chosen adequately with respect to matrix and analyte content, see Sect. 2.2.1.2). With this approach, no separate AQC measurement is required. Apart from the efforts which are saved, the uncertainty in the determined content may be smaller (compare Sect. 2.4.2.4). *So, with the solid sampling method, a direct calibration using a suitable (appropriate) CRM suggests itself, because AQC is "built-in" the analytical procedure.*

---

[8] Additional demands of AQC such as representativity (field or bulk sampling procedure) and tracebility (e.g. documentation) are not discussed here since the respective principles are identical to these applied with conventional methods.

– If the determination of the analyte content of the unknown samples is based on a reference solution, the CRM must be analyzed under the fixed analytical and instrumental conditions and the *result of the AQC measurement c_q* is to compare with its *certified value c_r*. The presence or absence of statistical *evidence for bias* can be estimated by a statistical test.

The bias is given by the difference of the certified and the determined content of the CRM. The *detection of no bias* may be regarded as evident, if the AQC result lies in the interval given by the *standard deviation s_D*, which is "associated with the measurement process" according to [146]

$$-a_2 - 2\ s_D \leq \bar{c}_q - c_r \leq 2\ s_D + a_1 \qquad\qquad \text{Eq. 2.29}$$

(a_1, a_2 are adjustment values which represent an acceptable limit for a bias in the analysis that is not statistically defined, but chosen by the user according to economic or technical limitations or stipulations; preferably the analytical method should be as accurate that these values can be chosen to zero.)

An estimation for s_D is given as the combination of the *between-laboratory standard deviation $s_L(c_{r,i})$* of the certified content and the *short-term standard deviation $s(c_{q,i})$* of test sample from the CRM with the applied method according to [146]

$$s_D = \sqrt{s_L^2(c_{r,i}) + \frac{s^2(c_{q,i})}{n_q}} \qquad\qquad \text{Eq. 2.30}$$

The assumption that the between-laboratory standard deviation is large compared to the short-term standard deviation [159], may not be realistic in direct solid sampling, because of the strong influence of the sampling error (heterogeneity) to the latter. Consequently the second term of Eq. 2.30 can only be neglected if the number of replicate measurements is large (e.g. $n_q \gg 10$).

If the conditions of Eq. 2.29 are fulfilled, there is no reason to believe – at the approx. 95% level of confidence – that the detected bias is significant. In other words, the observed difference between the determined and the certified content could be due to the variability of the measurement process.

Example (Evaluation of test data in Table 2.11)

For the AQC measurement of the BCR CRM 281 Rye Grass a Pb content of 2.51 mg/kg and a standard deviation of 0.39 mg/kg is determined by the 22 test samples using the calibration established by reference solutions. These values should be compared with the certified content of 2.38 mg/kg and the standard deviation of 0.21 mg/kg. It can be directly recognized that the bias of 0.13 mg/kg is not significant, which can be verified by Eq. 2.29, estimating $2s_D(c_{r,i}) = 0.46$ mg/kg ($\approx 2s_L(c_{r,i})$) (compare Eq. 2.30).

With another approach which is still often used, one tests whether or not the bias is smaller than the sum of the *confidence intervals CI* of the CRM-value and the result of the AQC-measurement respectively [160, 161]

$$\left|\overline{c}_q - c_r\right| \le CI(c_r) + CI(c_q)$$

Eq. 2.31

In contrast to Eq. 2.30, the "reference term" here will mostly be smaller than the "laboratory term". So, the paradox situation may appear that for two sets with identical means and standard deviations, the one at which a lower number of test samples is performed can be accepted, while the other must be rejected (or corrected) because with the larger number of replicates $CI(c_q)$ has been "shrunk", or, moreover, a laboratory result must be rejected, although its value is identical to an accepted result used for certification (compare Eqs. 2.38 and 2.39 and Fig. 2.26).[9]

Example (Evaluation of test data in Table 2.11)

The confidence interval for Pb in Rye Grass is certified as 2.38±0.11 mg/kg. The confidence interval (Eq. 2.39) for the determined content of this CRM is estimated as 2.51±0.17 mg/kg (neglecting the small contribution from the calibration based on reference solutions). Although with the addition of both confidence intervals (Eq. 2.31) it is also verified that a bias is not evident (0.13 mg/kg<0.28 mg/kg), the range of "accepted trueness" for the analysis is almost half compared that estimated by Eq. 2.29 (compare example above).

For overcoming this problem, NIST proposed adding the constant B to the right-hand side of Eq.2.31. "The value, B, is the user´s estimate of the magnitude of any uncorrected biases inherent in his measurement and is based on experience and professional judgement" [162]. Obviously, using the constant B that will be determined by most users by the analysis of CRMs, the procedure of trueness assessment is equivalent to the determination of a correction factor f_q (see below Eqs. 2.32 and 2.33).

If the test of Eq. 2.29 points to the appearance of a systematic error (bias), a *correction factor f_q* may be estimated from the AQC measurement according to

$$f_q = \frac{c_r}{c_q}$$

Eq. 2.32

The multiplication of f_q with the determined content gives the *corrected content of the unknown sample* which is then the best estimate for the true value

$$\overline{c}_{s,cor} = f_q \overline{c}_s$$

Eq. 2.33

---

9) This is particularily true if the certified confidence range is misinterpreted as the interval in which the determined value c_q of the CRM must lie.

Example (Evaluation of test data in Table 2.11)

From the certified content for Pb in BCR CRM 281 Rye Grass (2.38 mg/kg) and the determined content based on a calibration established with reference solutions (2.51 mg/kg) a correction factor of $f_q=0.95$ is calculated. However, because the bias is not significant (see examples above), no correction of the determined content of the unknown samples is required.

It should be stressed that the decision as to if such a correction is adequate cannot be met by a statistical test, but only by judgement based on profound analytical knowledge! This is the task of the analyst prior to the statistical treatment and involves mainly the choice of an appropriate CRM (comparable to the unknown sample with respect to matrix, analyte content, homogeneity), and the instrumental settings (e.g. the analytical line). Performing measurements and tests using a CRM the final result can be regarded as being "true". The trueness, however, is based on the quality of the certified content (i.e. the accepted reference value) [147].

Calibration based on a standard solution and an additional AQC-process may be recommended, either, if no appropriate CRM (with respect to matrix and/or analyte content) is available, or an otherwise suitable CRM show micro-heterogeneity. With such a material the establishment of a reliable calibration curve would require a very large number of calibration measurements or – in the case of a strong nugget effect – is even impossible. However the detection of a matrix effect, i.e. a systematic effect giving rise to a bias, should be possible with such a material if the number of test samples is sufficiently large and if necessary an appropriate outlier treatment is applied (see Sect. 2.4.3).

If a *laboratory test material* is used for calibration (see below) a reliable check for trueness using a CRM should be performed at least with the method development procedure.

2.4.2.2
Stability of Analytical Conditions

Beside the efforts connected with the method development, the analyst must guarantee stable analytical conditions during the entire analytical session. For this attempt at AQC, CRMs can also be used, using repeated analysis of these materials.

For visualizing the effects which indicate whether the analysis is stable or whether the measurand repeatedly reaches an unacceptable value, *control charts* may be used. Usually, the test sample measurements are recorded in a Shewart- or $\bar{x}$-chart, in which *control lines* are indicating the status of analysis. The purpose of these lines is explained by their names and their statistical implication [163, 164]:

- the *central or the target value (C$_r$-)line* gives the reference content,
- the *warning lines* indicate the range in between which a required portion of the repeated results should lie. Usually the control lines are chosen so that 95% of the results can be expected between the upper and lower warning lines, i.e.

$$C_r \pm \frac{2\sigma(c_{r,i})}{\sqrt{n_{qg}}}$$
Eq. 2.34

- The *action lines* are usually chosen so that between them 99.7% of the results can be expected. Virtually no result should be lie outside the region between the upper and lower action lines, i.e.

$$C_r \pm \frac{3\sigma(c_{r,i})}{\sqrt{n_{qg}}}$$
Eq. 2.35

where
$\sigma(c_{r,i})$=*standard deviation of the test sample population*
n_{qg}=*number of AQC test sample measurements of one group*

Usually each point of the $\bar{x}$-chart represents the mean of n_{qg} control measurements (subgroups). The additional use of a chart for the precision may give important information, e.g. the standard deviation or the range of results for the subgroups can be recorded simultaneously (s-chart or R-chart).

The *establishment of values for the target and control lines* depends on the procedures of calibration, the materials used, and the data available:

It is obvious that if a *CRM is used for the control of repeatability*, the target value must be the certified content, (particularly, if this material has been used for calibration).

The standard deviation for the establishment of the control lines can be estimated by the long term standard deviation $s_L(c_{q,i})$ achieved with the solid sampling method by the analysis of the CRM (within-laboratory) over a longer period, based on a large number of test samples. Additionally it must be considered that this value may be dominated by the sampling error, so it also depends on the (average) test sample mass used for its determination, so – if a significantly different test sample mass is used in the actual analysis – it must be corrected by Eq. 2.17.

Example (Evaluation of test data in Table 2.11)

The test sample measurements of the BCR CRM 281 Rye Grass are used for the control of stability (calibration established with reference solutions). The measurements are performed in groups of three (n_{qg}=3) over several hours (Table 2.11 lists the values in ascending order whereas here these have been combined in chonological order). For the establishment of the control values (Eqs. 2.34

and 2.35) a long term standard deviation of $s_L(c_{q,i})$=0.22 mg/kg is used as an esti-mate of $\sigma(c_{r,i})$ (this value is significantly smaller than the standard deviation $s(c_{q,i})$=0.39 mg/kg from the test data). Thus for groups of three test samples ("AQC samples") the warning lines are calculated as ±0.25 mg/kg and the action lines as ±0.38 mg/kg on every side of the mean 2.51 mg/kg. From Fig. 2.24 a, it can be rec-ognized that the means of AQC subgroup 4 and 5 lie outside the "warning band" (not entered in Fig. 2.24 a), however, they are still inside the "action band". The "s-chart" (Fig. 2.24 b) shows that the respective standard deviations are nearly twice those of the others, indicating that in both cases one extreme value is responsible for the increased mean (compare Table 2.11, CRM value no. 5 and no. 6).

At first glance, it would seem to be reasonable to use the standard deviation of the certified content $s(c_{r,i})$ to establish the control limits. However, the standard deviation of the certified content does not reflect the distribution of the results of the applied method, especially not the uncertainty due to analyte heterogeneity using microgram test samples. So, using the certified standard deviation for the calculation of the control limits, values may lie outside these limits without indi-cating a situation out of control. On the other hand, the use of the certified stan-dard deviation may be meaningful if the long term standard deviation for the paticular element of the CRM is not available. If the plotted points in the chart represent the mean of 3 or more replicate solid test sample measurements, the respective standard deviation of the mean (Eq. 2.38) should be smaller than the certified standard deviation, i.e. outlying points would be improbable. If the con-trol lines are calculated using $s(c_{r,i})$ not divided by the square root of n_{qg}, these may also indicate that the AQC measurement is in the "accepted" range.

Fig. 2.24. a Shewart chart from the control measure-ments (AQC) for the deter-mination of Pb; the target value und action lines *(straight lines)* are establis-hed from experimental values, each mean (point) based on three test samples of the CRM Rye Grass (see test example Table 2.12). Additionally the certified content and action lines cal-culated from the certified standard deviation are ente-red *(broken lines);* **b** corre-sponding s-chart.

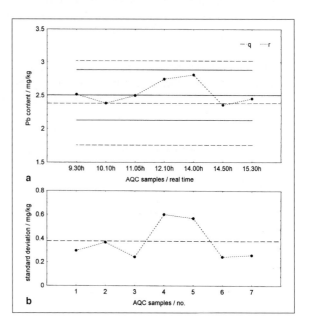

Example (Evaluation of test data in Table 2.11)

Additionally in Fig. 2.24 a action lines calculated from the reference standard deviation $3s(c_{r,i})=0.63$ mg/kg on each side of the certified content 2.38 mg/kg are given. Because the standard deviation for Pb in this material (8.8%) is comparatively large, in this case a broader band is outlined.

The check for stability can be performed by the repeated analysis of a *reference solution* as well (whether the calibration is established by a reference solution or by a CRM). This can be advantageous because pipetting of a solution is faster than the sampling procedure of a solid material. The target value and the standard deviation used for the control lines should be calculated from the subgroup measurements which are performed during the entire analytical session. Thus a useful control for no drift- or step-effects may be performed. However, no information about a bias (trueness) are available, so an additional independent assessment of trueness may be required if the calibration is based on a reference solution (see Sect. 2.4.2.1).

If a *non certified reference material (test material)* is used for the repeatability test, the same procedure as with the use of a standard solution can be performed. Such materials are often available from interlaboratory comparison campaigns or are laboratory-made [120, 128]. Reasons for the use of a test material can be given, if no suitable CRM is available, only a small amount of the CRM is at hand in the laboratory, or costs must be reduced.

Even if the results are all in between the inner range, the tendency of consecutive measurements may give hints that the analysis may run into an out-of-control situation or at least has reached an unsatisfactory analytical situation. According to the ISO-standard, pattern tests should be performed. "These tests should be viewed as simply practical rules for action whenever the presence of assignable causes is indicated. An indication of any of the conditions stipulated in these tests is an indication of the presence of assignable causes of variation that must be diagnosed and corrected" [164].

If a drift or a sudden shift (step) can be read from the chart, it may be possible to correct the analytical results for the unknown samples by a factor which can be evaluated e.g. by a trend analysis for the plotted points. Such a procedure is equivalent to a continuous recalibration for correcting a change in the instrument response e.g. due to lamp, furnace or electronic instabilities. In combination with a conscientious auditing of the analytical conditions, there is no reason to reject such a procedure[9]. For this purpose, the control measurements should be recorded, not just over the number of observations, but over the real

---

9) From the author's experience, such a procedure is more or less common in laboratories which carry out routine work (for all kinds of analytical methods), although these techniques of "internal AQC" are usually not revealed.

time of analysis (compare Fig. 2.24 a). Thus the character and the course of the effect comes out more pronouncedly.

It needs some experience to read the control charts pattern and a profound knowledge of the solid sampling method is required to draw adequate conclusions during or after an analytical session. However, applying the described methods of AQC, it guarantees that accurate results leave the laboratory.

2.4.2.3
Estimation of Uncertainty in the Analytical Result

The advanced concept of estimating the uncertainty in measurements due to different effects has not, up till now, been applied consistently in analytical atomic spectrometry, therefore it should be outlined in general before specific derivations for the solid sampling method are carried out.

The ISO-concept *"Guide to the expression of uncertainty in measurement"* [60] can be characterized by the following key points:

- Imperfection in a measurement gives rise to an error in the measurement result. However, because a (remaining) error cannot be quantified (if it were known, the measurement result could be corrected for this error!), instead the *uncertainty in the measurement* result due to uncontrolled effects in the measurement procedure should be estimated. The guide gives the statistical tools to consider all *standard uncertainties* of a complex measurement process in a *combined uncertainty.*
- Uncertainties due to *repeated measurements* are evaluated by statistical analysis (*type A evaluation*). This kind of uncertainty only is estimated with the classical approach of a "confidence interval" appying the standard deviation of the mean as the standard uncertainty.
- The standard uncertainties which can only be estimated by means *other than statistical analysis (type B evaluation)*, may be obtained by scientific judgement based on available information on the behavior and properties of relevant materials, instruments or methods; for that purpose often *a priori* distributions (e.g. rectangular) for the "input variables" must be used.
- Uncertainties due to *random effects* (giving rise to random errors) can be achieved by typ A evaluation if own measurements are performed, or must be estimated due to typ B evaluation, if the range and/or the distribution of the measurand can not be experimentally obtained.
- The uncertainty due to a *systematic effects* (giving rise to a systematic error) can be obtained by appying a *correction factor* f to compensate the measurement result for the respective error (for recognized systematic effects f will be unity) and than evaluating the uncertainty that is connected with this factor.
- All evaluated standard uncertainties are combined by addition of the respective square values (variances). The *combined uncertainty* is given by the

square root of this value. If the input variables are correlated, covariance terms must be considered with the evaluation of the combined uncertainty.
- The *extended uncertainty* specifies an interval around the measurement result that may be expected to encompass a large, specified fraction of the measurand value (that here is the mean content). For that purpose a *coverage faktor* k_p must be chosen which is multiplied with the combined uncertainty and thus provides an interval corresponding to an approximate level of confidence (e.g. 95 percent).

Obviously, with this evaluation of standard uncertainties and the treatment in their combination it is no longer distinguished between uncertainties of type A and of type B or due to random and systematic effects as it is in the classical "error concept" that is not qualified for the evaluation of *only one value* that includes the overall uncertainty in the entire measurement process. In the consequence important sources of uncertainty couldn't be considered and thus uncertainties generally are more or less underestimated.

This advanced concept released from ISO is concreted and exemplarily applied to some analytical measurements from Eurachem [165] and for the production and certification of reference materials from NIST [166]. Williams [167] gives a comprehensive introduction for the principles of this concept in analytical chemistry.

Generally it is the aim of an analysis to determine the *"true"* content C_s for the laboratory sample (measurand of the analysis). From the performed analysis the *content* c_s is derived which, however, is only an imperfect estimate for C_s. This result and its relation to the "true content" can be conveniently expressed using the *extended uncertainty* $U(c_s)$

$$C_s = c_s \pm U(c_s)$$

Eq. 2.36a

Alternatively, this interval is expressed as

$$c_s - U(c_s) \leq C_s \leq c_s + U(c_s)$$

Eq. 2.36b

The uncertainty limits given by $U(c_s)$ are set by the *combined uncertainty* $u(c_s)$ according to:

$$U(c_s) = k_p u(c_s)$$

Eq. 2.37

The size of the range is set by the *coverage factor* k_p. The value of k_p must be chosen on the basis of the level of confidence required and will be in the range of 2 to 3. The expressions in Eqs. 2.36 a and b mean that the best estimate for the unknown content C_s is given by c_s, and that the defined interval may be expected to encompass a required fraction of the distribution of the determinable contents.

In most cases, the best estimate for C_s is the arithmetic mean $\bar{c}_s$ of n_s independent test sample results. Then the *standard uncertainty* $u_s(c_s)$ in the estimated value can be expressed by the *standard deviation of the mean content* $\bar{c}_s$ so [11]

$$u_s\left(c_s\right)=s\left(\bar{c}_s\right)=\frac{s\left(c_{s,i}\right)}{\sqrt{n_s}} \qquad\qquad \text{Eq. 2.38}$$

(The basic statistics used in this chapter are listed in Table 2.9)

If the measurement result can be supposed to be normally distributed, the *coverage factor* k_p can be chosen as the *Student factor t* from the "t-table" according to the required level of confidence ("coverage probability", usually 95% chosen in analytical chemistry) and the *degree of freedom df* of the measurement (for a single mean value df=n-1 and df=n-2 for a linear regression).

Table 2.10 shows values of t for different level of confidence. For df=n-1 the values of the term t/n^{-2} which is the factor to multiply the experimental standard deviation with (see Eqs. 2.37, 2.38 and 2.39) are also tabulated, from these, the strong influence on the interval width with low numbers of test measurements can be recognized.

However, the choice of k_p from the t-table has the precondition that the distribution of the measurant values is *approximately normal*. As discussed below, this precondition may be fulfilled for solid sampling analysis. Type B uncertainties in other "input quantities", often do not have to be considered. Generally, it can be supposed that the analyte content in the test samples follows a normal distribution. Only in the case of a nugget effect may more or less skewed distributions appear. Moreover, because the number of test samples is large (compared to the analysis of decomposed samples), the distribution of the meaasurant, i.e. the measurement results $\bar{c}_s$ approaches normallity, whatever may be the probability distribution of the test sample results $c_{s,i}$ (Central Limit Theorem).

If only the sample measurement must be considered, the interval given by the extended uncertainty $U(c_s)$ specified by Eqs. 2.36 is identical with the classical approach of the *confidence interval CI* that is given by

$$CI=\bar{c}_s\pm tu_s\left(c_s\right) \qquad\qquad \text{Eq. 2.39}$$

Example

An RSD of 10% is typical for the precision of solid sampling measurements. If the number of replicate test measurements is around 10, the 95%-uncertainty

---

[11] Often only the precision of the analytical measurements, mostly expressed as the RSD, has been documented in the literature. However, without information about the number of test measurements, the influence of random effects to the uncertainty in the analytical result cannot be estimated.

Table 2.9 List of terms and statistics used for the estimation of uncertainty in the analytical result

Terms

a	mass of analyte in a test sample
b	slope of a regression line
b_0	intercept of a regression line
c	content, estimated from an anlytical measurement (mass fraction or concentration of the analyte)
Cl	confidence interval
f	correction factor for systematic effects
k_p	coverage factor
m	mass of test samples used for analytical measurement
n	number or replicate measurements (number of test samples)
R	instrument response value (e.g. absorbance, counts)
R'	mass specific response value (R devided by m)
R*	corrected response value for the average test sample mass
$\hat{R}$	response value on the estimated regression line
$s(x_i)$	standard deviation of test samples
$s(\bar{x})$	standard deviation of a mean
$s(R_i / x)$	standard deviation R on x of a regression estimate
S_{xx}	sum of squares about the mean
t	Students factor
u(x)	combined standard uncertainty in the measurand
$u_p(x)$	standard uncertainty in the process step p (see below)
$U(c_s)$	extended uncertainty in the estimated content of the unknown sample
V	volume of a solution or a slurry

The *measurands are indicated in parenthesis* of the respective statistic terms.
The *first subscripts on a variable* indicate the measurement process considered:
c for the calibration process, s for the unknown sample, r for the CRM
q for the the AQC-measurement, cs for calibration and sample measurements.
The *second subscript* indicate the successive test sample measurements (i = 1. .... n).
As usual the *upper bar* is used for indicating the *arithmetic mean* of the respective term.

Basic Statistics

Mean value of test samples for the measurand x

$$\bar{x} = \frac{1}{n}\sum_{i=1}^{n} x_i$$

Standard deviation of n test samples

$$s(x_i) = \sqrt{\frac{1}{n-1}\sum_{i=1}^{n}(x_i - \bar{x})^2}$$

Standard deviation of means

$$s(\bar{x}) = \frac{s(x_i)}{\sqrt{n}} = \sqrt{\frac{1}{n(n-1)}\sum_{i=1}^{n}(x_i - \bar{x})^2}$$

Standard deviation of a linear regression y on x

$$s\left(y_i / x\right) = \sqrt{\frac{1}{n-2}\sum_{i=1}^{n}(\hat{y}_i - y_i)^2}$$

Sum of squares about the mean of the variable x

$$S_{xx} = \sum_{i=1}^{n}(x_i - \bar{x})^2, \quad S_x = \sum_{i=1}^{n} x_i^2$$

Table 2.10 Students factor t for specified level of condidence for different degrees of freedom

degree of freedom df	Students factor t			$\dfrac{t}{\sqrt{n}}$
	level of confidence			for
	90%	95%	99%	n=df+1, 95%
1	6.314	12.71	63.66	8.99
2	2.920	4.303	9.925	2.48
3	2.353	3.182	5.841	1.59
4	2.132	2.776	4.604	1.24
5	2.015	2.571	4.032	1.05
6	1.943	2.447	3.707	0.925
7	1.895	2.365	3.499	0.836
8	1.860	2.306	3.355	0.769
9	1.833	2.262	3.250	0.715
10	1.812	2.228	3.169	0.672
11	1.796	2.201	3.106	0.635
12	1.782	2.179	3.055	0.604
15	1.753	2.131	2.947	0.533
16	1.746	2.120	2.921	0.514
20	1.725	2.086	2.845	0.455
25	1.708	2.060	2.787	0.404
30	1.697	2.042	2.750	0.367
40	1.684	2.021	2.704	0.316
50	1.676	2.009	2.678	0.284
100	1.660	1.984	2.626	0.198

interval is calculated to be ±7,5%. The analysis with prior decomposition of the sample may typically give 5% RSD, but mostly not more than 3 test samples (i.e. decomposition solutions!) are prepared for analysis. In this case, the 95%-uncertainty interval is increased to ±12,4%.

This example illustrates the statement that for the solid sampling technique often the worse precision due to sample heterogeneity is compensated by the generally large(r) set of replicates!

It is often overlooked that the uncertainty in an analysis is not only determined by the uncertainty in the test measurements of the unknown sample, but additionally by other uncertainty components which may contribute e.g. the uncertainties in the calibration measurement [168], in the AQC measurement [60], and in the reference content of the calibration material [95]. Figures 2.22 a, b visualize how the single uncertainty components in the entire analytical procedure are propagated to the uncertainty in the final result.

The dominating systematic effect in analytical atomic spectrometry is connected with interferences (spectral or by concomittants) occurring in the instrumental measurement. Mainly for the detection of bias by this origin, the AQC-process is required. Other effects are either physically propagated so that these are included in the instrumental precision (e.g. sampling effects due to sample heterogeneity, effects in the slurry preparation process, or random effects of pipettes or balances used for dosing the test samples) or are compensated in the entire process (e.g. systematic effects of pipettes and balances), or are significantly lower (e.g. random and systematic volumetric effects in the preparation of reference solutions).

The uncertainty in the calibration measurement depends from the method of calibration. These are derived in the next section (listed below in Table 2.12).

The *uncertainty in the reference value* of the material which is used for calibration or for the AQC measurement can contribute to the combined uncertainty as well. While the uncertainty in the concentration of a reference solution is comparatively small, with the use of a CRM the uncertainty in the certified content, must be taken into account. Because the uncertainty in the analyte content of a CRM is based on analytical measurements and statistical evaluation, it is basically of type A as well.

In the certification reports of the BCR, all the required information is given ([169, 159], compare Fig. 2.26). However, some producers release CRMs, for which only the extended uncertainty or the confidence interval is documented in the certificate, i.e. $s(c_{r,i})$ and/or n_r are not given. In this case the user must treat the uncertainty as to be of type B. In order to evaluate the required uncertainty component "... those who use the value must employ their own knowledge about the likely magnitude of the uncertainty, given the nature of the quantity, the reliablility of the source, the uncertainties obtained in practice for such quantities, etc." [60].

Example

NIST state the uncertainty in the certified value of their SRMs in the form $c_r \pm U(c_r)$, e.g. for Pb in SRM 1570 Spinach 1.2 ± 0.2 mg/kg. "The uncertainty of the values of the constituents ... include allowances for material inhomogeneity, method imprecision, and an estimate of possible biases of the analytical methods used" [Certificate SRM1570]. Additionally it is mentioned that the certified values "... are based on results obtained either by reference methods of known accuracy or by two or more independent, reliable analytical methods". From this information one may assume that the uncertainty value is estimated as the "extended combined uncertainty". Because NIST follows the concept of "reference or absolute" methods in the assessment of certified values, the number of results may be relatively small, so that $k_p = 2.6$ may reasonably be chosen. With this "judgement" the uncertainty in the certified content of Pb in "Spinach" is estimated as ± 0.077 mg/kg (Eq. 2.38 and Eq. 2.39).

It is a consequence of the concept of combined uncertainty [60, 165, 167] that the *uncertainty in the AQC process* must be considered with the uncertainty in the unknown sample result – even in the case where the correction factor can be chosen as $f_q = 1$, i.e. no evidence for a systematic effect (i.e. bias) has been detected (see Sect. 2.4.2.1).

The uncertainty in the correction factor follows from the combination of the uncertainties in the certified content of the CRM (which is used for AQC) and the AQC process – considering that the latter uncertainty component is still a *combined uncertainty* u_{cs} from the CRM test sample measurements and the calibration measurements (see below). The application of the laws of uncertainty propagation to Eq. 2.32 yields (addition of the squares of the relative uncertainties, covariances not considered):

$$\left(\frac{u(f_q)}{f_q}\right)^2 = \left(\frac{u(c_r)}{c_r}\right)^2 + \left(\frac{u_{cs}(\bar{c}_q)}{\bar{c}_q}\right)^2 \qquad \text{Eq. 2.40}$$

Correspondingly, the total uncertainty for the determined content of the unknown sample is achieved by the combination of the individual uncertainties according to Eq. 2.33

$$\left(\frac{u(c_{s,cor})}{\bar{c}_{s,cor}}\right)^2 = \left(\frac{u_{cs}(c_s)}{\bar{c}_s}\right)^2 + \left(\frac{u(f_q)}{f_q}\right)^2 \qquad \text{Eq. 2.41}$$

Actually, the validity of these propagations of uncertainties is independent of whether the final result must be corrected or can be used uncorrected. It reflects the uncertainties that are inherent in the process of AQC and the respective measurements and estimates.

In the section below the concrete expressions for those terms which depend on the applied calibration method are derived and the above discussed uncertainty contributions are considered for the estimation of the combined uncertainty. Because these are all based on experimental standard deviations, the coverage factor k_p can be chosen according to the t-statistic.

It should be mentioned that for the correction of systematic effects other than the analytical "matrix-effect", further corrections may be required:

A significant contribution in the solid sample analysis might be the *uncertainty in of the moisture content* (compare Eq. 2.27). Because generally only one determination is performed (e.g. drying at 105 °C to constant weight) no information may be available about the random effects connected with this measurement. Furthermore, systematic effects by the different volatility of various water components (e.g. surface water, capillary water, crystal water) may occur. Hence, the uncertainty in the required correction may be of type B[12].

---

12) Even if uncertainties of type B have to be considered, the coverage factor can be orientated by the values in the "t-table" as long as the convoluted distribution can be assumed to deviate not to much from normal. The expanded uncertainty then provides a range having an approximate level of confidence that may differ from the level given by the t-distribution; consequently, the term "confidence interval" should by no means be used.

2.4.2.4
Combined Uncertainty in Solid Sampling Results

Each significant effect delivers an *uncertainty contribution $u_i(c_s)$* to the *combined uncertainty $u(c_s)$* in the determined content of the unknown sample c_s. As discussed above, for the analysis of solid materials four contributions generally must be considered[13)]

$$u(c_s) = \sqrt{u_s^2(c_s) + u_c^2(c_s) + u_q^2(c_s) + u_r^2(c_s)}$$

Eq. 2.42

For the estimation of the extended uncertainty $U(c_s)$ (Eq. 2.37) the coverage factor k_p may be chosen from the "t-table" based on the *effective degree of freedom df_{eff}* that consideres the different numbers of test samples and different degrees of freedom in the measurements and which is defined as [60]

$$df_{eff} = \frac{u^4(c_s)}{\dfrac{u_s^4(c_s)}{df_s} + \dfrac{u_c^4(c_s)}{df_c} + \dfrac{u_q^4(c_s)}{df_q} + \dfrac{u_r^4(c_s)}{df_r}}$$

Eq. 2.43

Because of the fourth power (square of variances) in the uncertainty terms, df_{eff} approaches the sum of the values of df_i only if all uncertainty components are of (almost) the same value. However, if one uncertainty component dominates, df_{eff} approaches the df_i-value of the respective term. Hence a "balanced" design of the experiment in all uncertainty components is honored by a larger df_{eff} value and thus by a small value for the Student factor t.

The expressions for the uncertainty components listed in Table 2.12 following to the frequently applied calibration methods are derived and discussed below in detail. Sets of solid sampling data which are used as test examples for the calculation procedures described below are listed in Table 2.11.

a. Direct Comparison of Response Data

This method is commonly performed by using a CRM as the reference material (see Sect. 2.2.2.1). The unknown content is estimated directly by the mean of the *mass-specific response data* (Eq. 2.1) from the unknown sample and calibration measurements. The combined uncertainty in the finally calculated mean content can be estimated easily by the addition of the squares of the relative uncertainties [14, 171] in the terms of Eq. 2.2

---

13) For the analyis of solid materials after prior decomposition of the sample as well [660].

Table 2.11 Test data set for exemplary evaluation of solid sampling analyses
(direct solid sampling, graphite furnace AAS, Pb 283.3 nm, peak area)

Certified values for Pb of BCR CRM 281 Rye Grass (see Fig. 2.26) [from 159]:
Number of accepted sets of results: $n_r=16$
Mean of laboratory means: $c_r=2.382$ mg/kg
Std. dev. of means: $s_r=0.213$ mg/kg
Confidence interval : 2.382 ± 0.114 mg/kg

test sample no.	reference solution			BCR CRM 281 Rye Grass				unknown sample Maize Leaves		
	c_c (mg/l)	a_c (ng)	R_c (s)	m_r (mg)	a_r (ng)	R_r (s)	R_r' (s/mg)	m_s (mg)	R_s (s)	R_s' (s/mg)
1	0.05	1	0.050	0.276	0.657	0.027	0.099	0.096	0.076	0.791
2	0.05	1	0.051	0.316	0.753	0.036	0.115	0.114	0.079	0.697
3	0.05	1	0.051	0.384	0.915	0.039	0.101	0.122	0.110	0.902
4	0.05	1	0.052	0.409	0.974	0.052	0.127	0.125	0.105	0.840
5	0.10	2	0.103	0.459	1.093	0.082	0.178	0.128	0.122	0.953
6	0.10	2	0.104	0.470	1.120	0.082	0.174	0.142	0.106	0.747
7	0.10	2	0.104	0.475	1.131	0.051	0.107	0.143	0.128	0.895
8	0.10	2	0.107	0.527	1.255	0.061	0.115	0.158	0.156	0.987
9	0.15	3	0.152	0.547	1.303	0.073	0.134	0.158	0.145	0.918
10	0.15	3	0.155	0.618	1.472	0.082	0.132	0.160	0.111	0.694
11	0.15	3	0.156	0.697	1.660	0.100	0.144	0.169	0.146	0.864
12	0.15	3	0.159	0.780	1.858	0.095	0.122	0.174	0.157	0.902
13				0.798	1.901	0.092	0.115	0.184	0.151	0.821
14				0.800	1.906	0.103	0.129	0.185	0.160	0.865
15				0.864	2.058	0.107	0.124	0.201	0.137	0.682
16				0.878	2.091	0.127	0.145	0.250	0.210	0.840
17				0.879	2.094	0.120	0.137	0.255	0.186	0.729
18				1.046	2.491	0.124	0.118	0.255	0.210	0.823
19				1.217	2.899	0.155	0.127			
20				1.347	3.209	0.199	0.148			
21				1.359	3.237	0.177	0.130			
22				1.422	3.387	0.202	0.142			
mean		2.00	0.103	0.753	1.854	0.099	0.130	0.167	0.138	0.830
s				0.348			0.020	0.050		0.091

$$\frac{u(c_s)}{\bar{c}_s}=\sqrt{\left(\frac{u(R'_s)}{\bar{R}'_s}\right)^2+\left(\frac{u(R'_c)}{\bar{R}'_c}\right)^2+\left(\frac{u(c_r)}{c_r}\right)^2} \qquad\qquad \text{Eq. 2.44}$$

Resolving Eq. 2.44 to the combined uncertainty and tracing it back to the standard
deviations of the respective measurements yields the uncertainty components in
the finally calculated analytical result for the unknown sample (see Table 2.12).

At that point it should be noted again that for a calibration based directly on
a CRM, no separate AQC is required because the AQC is "built-in" the calibration
measurements. Accordingly no separate AQC-term appears in Eq. 2.44 (see the
discussion in the next section).

Table 2.12 Uncertainty components (variances) in solid sampling analysis for different methods of calibration

method of calibration	contribution of measurements to the combined uncertainty			
	unknown sample $u_s^2(c_s)$	calibration $u_c^2(c_s)$	reference material (CRM) $u_r^2(c_s)$	AQC-process $u_q^2(c_s)$
linear regression (CRM as calibrant material)	$\dfrac{s^2(R'_{s,i})}{b^2 n_s}$	$\dfrac{s^2(R_{c,i}/a)}{b^2 \overline{m}_s^2}\left(\dfrac{1}{n_c}+\dfrac{(\overline{R}_s-\overline{R}_c)^2}{b^2 S_{aa}}\right)$ regression line through the origin: $\dfrac{s^2(R_{c,i}/a)}{b^2 \overline{m}_s^2}\dfrac{\overline{R}_s^2}{b^2 S_a}$	$\left(\dfrac{\overline{a}_c}{\overline{m}_s \overline{c}_r}\right)^2\dfrac{s^2(c_{r,i})}{n_r}$	no AQC measurements required
linear regression (reference solution as calibrant material)	$\dfrac{s^2(R'_{s,i})}{b^2 n_s}$	$\left(\dfrac{s(R_{c,i}/a)}{\overline{m}_s b}\right)^2\left(\dfrac{1}{n_c}+\dfrac{(\overline{R}_s-\overline{R}_c)^2}{b^2 S_{aa}}\right)+$ $\left(\dfrac{\overline{c}_s}{\overline{c}_q}\dfrac{s(R_{c,i}/a)}{\overline{m}_s b}\right)^2\dfrac{(\overline{R}_q-\overline{R}_c)^2}{b^2 S_{aa}}$	$\left(\dfrac{\overline{c}_s}{\overline{c}_r}\right)^2\dfrac{s^2(c_{r,i})}{n_r}$	$\left(\dfrac{\overline{c}_s}{\overline{c}_q}\right)^2\dfrac{s^2(R'_{q,i})}{b^2 n_q}$
direct comparison of response data (CRM as calibrant material)	$\left(\dfrac{\overline{c}_s}{\overline{R}_s}\right)^2\dfrac{s^2(R'_{s,i})}{n_s}$	$\left(\dfrac{\overline{c}_s}{\overline{R}_c}\right)^2\dfrac{s^2(R'_{c,i})}{n_c}$	$\left(\dfrac{\overline{c}_s}{\overline{c}_r}\right)^2\dfrac{s^2(c_{r,i})}{n_r}$	no AQC measurements required
slurry analysis, linear regression (reference solution as calibrant material)	$\left(\dfrac{V_s}{m_s}\right)^2\dfrac{s^2(R^*_{s,i})}{b^2 n_s}$	$\left(\dfrac{V_s}{m_s}\dfrac{s(R_{c,i}/c)}{b}\right)^2\left(\dfrac{1}{n_c}+\dfrac{(\overline{R}^*_s-\overline{R}_c)^2}{b^2 S_{cc}}\right)+$ $\left(\dfrac{V_q\overline{c}_s}{m_q\overline{c}_q}\dfrac{s(R_{c,i}/c)}{b}\right)^2\dfrac{(\overline{R}_q-\overline{R}_c)^2}{b^2 S_{cc}}$	$\left(\dfrac{\overline{c}_s}{\overline{c}_r}\right)^2\dfrac{s^2(c_{r,i})}{n_r}$	$\left(\dfrac{V_q}{m_q}\dfrac{\overline{c}_s}{\overline{c}_q}\right)^2\dfrac{s^2(R^*_{q,i})}{b^2 n_q}$
all calibration methods	$\dfrac{s^2(c_{r,i})}{n_r}$	uncertainty in calibration small compared to the other contributions	$\left(\dfrac{\overline{c}_s}{\overline{c}_r}\right)^2\dfrac{s^2(c_{r,i})}{n_r}$	$\left(\dfrac{\overline{c}_s}{\overline{c}_q}\right)^2\dfrac{s^2(c_{q,i})}{n_q}$

Example (Evaluation of test data in Table 2.11)

For the content of the Maize Leaves sample 15.2 mg/kg was estimated (see example in Sect. 2.2.2.1). The uncertainty contributions of the measurements are calculated according to Eq. 2.44 (compare Table 2.12) to u_s=0.39 mg/kg and u_c=0.50 mg/kg. The transformation of the uncertainty in the certified content into its contribution in the unknown sample content yields u_r=0.34 mg/kg. Using these components, the combined uncertainty in the unknown sample content is calculated to u=0.72 mg/kg (Eq. 2.42). Because of the large value of the effective degree of freedom (df_{eff}=51 calculated from Eq. 2.42) a coverage factor of 2 can be chosen. Thus the extented uncertainty for the sample content is estimated as ± 1.4 mg/kg($\pm 9.5\%$).

b. Calibration Based on Linear Regression, Reference Solution as the Calibrant Material

If a *linear regression* curve for the instrument response R_c on the *analyte amount a* is calculated from a set of n_c calibration measurements, two parameters are estimated (e.g. slope and intercept or slope and mean response value) each connected with an uncertainty. Accordingly, the combined uncertainty consists of two terms [172]. The estimated uncertainty is a function of the analyte amount a and shows the smallest value at the centroid $(\bar{R}_c, \bar{a}_c)$. The "extended uncertainty band" for the regression line estimate is given by [173] (for the statistics used see Table 2.9)

$$U(\hat{R}) = \hat{R} + k_p \sqrt{s^2 (R_{c,i}/a) \left(\frac{1}{n_c} + \frac{(\bar{a}_c - a)^2}{S_{aa}} \right)}$$

Eq. 2.45

where
$\hat{R}$=*value of R on the regression line*

As a consequence of the fact that both slope and intercept can show an error, the uncertainty for the regression line shows the characteristic form (as can be recognized in Fig. 2.25 for the calibration based on the CRM test data set). Sample measurements near the centroid give the most "certain" results, while measurements at the ends of the regression curve can show a significantly larger uncertainty. If a small uncertainty for the sample measurement is of importance, this property must be taken into account in the *experimental design* (e.g. additional calibration points at the ends or and sample measurements near the centroid).

The combined uncertainty in the unknown sample content can be approximated by propagating the uncertainties from calibration and sample measurements according to the "analytical function" (inverse calibration function) expressed in the mean of the response values and the slope [168], considering

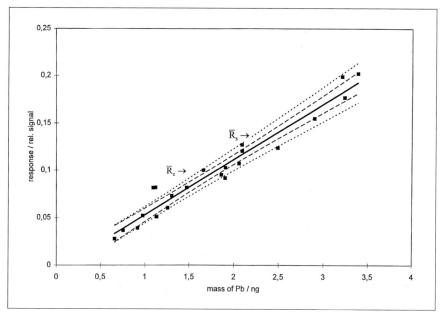

Fig. 2.25. Regression lines estimated from the test data of Table 2.12, using the CRM Rye Grass for calibration. The inner uncertainty bands consider only the uncertainty in the calibration measurement, while the outer band includes the uncertainty in the certified content of the CRM. The "centroid" of the R_c data is marked, i.e. the section that would give lowest uncertainty for sample measurement, the mean of the R_s values from the "unknown" test example is indicated.

that the sample content is calculated from the analyte amount devided by the mass of the test samples (Eq. 2.5)

$$\bar{c}_s = \frac{1}{\overline{m}_s}\left(\frac{\overline{R}_s - \overline{R}_c}{b} + \bar{a}_c\right)$$
Eq. 2.46

The propagation of the uncertainties in the terms of Eq. 2.46 which show a pronounced uncertainty (a_c supposed to be "error-free") yields the combined uncertainty in the determined analyte content

$$u_{cs}(c_s) = \frac{1}{b}\sqrt{\frac{s^2(R'_{s,i})}{n_s} + \frac{s^2(R_{c,i}/a)}{\overline{m}_s^2}\left(\frac{1}{n_c} + \frac{(\overline{R}_s - \overline{R}_c)^2}{b^2 S_{aa}}\right)}$$
Eq. 2.47

The uncertainty in the unknown sample measurement is represented by the standard deviation of the mass-specific response data (Eq. 2.1) because the primary response values do not represent the "response distribution" due to random effects (test sample mass arbitrarily chosen by the operator).

As discussed above, generally an AQC process is required, if the calibration curve is established by using standard solutions. The combined uncertainty in the determined content of the unknown sample can be derived as discussed above, by propagating the uncertainties of the terms according to Eqs. 2.40 and 2.41. The resulting four uncertainty components are listed in Table 2.12.

It should be noted that the uncertainty in the calibration line estimate contributes to both measurements, for the unknown sample and the AQC measurement. However, if the uncertainty in the calibration measurement is small compared to the solid sample measurement (no analyte heterogeneity connected with the reference solution), its contribution may be neglected. In this case, the remaining terms can be related directly to the uncertainty in the content values (instead of the response values), thus the calculation becomes simpler (see Table 2.12). Note: The analysis of the contents for the unknown sample and the CRM, i.e. the AQC measurement, are based on the same calibration line, thus these values are not totally independent. Consequently covariances must be calculated for a complete estimate of the combined uncertainty. For two reasons this extensive treatment is not performed here: As mentioned above, the uncertainty contribution of reference solutions will be very small, so that these can be neglected in any case. Second, the respective *covariances will lessen* the combined uncertainty further so that the given estimates deliver only a very slight overestimation (that may be compensated by slight underestimations due to unconsidered small effects).

Example (Evaluation of test data in Table 2.11)

From the linear regression using the reference solution measurements the required values are calculated to b=0.0522 s/mg, b_0=-0.0008 s, $s(R_{c,i}/a)$=0.002 s, S_{aa}=8 ng^2. Using this regression line for the Maize Leaves measurements a content of 16.0 mg/kg is calculated. The measurement of the CRM (AQC) has shown no significant bias to the certified value (see example in Sect. 2.4.2.1). The uncertainty contributions are estimated as being u_c=0.14 mg/kg, u_s=0.41 mg/kg, u_q=0.40 mg/kg, u_r=0.35 mg/kg. Thus, the combined uncertainty is $u(c_s)$=0.686 mg/kg (0.671 mg/kg if the uncertainty in the calibration is not considered). The effective degree of freedom is calculated to 56, so a coverage factor k_p=2 may be chosen. The analytical result including the extended uncertainty should be given as 16.0±1.4 mg/kg (±8.6%). It might be inviting to state only the extended uncertainty in the sample measurement which amounts in this example to only ±5.6%, especially if the AQC shows that bias is not evident. However, the uncertainty in the assessment of trueness must be considered. Otherwise the uncertainty in the final result would be considerably underestimated.

c. Calibration Based on Linear Regression, CRM as the Calibrant Material

If a CRM is used as the calibrant material, the uncertainty in the certified content – which may be up to two orders of magnitude larger than that in the concentration of a reference solution – contributes an additional term to the uncertainty in the calibration curve estimate. With the propagation of the uncertainties, the value of a_c can no longer considered to be "error-free", instead is connected with the uncertainty in the certified content which can be estimated according to Eq. 2.4a as

$$u(a_c) = m_c\, u(c_r) = \frac{a_c}{c_r} u(c_r)$$

Eq. 2.48

Consequently the extended uncertainty for the regression line estimate is widened at the "upper end" (the CRM term is proportional to the analyte mass a) according to

$$U(\hat{R}) = \hat{R} \pm k_p \sqrt{s^2(R_{c,i}/a)\left(\frac{1}{n_c} + \frac{(a - \bar{a}_c)^2}{S_{aa}}\right) + b^2 \frac{a^2}{c_r^2}\frac{s^2(c_r)}{n_r}}$$

Eq. 2.49

Figure 2.25 shows also the influence of the CRM uncertainty term on the uncertainty band of the regression line estimate.

The combined uncertainty for the determined analyte content of the unknown sample can be approximated by applying the laws of uncertainty propagation to Eq. 2.46, considering now the uncertainty in $\bar{a}_c$. The three uncertainty components obtained are listed in Table 2.12. With the derivation, a simplifying approximation is made, i.e. the slope b is not influenced by the uncertainty in the a_c values. In [174] is shown that if the uncertainty in the independent variable is not larger than the uncertainty in the dependent variable, the influence on the estimated slope of a regression line is very small. This assumption will mostly be given because the uncertainty in the certified content are generally smaller than in the determined content (if not, orthogonal regression procedures are required).

Example (Evaluation of test data in Table 2.11)

From the linear regression using the measurements of the CRM Rye Grass the required values are calculated as b=0.059 s/mg, b_0=-0.0055 s, s($R_{c,i}/a$)=0.0011 s, S_{aa}=14 ng². Using this regression line the sample content of the Maize Leaves sample is estimated to 14.8 mg/kg. The uncertainty contributions are estimated to u_s=0.37 mg/kg, u_c=0.31 mg/kg and u_r=0.24 mg/kg. The combination yields u=0.54 mg/kg. Using these values, Eq. 2.43 yields df_{eff}=47. The extended uncertainty is thus estimated as being 14.8±1.1 mg/kg (±7.3%).

If it can be verified that *no intercept* must be taken into consideration, the uncertainty in the calibration measurement is given only by the uncertainty for the slope of the regression line [172]. The first term in Eq. 2.49, which represents the uncertainty in the regression line estimate may be replaced by a simpler term (see Table 2.12). Thus the uncertainty toward the origin becomes very narrow and the value of k_p can be chosen for df=n-1 because only one parameter is estimated. In this case the estimated uncertainty in the content of an unknown sample may be significantly smaller if its mean response value is below the regression line centroid. (As can recognized from the test sample data and Fig. 2.25, the latter condition is not given with the test example, so in this case no smaller uncertainty could be achieved forcing the regression line through the origin.)

d. Calibration Based on Analyte Addition

The laborious AAM is generally applied in order to overcome a matrix effect. If this assumption is proven to be valid (or at least reasonable from analytical knowledge), *no AQC measurement* is required, even if a reference solution is used for analyte addition.

However, because the analytical result is based on an *extrapolation process* – i.e. the estimated value may be far outside the narrow uncertainty interval at the centroid – the uncertainty in the content of the unknown sample will be significantly enlarged. This points to the importance of a suitable experimental design, i.e. a larger number of test samples without addition should be performed in order to get a narrow uncertainty range at that end over which the extrapolation is performed (compare Figs. 2.9 a, b).

If the *"sample variation-AAM"* is applied, i.e. the added analyte amount is kept constant, the uncertainty in the sample mass m_a, which corresponds to the added analyte amount, can be estimated by the uncertainty in the regression estimate at R=0. Considering that the final content is calculated according to Eq. 2.6, the uncertainty in the unknown sample content is given by [163]

$$u(c_s) = \frac{a_a}{m_a^2} u(m_a) = \frac{c_s}{m_a} \frac{s(R/m)}{b} \sqrt{\frac{1}{n} + \frac{\overline{R}^2}{b^2 S_{mm}}} \qquad \text{Eq. 2.50}$$

Example

In Sect. 2.2.2.3, an AAM analysis of Ag in a silicate rock material [96] is discussed (see Fig. 2.9 a). The required statistics of the linear regression of the n=17 test samples are calculated as $\overline{R}$=0.367, s(R/m)=0.013 and S_{mm}=1.68 mg². Considering these values, as the result of the analysis the 95%-uncertainty interval can be estimated as 3.3±0.48 mg/kg (Eq. 2.50).

If *"sample & analyte variation-AAM"* is applied, i.e. mass of the unknown sample and the added analyte amount are varied, a linear regression "specific response on the added analyte amount" is performed. As mentioned with the discussion of this approach of AAM in Sect. 2.2.2.3, the uncertainty in R_0 must additionally considered in the combined uncertainty. With the simplifying (conservative) assumptions that all response data are connected with the uncertainty in the mean of the R_0-values (instead of only the R_a-values) and the variables are not correlated, the propagation of the uncertainties yields

$$u(c_s) = \frac{1}{m_0\,b}\sqrt{(\overline{m}-m_0)^2\,\frac{s^2(R'_{0,i})}{n_0} + s^2(R'_i/a)\left(\frac{1}{n} + \frac{\overline{R}'^2}{b^2 S_{aa}}\right)}$$ Eq. 2.51

Example

In Table 2.3, the analytical data and the respective mean values from the determination of Se in Bovine Liver [100] are listed (see example in Sect. 2.2.2.3 a). The additionally required statistics for applying Eq. 2.51 are calculated as: b = 0.000119 s/mg · ng, $s(R'_{0i})$ = 0.0182s/mg, $s(R'_i/a)$ = 0.0133s/mg, S_{aa} = 1.600.000 ng^2. From these data the uncertainty in the calculated content is estimated as 107 mg/kg, and the uncertainty interval (df=13) as 778± 231 mg/kg. The uncertainty component from the regression is the dominating contribution (104 mg/kg), while the propagated uncertainty in the response without addition (Eq. 2.7) only contributes a small part.

A derivation of the uncertainty limits for the unknown sample content applying the *generalized analyte addition method GAAM*, has been published by Berglund and Baxter. The authors state: "Considering the GAAM, ... the terms D and S required to calculate the concentration are correlated and the covariance COV(D,S) must be included. This additional source of error will, of course, increase the uncertainty in the estimated concentration. Uncertainty limits for the concentration are constructed using Fieller´s theorem which yields a rather complicated expression. It may be seen that the distribution of C is generally skewed about the point estimate, which tends to be subject to positive bias when the experimental data exhibit a high level of random error (as may be expected due to inherent material inhomogeneity). Minimizing the extent of the bias may be achieved by increasing the number of data points as C then converges on its true value. This will also narrow the practical uncertainty limits" [95].

In combination with [95] a PC-program is delivered for the calculation of the uncertainty interval. So the very complex expressions are not given here.

For the discussion of the particularities of the estimation of the uncertainty contributions in *slurry analysis* see Sect. 5.2.2.4.

2.4.2.5
Are the Effects Worth the Effort?

The above derived expressions for the uncertainty in the determined analyte content of the unknown sample look rather complex. However, modern programs for table calculation offer programming by simple "macros" or program languages. Because today all data are normally available from a file of a PC-program, it does not requires much more effort to calculate the different statistical terms. So it is no longer a complicated and tedious calculation procedure to get an reliable estimate for the uncertainty in the analytical result[14].

The calculation of the combined uncertainty due to all random and (revealed) systematic effects often leads to a disillusioning large uncertainty in the analytical results. Especially, if this range is compared with those from other laboratories, where no rigorous statistical treatment and perhaps even no AQC measurements are performed, this trustworthiness may be a disadvantage.

So, actually, the question may arise as to whether such a rigorous treatment is worth the efforts.

It is a fact that in analytical chemistry generally often only the uncertainty in the test sample measurement is stated (or even only the precision). Moreover, in the literature it has even been judged as a disadvantage of the calibration using a CRM that the uncertainty in the certified content of the CRM increases more or less significantly the uncertainty interval of solid sampling results [95, 175]. However, the above rigorous statistical derivation shows that the direct calibration by means of a CRM more likely may result in a smaller uncertainty interval [660] compared to an analysis after decomposition and calibration by standard solutions because the additionally required AQC-measurement contributes significantly to the combined uncertainty.

Thus the consideration of the uncertainties in the AQC-measurement and the CRM content with solid sampling results would often give *no comparable values* for the uncertainty compared to other data and would lead to unjust impressions about the quality of solid sampling results. Actually, the solid sample analysis with the graphite furnace yields results whose accuracy are fully comparable with these from the established methods – or is sometimes even better if all souces of uncertainty connected with the entire procedure of the method are considered.

Additionally, it should be considered that the uncertainty in a CRM value has a specific character. The certified content of a CRM results from different sets of measurements performed by highly reputable laboratories using the established analytical techniques. The evaluation of the analytical data in the certification of the BCR is visualized in Fig. 2.26 for Pb in the CRM 281 Rye Grass. With the statistical evaluation of these data still unknown, systematic effects inherent in a particular method and from the particular laboratory are "randomized" [176, 160] when the certified content is estimated by the mean of mean values from the laboratory results. The final uncertainty in the certified content therefore is an expression of today´s "limits of attainable trueness". This is the basic meaning if the term "accepted" content [145] or "conventional true" content [60], is recommended to use instead of the term "true" content.

The use of (suitable) CRMs for calibration gives the best certainty in the determined analyte content of the unknown sample which can be reached today in rou-

---

14) A recently published software tool gives a practical guidance for the raw data processing and facilitates the uncertainty evaluation considerably [709].

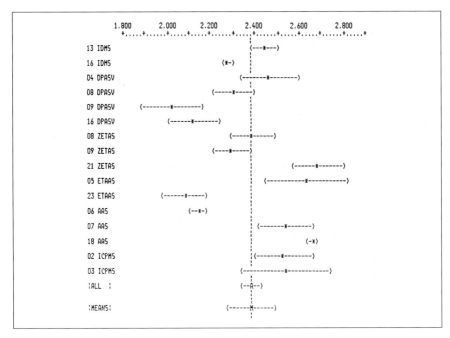

Fig. 2.26. Bar-graphs for the laboratory means and standard deviations considered in the certification of the Pb content in the BCR CRM 281 Rye Gras. The applied analytical methods are indicated by the well known abbreviations (following the laboratory code number). The *bar* "MEANS" gives the certified value, estimated from the mean of the mean values from all laboratories (n_r=16) and the respective confidence interval that is 2.382 ±0.114 mg/kg, see Table 2.11 (the bar "ALL" gives the mean of all single 101 values and the respective very narrow confidence interval in which the existing systematic effects between the laboratories are still totally unrecognizable) (from: [159], with permission).

tine analysis. The uncertainty in the certification may be judged at first as a highly valid information on the "state of the art" of the analytical methods and techniques of today!

The excellent "Guide to the expression of uncertainty in measurements" [60] released by ISO and the concretization "Quanatifying uncertainty in analytical measurements" [165] released by EURACHEM present the required statistical tools and are clearing up problems in the treatment and the quantificaton of all effects which contribute to the uncertainty. A comprehensive overview of the evaluation process for measurement uncertainty in analytical chemistry has recently been given by Williams in [167]. Today, it is still within the responsibility of the analyst to decide which contribution of errors is included in the reported uncertainty. But it is to be hoped that, in the future, generally more realistic estimations for the uncertainty are given. At least a *complete report* of the considered uncertainty components and a full documentation of how they were evaluated are an important demand (with respect to all methods) for the presentation of analytical data.

2.4.3
Treatment of Outliers

A single result in a set of several test measurements is regarded as an outlier if it appears to differ unreasonably from the others in the set.

Outliers may be attributed to gross errors, such as "human errors" (e.g. miscalculations, use of wrong units, misidentification, erroneous transcription) or gross errors in the analytical procedure (e.g. instrument malfunction, contamination). Outliers caused by these kinds of errors undoubtedly must be rejected. It is a basic task for the analyst to identify the causes for the occurrence of gross errors.

However, in a limited set of test measurements, outliers may occur even if gross errors are not present. A single result can have an outlying value although it is part of the population, but is overrepresented in relation to the size of the test samples being determined.

It is the aim of outlier tests (e.g. Grubb's test) to recognize outliers by an estimation for the representativity of extreme values assuming that the population is normally distributed. If a single value cannot be expected in a set, it is recognized as an outlier and it should not be included in the statistical evaluation (outlier rejection).

It was shown in Sect. 2.3.2 that solid sampling test results – more than with other methods – may show distributions which are more or less rightsided (tail to larger contents) due to the influence of "analyte nuggets" with small test sample masses. In a small set of replicate measurements a high value – which has a comparatively low probability, but is part of the population – appears as an outlier (compare e.g. Figs. 2.16 and 2.21).

In this situation, an outlier test based on the assumption of a normal distribution, would force the set of test results to "normality". However, if the true distribution is not known and the number of replicate measurements is too low for a reliable estimation, it cannot be decided if "the god of chance" has given the extreme values representatively with respect to the population (i.e. laboratory sample). However, without rejection of distinct "outliers" the bias (to higher values) would be larger than the bias (to lower values) with its rejection! So, although "outliers" may be part of the real (skewed) distibution, they should be rejected in order to limit a bias [61].

Example

Anzano et al. [56] reported on outliers in a series of 76 test samples (Cu in "animal feed"). The mean of all determined values correspond exactly to the reference value of 21µg/g with a RSD of 46%. If 19 outliers (outside the ±2s interval) were rejected, the determined mean declined to 17.8 µg/g and the RSD to 20%. A series of 6 test samples yielded a content of 17.6 µg/g with a RSD of only 6.7%.

A practical treatment of outliers in a small set of test samples (e.g. <10) for part-ly overcoming the described dilemma was proposed by Winsor [126]: The value of the supposed outlier should be reduced to the highest but one value (second high-est value). Thus "winsoring" reduces the strong influence of outlying values, but the mean is shifted more or less modestly into the direction of the "true" value.

Example

Table 2.13 shows 8 test sample results from a mercury determination in a homog-enized waste sample. The mean is calculated to 1.49 mg/kg (RSD 78%). Obious-ly two values are outliers, if a normal distribution is assumed (Grubb's test). The rejection of these values results in a mean of 1.01 mg/kg (RSD 10%). If it is sup-posed that the distribution may be not normal, only the largest value should be reduced to the highest but one value. Winsoring yields a mean of 1.13 mg/kg (RSD 21%), which is indeed a better estimation of the accepted Hg content (in labourios control measurements a content of 1.18 mg/kg was found).

So this procedure is recommended, if outliers occur which are suspected as being caused by the nugget effect. However, it must be pointed out that winsor-ing is only adequate if contamination effects (or other gross errors) can be excluded to be the source of the observed outliers!

It is obvious that at first it is recommended to increase the number of repli-cate measurements, in order to get a better estimation for the analyte distribu-tion. However, in routine analysis, time and costs are limited. So "winsoring" of solid sampling data is comparable to the situation where with the conventional analysis only 2 (or even 1) samples are decomposed or one of three measure-ments are rejected after an outlier test. All these are suitable actions for achiev-ing sufficiently accurate results despite limited expenditures.

The potential of the SS-GF-AAS method for getting reliable results even in the case of severe heterogeneity effects was proven with the analysis of the – above documented – Bovine Muscle sample and other micro-heterogeneous animal tissues during the BCR certification campaign, documented in the certification report [177].

Table 2.13 Outlier treatment in a set of test sample results form a determiantion of Hg in the sample "homogenized waste"

no.	all data	outl. rejected	winsorised
1	0.91	0.91	0.91
2	1.49	1.49	1.49
3	1.12	1.12	1.12
4	1.13	1.13	1.13
5	0.93	0.93	0.93
6	1.04	1.04	1.04
7	4.36	-	1.49
8	0.94	0.94	0.94
means	1.49	1.08	1.13
RSD	78%	18%	21%

2.5
Accomplishment of the Analytical Process

As discussed in Chapter 1, a successful application of the solid sampling method needs a "specific manner of thinking" in each step of the analytical process. The sections of this chapter describes these differences in detail. The following section outlines the process of finding the appropriate analytical conditions. Of course the procedures are different in detail with the analytical techniques under discussion, and the specific procedures (AAS, ETV-ICP, slurry introduction) are discussed in the respective chapters. However, it might be useful to get a general idea of the *complex interrelation of the numerous analytical parameters* characteristic for the solid sampling method already here.

A general and important tool common in the graphite furnace techniques is the examination of *transient signals*. The comparison of signals shapes has proven to be suitable for characterization and identification of differences in vaporization/atomization behavior and is also documented in various figures in the different chapters of this book (see Fig. 2.28 and compare also Table. 3.5). The influence of the measures discussed in this section can be studied due to the change of the transient peaks (e.g. the mixture with graphite powder). Attempts to classify materials due to the signal shape are made in order to choose the most adequate reference material [178, 179].

2.5.1
Adequate Test Sample Mass

Running a solid sampling analysis, the choice of the test sample quantity is an important and continuous task for the operator.

The favorable test sample mass for the actual analysis depends in a complex manner on the *sample properties* (e.g. analyte content, matrix composition, bulk density, particle size, homogeneity), the *spectroscopic and instrumental features* (e.g. range of analysis, efficiency of background correction, sample loading capacity), and the *demands on the analytical results* (e.g. accuracy, range of uncertainty, AQC).

These various limiting and occasionally contrary factors must be considered e.g. the procedure is as described below:

If the analyte content c of the sample is approximately known, and the upper and lower limits of detection a_{max} and a_{min} ($=a_L$) are fixed with the spectroscopic and instrumental parameters, a *range for the practicable mass of test sample* can be calculated by (unit examples in the "ppm-range" in parenthesis)

$$m_{max/min}\left[1\ mg\right] = \frac{a_{max/min}}{c}\frac{\left[1\ ng\right]}{\left[1\ \mu g/g\right]}$$

Eq. 2.52

Mostly the "instrumental mass range" calculated according to Eq. 2.52 does not fit completely with the analytically required mass range e.g. with respect to sample content or sample heterogeneity. Additionally the "operational mass range" must be considered (sample amount which can be handled). So the operator must meet the decision as to if the reduced overlapping range for the test sample mass range is still appropriate for the analysis. These criteria can be outlined as follows:

If micro-heterogeneity for the sample is supposed, the sample mass should be as large as possible in order to get low sampling uncertainty. However, even if the calculation using Eq. 2.52 allows a *large test sample mass*, these often cannot be used because of further limitations which may be given:

- by the maximal loading capacity of the graphite tube or of the sample carrier (see the survey of sample carrier systems, Sect. 2.1.3.2),
- by the behavior of the matrix (e.g. building of droplets, fizzling effects), or
- if the background in the furnace (AAS) or the matrix vapor loading for the ICP is too intense.

While for the enlargement of the test sample mass the pellet technique can be used (see Sect. 3.2.2), both the latter problems often can be reduced by thermal pretreatment or by matrix modification.

If no micro-heterogeneity is supposed, it can be advantageous to run small test samples, e.g. because a thermal pretreatment can be avoided and thus the time of analysis and the danger of losses of volatile analyte compounds are reduced.

The *lower limit of the sample mass* is primarily given by the problems of dosing extreme small sample quantities (if the instrumental limit of detection is sufficient for the given sample). For sufficiently ground materials a separation of 0.02–0.05 mg may be feasible with some care. The use of an extremely small sample mass can be compulsory if the sample analyte content is so high that only with small test samples can the upper limit of detection be met (Eq. 2.52).

Although generally, three orders of magnitude for the test sample mass introduced into the furnace are feasible (e.g. direct: 20 µg-20 mg, slurry: 2 µg-2 mg), in practice only a smaller range can be used. With the direct technique using test sample masses between 0.3 mg and 3.0 mg often acceptable compromise conditions are given with respect to sample handling, precision of the analytical results, and the background during atomization.

If no appropriate test sample mass can be used, considering these "boundery conditions", other measures must be taken into account, e.g. alternative lines, chance of instrumental conditions.

2.5.2
Limit of Detection and Limit of Analysis

In Sect. 4.5.1.4, the problem of obtaining a reliable value for the minimum detectable analyte mass a_L (limit of detection) for the direct analysis of solid samples are described. However, in most cases – particularily in GF-AAS , where often the calibration with reference solutions yields good results – it will be a good estimation to use the conventional approach, using the "*zero-mass response*" (empty sample carriers inserted into the furnace) as the measure of the blank value R_B. Thus, the standard deviation of blank value measurements s_B is affected primarily by the base line scatter ("noise"), in the case of the determination of ubiquitous elements occuring on sensitive lines also by random contamination [180,181]. The smallest detectable response R_L is estimated by Eq. 4.5. The transformation into the corresponding analyte mass a_L is carried out by using reference solutions (via direct comparison of mass-specific response data or a calibration function).

Having a value of the limit of detection at disposal the *limit of analysis* can be obtained, expressed as the *lowest content c_L* determinable in the given material according to

$$c_L = \frac{a_L}{m_{max}}$$

Eq. 2.53

This figure reflects the "detection power" of the present technique for a certain element and a certain matrix composition.

It is the value of c_L that is used to evaluate if different techniques and/or methods are comparable, because it is related to the real sample. Using comparable GF-systems (graphite tubes) the limits of analysis with the solid sampling method are significantly smaller – often more than one order of magnitude than these achieved with "sample decomposition methods" (see e.g. [26, 182, 180, 181]).

Example

Assuming a detection limit of 1 pg for Cd with a given furnace/spectrometer system and a maximum test sample mass for biological samples of only 2 mg because of background problems; thus unknown samples down to 0.5 ng/g are analyzable, while for geological samples, the maximum mass will be larger, e.g. 15 mg, yielding a analysis limit of below 0.1 ng/g.

If a *slurry* is analyzed, the original sample is diluted by the "m/V-factor" (mass of solid material/volume of the slurry solution, compare Eq. 5.5). The estimated limit of analysis than must be related to the concentration of the slurry solution, replacing m_{max} by the maximum injectable volume $V_{ss,max}$ of the test samples. Assuming a sample mass of 10 mg for the above example, a slurry volume of 1 ml, and an injected volume of 20 µl, the limit of analysis related to the solid sample can be calculate as 5 ng/g. This exemplary calculation illustrates the fact that the "detection power" of direct solid sampling is generally larger than that of slurry analysis (and of the analysis of decomposed samples as well). Using comparable GF-systems the limit of analysis with the direct solid sampling is significantly lower – often more than one order of magnitude [180].

2.5.3
Thermal Sample Pretreatment in the Graphite Furnace

An in-situ ashing/pyrolysis[15] step in the furnace may be helpful, if the background absorption or the loading of the ICP by matrix vapor is too large when using an otherwise favorable test sample mass.

In order to cause no loss of analyte with the thermal pretreatment, the temperature which is applied must be chosen carefully. The experimental design for achieving the optimum temperature adjustment with the particular furnace/sample carrier equipment is well conceived in the analysis of decomposed sample (solutions) and of solids (direct and slurry sampling) as well.

The experimental data are evaluated and presented in a *pyrolysis diagram* which gives the relation between the analyte signals in the measurement step (atomization or vaporization) depending from the pretreatment (pyrolysis) temperature. Figure 2.27 a shows the interpretation of the plotted curves [10]:

background absorbance as a function of the pyrolysis temperature at the optimum atomization temperature (curve a);

analyte absorbance as a function of the pyrolysis temperature at the optimum atomization temperature (curve b);

analyte absorbance as a function of the atomization temperature at the optimum pyrolysis temperature (curve c).

T_1 is the minimum temperature for rapid pyrolysis of the matrix, T_2 is the minimum pyrolysis temperature for effective use of the background correction technique, T_3 is the optimum pyrolysis temperature, T_4 is the optimum atomization temperature.

Representative pyrolysis diagrams for a variety of elements for Bovine Liver are presented by Chakrabarti et al. [10, 11]. Figures 2.27 b and c show examples from these investigations which are particularly instructive, because the background signals are also presented for samples atomized from a L'vov platform. For Zn in Bovine Liver there is a sufficiently large pretreatment temperature range where the background is reduced drastically (>600 °C) before loss of Zn appears (>900 °C). Pretreatment of Cd in Bovine Liver shows a significant loss if the temperature exceeds 400 °C at which temperature the background absorbance is reduced only from 2.4 to 2.1. In these diagrams additional atomization curves are included (circles) from which the recommended minimum temperature for the measurement step can be read (e.g. for Zn and Cd>1800 °C).

Experimental data obtained with direct solid sampling for some elements in different matrices are summarized in Table 2.14. However, it must be pointed out that the temperatures listed are not absolutely reliable for the following reasons:

---

15) In the literature other terms are used for this process, e.g. ashing, thermal decomposition, charring. Because of the furnace shielding gas (argon or nitrogen), not enough oxygen is available this process should be termed pyrolysis; ashing should be used only if oxygen is added with the pretreatment step.

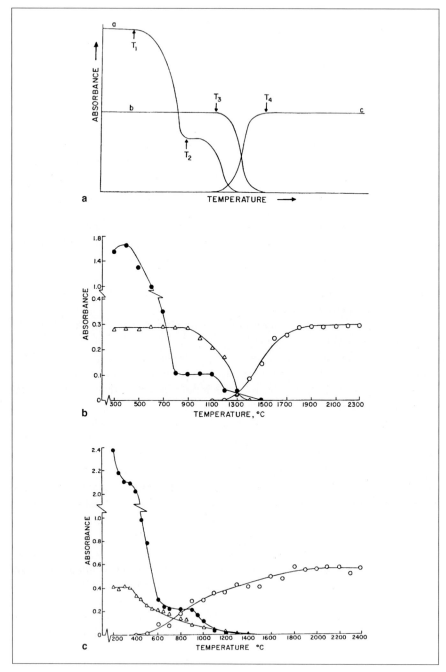

Fig. 2.27. Pyrolysis curves for finding out the optimum furnace program (compromise conditions) for a bovine muscle sample (from:[10], with permission):
a Principal representation (see text); **b** Zn absorbance over temperature for test sample mass of 1.0 mg; **c** Cd absorbance over temperature for test sample mass of 0.5 mg

Table 2.14 Survey of maximum pretreatment temperatures from various recovery experiments

element material	matrix/	modifier	max. pyrol. temperature (°C)	references
Al	biopsy	-	1200	
		$MgNO_3$	1800	[155]
As	biological	$6\mu g$ Ni $+3m\ H_2SO_4$	800	[208]
		$+ 3M\ HNO_3$ 10%HNO_3 $+ 0.2\%\ NiNO_3$ $(10\mu l)$	1100	[151]
Cd	animal	-	400	[86],[19],[10],
		$(NH_4)_2SO_4$	650	[11], [318]
	milk	"cocktail"	800	
Co	animal	-	1200	[10],[11]
Cr	plant	-	550	
	animal	-	1050	[331]
	milk	"cocktail"	1600	[318]
Cu	animal	-	1000	[10],[11]
	leaves	-	1300	[46]
			750	[150],[88]
Fe	animal	-	1300	[10], [11]
In	synth.std	-	900	[343]
Mn	leaves	-	1000	[86]
	coal	-	1600	[51]
Pb	animal	-	900	[10],[11]
	leaves	-	500	[86],[296]
	PVC	-	800	[20]
	plastic	$(NH_4)2HPO_4$ $+ Mg(NO_3)$	750	[150]
	milk	"cocktail"	900	[318]
Se	biological	-	300	[284, 674]
		Ag	900	[284]
		Palladium		[674]
	milk	"cocktail"	1600	[318]
	ash	-	700	[283]
		Ni-powder	2000	[282]
Zn	animal	-	900	[10],[11]
	leaves	-	700	[86]
	milk	"cocktail"	1100	[318]

*   the "cocktail" was composed from: 0.4% $Rh(NO_3)_3$ + 0.25% $Mg(NO_3)_2$ + 0.4% HNO3 + 3% Triton X-100

The chemical compounds of the analyte element in real samples are mostly not known, thus the volatility cannot be predicted and the findings for a particular matrix is not transferable to a different matrix, even if it is a similar type (e.g. the maximum pretreatment temperature for the determination of Pb in Bovine Liver was found to be 800 °C, whereas for Orchard Leaves, losses of Pb take place if the pretreatment temperature exceeds 500 °C [86])!

Furthermore, the temperature adjustment is related to the tube type. Differences of the cuvette design and the sample deposition area (e.g. a platform carrier), can cause temperature differences of around 100 °C or more (10, 11).

Besides the fact that it is extremely difficult to measure and control the true temperature exactly, not for the sample deposition area and not at all for the sample material. The adjustment and controlling facilities of commercial intruments will not give values which are sufficiently reliable for setting the maximal possible pretreatment temperture.

Thus the performance of a recovery study is mandatory, if the temperature must be increased to the maximum possible in order to separate the matrix to the largest extent.

The duration of the thermal pretreatment step depends at first on the heat transfer characteristic of the furnace configuration. Whereas with tube wall deposition the heat is directly transferred to the sample, a platform is needed at the low temperature level up to 1 min for reaching the final temperature. Second, the duration depends on the sample material and the introduced test sample mass. Mostly pyrolysis steps of 30 s to 90 s are applied, but for critical determinations (e.g. Cd in milk powder without a modifier) a pyrolysis step of some minutes may be required to reduce the background of large test samples to an acceptable level.

From the data presented in Table 2.14 it can be generally concluded that only for some volatile elements (or their volatile compounds) can a thermal pretreatment with temperatures below 700 °C lead to problems due to analyte loss. For these elements, the matrix modification method can be applied (see below), where the addition of a reagent to the test samples will transform the analyte compounds to a thermally more stable compound.

It has been proven with the analysis of decomposed samples that with *oxygen ashing* (oxygen as an alternative gas with the thermal pretreatment step), the matrix of biological materials can be separated more effectively and at lower temperatures (see also Sect.5.4.2). This technique was used for the solid sample analysis of Pb in Bovine Liver, where the background was reduced from an absorption of 1.4 to 0.25 below 900 °C [183]. In [155] it was pointed out that interference with the light beam by the build-up of carbon skeletons and swelling were reduced by oxygen ashing. In [184] argon +20% oxygen was used during preheating of graphite samples in order to converting them into carbon dioxide; accordingly no residues had to be removed from the furnace after atomization. The oxygen ashing technique needs some additional instrumental equipment for the gas sequence control (valves, electric/electronic control circuit). Prior to the initiation of the atomization cycle, the oxygen must be flushed out of the tube with the shielding gas in order to minimize oxidation of the graphite at high temperatures.

Atsuya et al. [185, 87, 186] popose *pre-ashing* biological samples in a muffle furnace prior to the solid sample analysis if the limit of analysis is not sufficient. They found that this sample pretreatment step yields a "concentration factor" (sample weight before devided by sample weight after pre-ashing). The procedure was successfully applied for the direct determination of Co, Ni, Mn, and Pb, where otherwise the content of the original sample on the ng/g level would be near or below the limit of analysis. The background is reduced drastically by the pre-heating, however, the inorganic matrix components are also concentrated, which may cause strong matrix effects. For this reason, the authors propose for calibrating the pre-ashed biological samples, the application of the analyte addition method (AAM) [186] or the preparation of synthetic reference materials prepared e.g. by using magnesium oxinate coprecipitates [87] (compare Sect. 2.2.1.3). A good correlation was found between the total content of Na, K, Mg, and Ca in biological samples (type. 1-10%) and the achieved concentration factor (2 to 50) for these materials. A significant increase for the analyte content is generally gained with pre-ashing temperatures >500–600 °C (for 30 min), however, if more volatile elements are to be determined, a pre-ashing temperature must be lowered sufficiently whicht will result in lower concentration factors (see Table 2.14).

The *second surface technique* was proposed by Rettberg and Holcombe [24]. Prior to the atomization, the analyte is separated from the solid matrix inside the tube in order to reduce the background absorption. The analyte is vaporized followed by a recondensation on a cooled Ta "plug" in a transfer heating cycle below the typical appearence temperature of the element under determination. The inventors decribe the advantages as follows: "Since the analyte is ´destilled´ from the sample to the cooled Ta insert and the remaining solid residue is physically removed from the furnace, the background levels during atomization are significantly lower than would be obtained by complete volatilization of the entire sample. The thin, uniform layer of condensate on the plug is devoid of most of the original matrix, and the analyte supply function is less dependent on the original sample. Thus, peak hights have been used successfully".

2.5.4
In-Situ Matrix Modification

2.5.4.1
Chemical Matrix Modification

With the analysis of decomposed samples the method of matrix modification by the addition of liquid reagents to the test samples is well established (e.g. see [276]). Besides the transformation to analyte-compounds which are thermally more stable (see discussion above) the vaporization/atomization characteristic can be influenced with the aim of overcoming interferences.

Although these methods of matrix modification can be applied in direct solid sampling it is not in widespread use for various reasons. [16].

The chemical mechanisms only can take place, if the particles of test sample are more or less complete moistened by the liquid modifier. It is easy to imagine that this is difficult to achieve for a powdered material [150].

Moreover one of the main advantage of direct solid sampling can be lost. With the addition of a reagent the possibility of a blank value is introduced into the analysis, besides the additional handling step (which is automized in the analysis of liquids).

So if a problem with the method development appears – it is recommended that one first looks for an alternative way of overcoming it before the approach of modifying the matrix by adding liquid reagents to the sample is considered.

However, for the determination of As and Se in certain typs of samples, matrix modification seems to be mandatory for direct solid sampling as well, because of the high volatility of a variety of their compounds (compare Table 2.14).

Esser used the "classical" modifier NH_4NO_3 as a solid powder for elimination of cloride during the thermal pretreatment [648]. He ground a mixture of the sample and a surplus of ammonium nitrate 1+10 down to particles of <90μm. After reatment at 350 °C the depressing effect (–50%) of added NaCl in the determination of Tl in a dust sample could be overcome; also the certified value of Tl in City Waste Incineration Ash (BCR CRM 176) (content of cloride 5%) was determined, while without this matrix modifier only 20% could be found.

For *slurry sampling* the use of liquid matrix modifiers is more common, therefore its effects are discussed in Sect. 5.4.1. Apart from the problems mentioned above the experiences can be generally applied to direct solid sampling.

2.5.4.2
The Use of Graphite Powder

Since direct solid sampling experiments with the graphite furnace has been performed, investigators have reported repeatedly about the advantageous analytical effects of mixing pulverized samples with graphite powder (e.g. [188]). This technique must be classified as a kind of matrix modifying of solid samples.

With the analysis of inorganic samples often distinct *double or multiple peaks* can be observed. Such transient signals indicate that the vaporization of the analyte element is hampered by the refractory matrix. Simply described, the atoms on the surface of the particles are freed more or less with the normal appearance temperature, while migration and diffusion of the atoms deeper inside the melt-

---

16) There is a discrepancy in the literature about the necessity of matrix modifying. Some of workers, specially those who are familier with the classical approaches, have mentioned the use of matrix modifiers without showing or giving the reason for its need.

ed particles are delayed. This effect can be superimposed by the building of larger droplets and chemical reactions.

After mixing a pulverized sample with graphite powder such transient signals are often changed to signals which show a shape which is typical for electrothermal vaporization/atomization and which is comparable to the calibrant material (even to that of a solution). Figures 2.28 a and b show examples of this impressive effect.

From these experiences with sample/graphite mixtures it can be concluded that particularly for silicate materials or materials which contain silicates (rocks, minerals, soils, sediments, mud, sludge) this technique is recommended. De Kersabiec et al. [122] pointed out that for the determination of particular elements in *geological materials* such as Pb [189, 154, 4, 190, 260] As and Zn [191] it is mandatory in order to get signals which can be evaluated. Other workers have mentioned the application of this technique with Bi [81] and Ag [81, 96, 192]. Esser has documented comparable observations for the surpression of double peaks

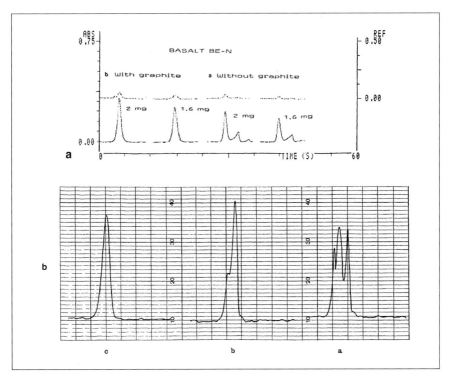

Fig. 2.28. Transient atomization signals, which show the influence of graphite podwer to the atomisation behaviour of the analyte:
a Signals for Pb in a pulverized basalt sample (**a**) with and (**b**) without a mixture with graphite powder (sample/graphite 9+1) (from: [122], with permission).
b Signals for Zn in cement with different amounts of graphite in the mixture (**a** sample only, **b** sample/graphite 1+1, **c** sample/graphite 3+1).

by the addition of graphite powder with the determination of Pb and Zn in *cement* [179]. Siemer et al. [189] showed that Pb was freed and atomized considerably faster if *glass* samples are mixed with graphite powder (1+1). Bley [193] found that for the determination of Ni in *soils* and *sediments* graphite addition is required in order to get "undisturbed" signals. For the determination of Sb a signal depression with a mixture of graphite (2+1) is reported [192], however very high amount of graphite was required to the sample (1+29) in order to get favorable signal shapes [122].

Similar observations have been documented by Nakamura et al. [130, 131] for the analysis of *carbonate and silicate rock* powders. They found an optimum mixture with graphite in the range (1+1)–(1+2) yielding transient signals for Li, Be, Co, Ni, Cu, Rb, Cs, Pb, Bi that are very similar to these from reference solutions, e.g. the slope for the calibration line for Cu was the same as that obtained for a reference solution, while it is significantly flatter below that range and declines again with increasing amounts of graphite [130]. For the analysis of *ceramic materials* for Pb and Cu these workers found that a mixture with graphite powder of 1+2 as a minimum is required in order to enhance the peak area signal to an constant level. The enhancement factor was approx. 2 for Si_3N_4 powder [194].

Takada [195] used a "bed" of graphite powder for the analysis of high purity *aluminium chips* (traces of Cu). He concluded that graphite powder acts like a buffer and may be called a matrix modifier as well, because of positive analytical effects namely, better precision, larger test sample mass possible, no sputtering effects of molten alluminium drops.

An exact and complete explanation of the chemical and physical mechanisms has not yet been given, however, some potential effects of the graphite are mentioned in the literature such as better heat properties (conduction and emissivity), forced reduction of oxides, no building of droplets (see Fig. 2.29), and eutectic effects (lowering of melting points). From X-ray diffraction patterns Nakamura et al. [194] deduced that there was a chemical conversion process of Si_3N_4 that was mixed with graphite powder during atomization.

Beside the described beneficial analytical effects a mixture with graphite powder gives additional advantages with the direct analysis of solid samples [96]:

It prevents melting or sintering beads or vitreous layers, hence sample residues as potential interfering compounds are easily to remove from the graphite surface (see Fig. 2.29, similar images are documented in [131]). The graphite deposition area (e.g. the sample carrier) is protected, because aggressive sample compounds find the dispersed graphite for effective reaction. Thus the sample carrier stands substantially longer. Our own observations with the analysis of *lime* samples show that the graphite bottom of the sample carrier was destroyed by very few runs, where, with a graphite-mixture it stands up to 50 runs.

Various workers have used graphite powder simply for *diluting pulverized samples* [196, 51, 117] in addition to the effects described above. This technique can be helpful if the sample contains a large content and the range of analysis is

Fig. 2.29. Residues from a test sample of a sediment sample in a platform boat after atomization.
a Without the mixture with graphite powder a compact melting drop has been formed; **b** with a mixture sample/graphite powder (1+1) the test sample remains as a pulver.

limited by instrumental adjustment or by spectral limitations (as it is with AAS, see Sect. 3.2). For this purpose also high purity urea [197] and solid naphthalene [46] were used. However, the analyst should note that the dilution is equivalent to a reduction of the weighed test sample mass. Thus a dilution is only required if the balance available has no sufficient resolving capability ("half-microgramm or worse), or the handling and dosing of extremely small test samples is regarded as too puzzling.

The proportions of the sample + graphite mixture range from 1+1 [189] to 1+999 [196], for the purpose of surpressing interfering effects mostly between 1+3 [154, 179] to 1+9 [96, 122, 260].

Some potential drawbacks of this techniques – which may be of importance only in exteme cases – should be mentioned:

At first glance it is somewhat astonishing that no increase of imprecision was reported after mixing samples with graphite powder. For getting an idea how such a mixture affects the sampling error an experiment was performed, where a large number of sewage sludge test samples with and without the addition of graphite (1+9) was analyzed. The results, documented in Figs. 2.30 a, b show that no significant additional sampling error was introduced by mixing the sample with graphite powder.

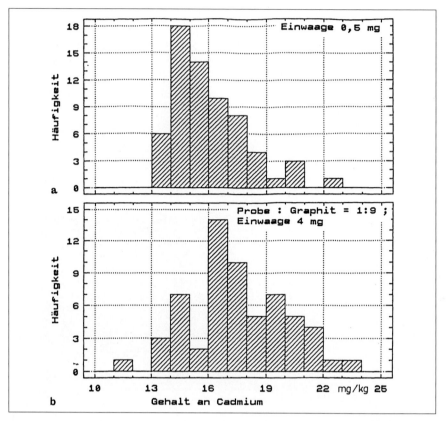

Fig. 2.30. Histograms from test sample determinations of Cd in Sewage Sludge (BCR CRM 146, c_r=18.0 mg/kg).
a The direct analysis of the material leads to 12.5% RSD (c=16.0 mg/kg, m=0.5 mg, n=65).
b Analysis of a mixture sample/graphite powder (1+9) leads to 13.7% RSD (c=17.5 mg/kg, m=4 mg, n=60).

Indeed, a rough estimation of the sampling error shows that only a very small additional uncertainty due to the heterogeneity of the mixture can be expected with a mixture with graphite powder (see example Sect. 2.3.2). Still, Pinta et al. [190] found that larger portions of granite powder + graphite may reduce the precision due to better representativity: 25 mg+25 mg yielded 10.8% RSD a stepwise increase up to 200 mg+200 mg yielded 6.7% RSD.

A homogeneous distribution of the sample particles in the mixture can be easily achieved with fine graphite powder, by shaking or grinding the components in a vessel/ball combination. Segregation of sample particles is highly improbable because of the strong adhesive property of graphite powder.

A precondition for accurate analytical results is the use of *spectrally pure graphite powder* e.g. RW-A (Ringsdorff, D-Bad Godesberg), Carbopur (Schunk,

D-Giessen), SP-1 (Nippon Carbon, Japan) or spectroscopic grade graphite (United Carbon, MI, USA). Some workers achieved graphite powder by grinding up (and sieving) pieces from spectrally pure graphite (e.g. from electrodes). Problems caused by blanks have still not been reported in the literature. However, because, even in these materials, traces of metal are stated from the manifacturer below $0.01-0.1$ µg/g, test measurements for blanks should be performed if low traces are to be determined.

A limitation of the applicability of this matrix modification technique may be given, if the volume of the test samples is enlaged by the "low density" graphite powder to such an extent that the loading capacity of the sample carrier or the furnace cross section is not sufficient. In this case the "pellet technique" [198] may be helpful (see Sect. 3.2.2).

However, in view of the clear positive effects that occur in the vaporization/atomization behavior of anorganic materials, the analyst should apply matrix modification using graphite powders whenever unfavorable signals or inferior results are suspected.

CHAPTER 3

Direct Solid Sampling with Graphite Furnace Atomic Absorption Spectrometry (GF-AAS)

ULRICH KURFÜRST

The basic experience with solid sampling using the graphite furnace were made with the detection by atomic absorption spectrometry. The methodological tools outlined in Chapter 2 were developed primarily with this analytical technique. Therefore, in this chapter, only t he specific conditions for certain analyte/matrix combinations will be discussed in light of reported experiences on solid sample analysis with GF-AAS. Considerations are focused on the specifics of the direct solid sampling method, many of which are also valid for the slurry technique. A more general treatment of the GF-AAS technique (e.g. see [23]) is beyond the scope of this book.

In solid sampling, increased attention must be paid to the elimination or at least the reduction of interferences. Thus it is important to begin with an outline of the significant instrumental parameters that address this key analytical need. The basic instrumental requirement for direct solid sampling – a suitable sample introduction system – is discussed in Sect. 2.1.3.

3.1
Instrumental Demands for the Elimination of Interferences

With spectroscopic techniques a distinction can be made between *spectral* and *non-spectral interferences* due to matrix concomitants [21]. Correspondingly, in GF-AAS interferences due to nonspecific absorption (background) and the occurrence of interferences connected with the vaporization/atomization process in the absorption volume are to be discussed.

For overcome the first type of interference an effective background measurement facility must be available. For the reduction of the latter, isothermal conditions should be reached with the vaporization/atomization of matrix and analyte.

3.1.1
Background Measurement and Correction Techniques

Obviously, with the direct introduction of solid materials into the furnace, background absorption due to concomitant compounds can rise drastically (it is just these problems that have led to the improvement of chemical sample treatment, in order to separate interfering concomitants from the analyte). Accordingly, the solid sampling technique has been closely connected with progress in the background measurement techniques.

Spectral interfering concomitants can be classified according to whether they absorb with a continuum or with a line characteristic. Scatter by particles from the matrix (e.g. from smoke or salt aerosols) or absorption due to photo-dissociation of gaseous molecules (e.g. from SO and SO_3 [212]) show continious absorption spectra, while interfering atoms only absorb on narrow lines. The absorption spectra of (small) molecules can show a continuum or a structured line characteristic, depending on the specific spectroscopic properies (electronic transition, vibration and rotation quantum states, broadening effects).

3.1.1.1
Background Measurement Using Line Sources

The background correction technique in atomic absorption spectrometry using the Zeeman effect (rather misleading when abbreviated as "Zeeman-AAS" or ZAAS) is the most advanced one today.

With this technique, a strong magnet is applied either to the spectral source i.e. the "lamp" (named "direct ZAAS") or to the absorption volume i.e. the graphite tube (named "inverse ZAAS"). Correspondingly, the emitting atomic lines or the absorbing atomic lines, respectively, are split by the Zeeman effect. Thus, in the case of direct ZAAS, reference lines for background measurement are created closely beside the analytical line ($\sim 10^{-2}$ nm at a magnetic field of 1 T). In the case of inverse ZAAS, the analyte absorption on the analytical line is "switched off" due to the shifted absorption lines, allowing background measurement "under the line". No further spectroscopic description of the Zeeman technique for background measurement will be given here, for details see e.g. [213, 214].

By taking advantage of the Zeeman effect, an excellent background correction can be achieved as a consequence of the line character of the reference lines and because analytical and reference beam are created in the same spectral light source. The line characteristic of the background measurement with Zeeman-AAS in principle allows us to measure and correct continuous and structured unspecific absorbances. Generally, it is stated that unspecific absorbance up to 2 is accurately corrected.

As usual, a "price" must be paid for a significant advance. In the case of direct ZAAS, a special spectral source that operates in a strong magnetic field is required, while in the case of inverse ZAAS, the heavy magnet restricts the spatial shape of the absorption volume. The last is unpropitious for the design of a special graphite furnace for direct solid sampling (large absorption volume in order to achieve ease of sample handling and adapted sensitivity, compare discussion below). Moreover, the analytical features (sensitivity, linearity) may be adversely affected some extent by uncomplete separation of the line pattern and the "anomalous Zeeman effect".

The development of the ZAAS-technique was a further breakthrough for the application of solid sampling with AAS to the analysis of samples which produce a strong background e.g. biological tissues and sludges. The first ZAAS-instrument was developed by Hadeishi in 1973 specially for direct solid sample analysis [6, 214].

However, in the last decade, even in ZAAS some *spectral interferences,* which are not completely corrected, have been observed. In the case of direct ZAAS, an overlapping of the lines of the analyte atom and that of other atoms or small molecules in the range of the Zeeman pattern (some 10^{-3} nm) may yield incorrect background measurement, while in the case of inverse ZAAS, a Zeeman splitting of a coincidental line from the interfering species may lead to erroneous background measurement. Accordingly, over- or undercorrection may take place, as Massmann has shown theoretically and by a first analytical example [215].

Such interferences observed up to the present are rare (compared to those occuring on D_2-correction, see e.g. [216]), however, the user must be aware that spectral interference may occur – for a digested sample as well – even using ZAAS. In Fig. 3.1 a, b, c, typical signals for overcorrection are documented. Table 3.1 lists the observed interferences with identification of the interfering species. It should be mentioned that an *over*correction is more easy to identify because the respective negative trace of the transient signal. *Under*correction has been observed up till now only in very few cases, e.g. in [217] a positive signal from Pt on the Fe line at 271.9 nm is documented. However, it can be assumed that such effects may be overlooked because the respective "positive" signal can be easily misinterpreted as a matrix effect that has led to a structured transient analyte signal [218].

Some of these interferences are of paticular relevance for solid sampling if the interfering compound is formed with the vaporization of the matrix. In Table 3.1 examples of matrix types are specified. It should be noted that the interferences from the PO molecule are observed especially with the use of large amounts of phospate based matrix modifier, however, an effect might also occur in phosphate-containing samples, e.g. bone, dust, sludge, fertilizer.

Background measurement according to *Smith and Hieftje (S/H-system)* utilizes the effect of self-reversal of atomic resonance lines occuring at a very high density of emitting atoms in the spectral source. By using a very high current pulse applied to the hollow cathode lamp (of special design), extreme self-absorption with a strong line broadening combined with a significant reduction in the intensity at the line center occurs, creating reference "lines" beside the analytical line.

The spectroscopic conditions with respect of reference and analytical lines are similar to that of direct ZAAS. Consequently this technique shows a comparable background correction capability which also allows solid sampling to be carried out for biological materials without background problems [140]. However, because self-absorption and still more the occurrence of self-reversal correlates

Fig. 3.1. Examples for overcorrection of background absorption with Zeeman-AAS (compare Table 3.1)

a Pb signal at 283.3 nm, from 2 mg of a grass sample. The *upper trace* shows the total transmission, the *lower trace* the Zeeman-corrected signal. The analytical (positive) Pb signal is overlapped by a (negative) signal which indicates an uncorrected interference.

b The Zeeman-corrected Cd signal at 326.1 nm, from 5 μl of 20%V/V H_3PO_4 (from [217], with permission).

c Zeeman-corrected Au signals at 242.8 nm. From right to left: standard solution (1, 2, 3 ng) and from a mineralogical material (solid sampling). The interfering compound from the solid material that leads to the negative signals is still not identified (possibly due to the NO molecule, which has a coincident absorbing band [297])

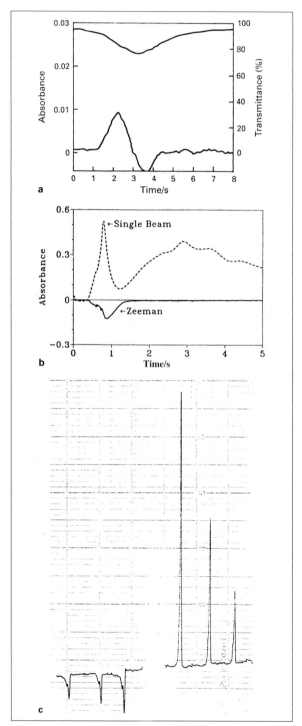

Table 3.1 Observed spectral interferences occuring even with Zeeman background correction, that may be of paticular relevance for solid sample analysis

Element	Analytical line (nm)	Interferent	Matrix type from which the interferant may occur	Ref.
Ag	328.1	PO	environmental, modifier	[298, 299]
As	193.7, 197.2	Pb	lead-base alloys	[300]
Au	242.8	(unidentified)	geological,	[this work]
	242.8, 267.6	Co	metallurgical	[301]
Bi	223.1	Fe/Ni	nickel-base alloy, steel	[302]
Cd	326.1	PO	environmental, modifier	[303, 299, 217]
Fe	246.3, 248.3 250.1, 252.3	PO	environmental, modifier	[299]
	358.1, 386.0	CN	biological	[304]
Ga	287.4	Fe	ore, steel	[305]
Hg	253.7	Co	metallurgical	[301, 217]
Cr	357.9	CN	biological	[304]
Pb	283.3	S_2	coal, biological	[296]
	217.0, 239.4, 283.3	PO	environm., biological, modifier	[306]
	261.4	Co	metallurgical	[218]
Pd	247.6	Pb	lead-base alloy	[307]
	244.3, 247.6	PO	environmental, modifier	[299, 303]
Sb	206.8	PO	environmental, modifier	[299]
Se	196.0/204.0	PO	environmental, biological	[308, 309]
		NO	modifier	[310]
Tl	276.8	CS	coal, biological	[311]
Zn	213.9	Fe	ore, steel	[305]
		PO	environmental, modifier	[299]
		NO	biological , modifier	[310]

with the sensitivity of the line, the calibration function show a significant unlinearity for the more insensitive lines. Moreover, for the very insensitive lines (e.g. non resonance lines) no self-reversal can be achieved at all, so that for these lines, no background measurement is possible by S/H-AAS[1] a disadvantage that may restrict the application to solid sampling (compare discussion below).

This pointing out of the limitations of the Zeeman technique should not lead to the erroneous impression that measurements using this background correction technique are prone to being adversely influenced by incorrect background measurement. On the contrary, this technique generally gives the best background correction that can be achieved today.

---

1) Because of this limitation, AAS-instruments which use the S/H-technique are also equipped with a D_2-system.

ontinuum Sources

Background measurement techniques using a second spectral source which emits a spectral continuum – a deuterium (D_2-) lamp is most commonly used – has been standard with commercial AAS-instruments for more than two decades. The introduction of this technique was a great advance in trace analysis by AAS. For flame-AAS, it is still common and usually adequate. However, the correction capability of this technique is limited by its spectroscopic properties. The spectral line background positioned in the range of the spectral bandpass of 0.1–1 nm (e.g. due to the absorption of small molecules) cannot be measured correctly and also the correction of a background level which is too high (absorbance >1) might be incorrect because of an uneven intensity distribution in the cross section of the absortion volume due to different caustic surfaces of the two lamps (HCL, D_2).

In the first period of solid sampling experiments, only this technique was available. Successful analyses have been reported for samples which cause no large or structured unspecific absorption, particularly of geological and metallurgical materials. It was proven at an early stage of the present technique that by using GF-AAS equipped with a deuterium background correction facility, reliable analytical results can be attained.

D_2-lamps of the hollow-cathode type that show an enhanced "overlapping" of the intensity distribution with the HCL that emits the analytical line, in combination with high transmission optical systems, have led to a higher acceptable background level [204, Jena Analytik]. Of course it is still more favorable, if an atomizer is used for the analysis that suppresses the appearance of the background level more than conventional furnaces tubes (e.g. side heated tube, see discussion below).

These technical improvements and the better understanding of analyte/concomitant chemistry inside the graphite furnace have widened the field for solid sampling using D_2-AAS. Consequently a great number of applications even for biological materials have been reported over the last decade [e.g. 219, 220, 204, 132]. However, it must be pointed out that the operator has the responsible task – much more than with the Zeeman effect background correction – to check if the actual D_2-background measurement may be assumed to be accurate (e.g. by the analysis of a CRM). Methods for the reduction of matrix interferences may help to overcome limitations of the D_2-technique for a concrete analytical problem (see below and Sect. 2.5.4).

3.1.2
Analyte Atomization Under Isothermal Conditions

Non-spectral interferences can be classified with respect to the place and stage of its occurrence [21]. For GF-AAS, volatilization and vapor phase interferences due to concomitants are of highest importance.

It is well established that these kinds of interference can be suppressed if the gas temperature in the atomization cell has reached a stable level when vaporization/atomization takes place. This condition has been described in the literature as "isothermal", whereas in the strict sense of the word (and of the analytical requirements) the cell should be void of temperature gradients as well i.e. show temporal and spatial isothermality [221].

If the given instrumentation does not allow us to apply the advantageous analytical conditions outlined below, or if interferences occur with the analysis despite background correction and isothermal atomization, alternate techniques may lead to success. Two kind of action may be considered for avoiding interference by the sample matrix: *Separation* of the interfering concomitants and/or *reduction* of their interfering potential. While in solid sampling the first can often only be achieved by thermal treatment, the latter needs the help of chemical reagents. These measures are discussed in Sect. 2.5.

3.1.2.1
Techniques for Providing Isothermal Conditions

Various techniques for achieving (temporal) isothermal conditions are described in the literature, however, the platform technique proposed and investigated by L´vov as long ago as 1978 [222] is today the one almost universally used. The platform effect may be described as follows:

If the test sample is deposited on a small rectangularly shaped platform inside the graphite tube, thermal conduction from the directly resistively-heated wall to the test sample is reduced almost completely. In this case, radiation from the wall is the main heating mechanism for the platform and the sample [223]. Because of the heat capacity of the platform and the characteristic of radiation energy transport, its heating is delayed compared to that of the tube wall. Because the temperature of the gas atmosphere inside the tube follows the temperature of the wall more closely [224, 225], sample deposition on the platform may lead to analyte vaporization/atomization into a hotter gas atmosphere.

In [206], a simple model was proposed for calculating the temperature courses of the platform by setting the heat transfer from radiation equal to the heat storage of the platform body. Thus the following relation is derived

$$\frac{dT_p}{dt} = \sigma C \left(T_w^4 - T_p^4 \right) \frac{A_p}{c \, m_p} \qquad \qquad \text{Eq. 3.1}$$

where
T_w, T_p =temperatures of wall and platform respectively
σ =radiation constant (5.67· 10^{-8} w/m^{2}K^4)
A_w, A_p =surface area of wall and platform respectively
m_p =mass of the platform
c =specific heat of the platform material
C =heat exchange constant

For the model of totally surrounded surfaces which includes the relation of the heat exchanging surfaces of wall and platform, and the respective *emissivities* ϵ_w and ϵ_p (e.g. near 1 for graphite and charred organic samples), C may be estimated according to

$$C = \left[\frac{1}{\varepsilon_p} + \frac{A_p}{A_w} \left(\frac{1}{\varepsilon_w} - 1 \right) \right]^{-1}$$

Eq. 3.2

Figure 3.2 shows how the the temperature of the platform follows that of the wall temperature, calculated by Eqs. 3.1 and 3.2. More sophisticated considerations, also based on this model, are derived and discussed in detail by Chakrabarti et al. [223].

If it can be assumed that the temperature of the sample is directly coupled with the deposition area – as it is with liquid samples placed and dried on the platform – the experimentally observed effects of sample vaporization/atomization from a platform can be understood:

- From the shifted signals, which appear from samples atomized form a platform, it is obvious that the platform temperature is significantly delayed compared to the tube wall temperature.
- The average gas atmosphere temperature inside the tube is also heated with delay and significantly lower than the tube wall temperature. Its heating progress can be assumed to be between the courses of tube wall and platform temperature [223, 225, 226]. Thus the sample is vaporised into a hotter gas atmosphere than with "wall atomization".
- From Fig. 3.2 it can be seen that the heating characteristic of the platform has three sections: In the first, only a very flat incline occurs (delay period). Then the temperature gradient is steeper than that of the wall, thus for elements with an appearance time in this region, the atomization process takes place over a shorter period and this leads to "slimmer" signals. In the last region, the temperature of the platform approaches that of the wall in a flattened

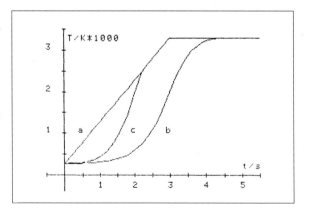

Fig. 3.2. Temperature course for a platform (*b*) in relation to the igniting wall temperature (*a*) calculated by Eqs. 3.1 and 3.2 (parameters: A_w=10 cm^2, A_p=1 cm^2, m_p=138 mg, ε_w=ε_p=0.95). (*c*) shows the hypothetical temperature of a test sample of m_s=2 mg (=m_p), A_s=0.1 cm^2 (=A_p)

incline. For the refractory elements with an appearance temperature in this region the atomization process takes place in a longer period, leading to broader signals than with wall atomization.

Since the introduction of the platform technique it was frequently shown that interferences by concomitants – mostly from a salt matrix – are drastically reduced, particularily for the volatile elements (e.g. Cd, Pb, Tl) [227, 228].

Although the vaporization/atomization is delayed when using a platform, the more volatile elements may appear in the gas phase when the wall temperature is still rising. So, complete isothermal conditions are not always reached with the platform technique. Unfortunately, for solid samples these may still be adversely affected (see discussion below). However, because of the simplicity of its technical realization, it constitutes such an obvious improvement that this technique is offered today by all manufacturers of AAS-instruments.

Nearly ideal temporal isothermal conditions can be reached if a second, separately heated graphite crucible is used to vaporize the sample outside the absorption cell (two-step atomizer). [2] If the vaporization products of the sample enter a *preheated absorption cell*, temporal isothermal conditions are reached [229, 230]. Solid sampling using such a unit has led to accurate results [14]. Lundberg et al. stated: "The furnace also easily lends itself to the direct analysis of solid materials, because of the controlled vaporization, the ability to standardize against an aqueous calibration curve and the fact that the light path is not at all obstructed by the sample" [231]. However, the mechanical and electrical connections of two independent heated graphite parts are technically complicated and require additional expenditure.

Preheating of the absorption volume is also possible using the probe technique [192, 211]. If the furnace reaches the atomization temperature before the test sample is introduced by the means of a sample probe, the atomization may take place under nearly isothermal condition. This technique can only be applied to previously dried samples because liquid droplets would "explode" when introduced into the hot tube. The introduction of the probe can be performed through an enlarged hole in the side of the tube. This technique has been applied to commercial AAS-instruments [211].

Strictly isothermal conditions have to be characterized not only temporally – as discussed above – but spatially as well, i.e. additionally a constant temperature profile over the absorption cell must be reached. With the graphite furnace of the Massmann type, a pronounced temperature gradient arises because of the cooled ends of the tube. This effect is responsible for the appearance of memory effects and other interferences. With the modern side heated tubes (integrat-

---

2) The electrode furnace developed by L`vov basically corresponds to a two step system, using moving graphite parts [3]. This partly explains why with this atomization unit, successful analysis of solid materials was achieved even in its very early days, in contrast to the simplified Massmann furnace system.

ed contact cuvette (ICC) the absorption cell is heated spatially nearly isothermally. Frech et al. achieved a significant reduction in non-spectral and spectral interferences with this invention [221, 232] (see Fig. 2.4f).

Of course it would be preferable to couple the two-step technique with a side-heated absorption tube [231], however technically it seems to be a very complex instrumentation. However, with a side-heated graphite tube and platform atomization, spatial and temporal isothermality is reached to a great extent [202] even at temperatures which are a compromise for multielement determination [233].

If the graphite tube is heated at a superfast heating rate of about 10^5K/s [234] by the power support of a capacitive discharge, the rise of the gas temperature in the absorption cell is so fast compared to the vaporisation of the sample, that the atomisation conditions are "quasi-isothermal". Because the necessary electrical/electronic control is technically difficult, this technique has not found wide acceptance.

3.1.2.2
Atomization Characteristics for Solid Samples

Transient signals from liquid and solid samples often show distinct differences in appearance time and shape. In the literature, differences in the chemical bonding of the analyte are considered in the first place when signals are interpreted. However, the atomization behavior is not only determined by e.g. the bonding energies, but also by the thermo-physical properties and conditions in the furnace. Because these effects are often overlooked, these will be discussed below with respect to sample deposition on the wall or on the platform.

With *wall deposition* of reference solutions, the temperature of the sample material is strongly coupled to the wall temperature because after the drying step, the analyte is enclosed in a thin salt layer on the graphite surface. Consequently the vaporization/atomization may start immediately when the wall reaches the respective temperature. Test samples introduced as solids have substantially less thermal contact with the wall so that the heating of the material is hampered. Thus the signals are delayed in comparison to analyte signals from reference solutions [183, 220, 235, 49]. This effect is shown for lead in biological material in Fig. 3.3 a (compare also Fig. 3.8).

With deposition of solid test samples from organic materials on a *platform* the appearence of the signals may behave in the opposite manner, i.e. these from reference solutions may be delayed and broadened compared to these from solid materials. Figure 3.4 show examples of this behavior. These effects can be understood by considering the different thermo-physical conditions of the liquid and solid samples on the platform: Again, it must be considered that the solid material is not directly coupled to the temperature of the platform. So, – described very simply – the solid material itself acts as a platform, i.e. heating by radiation becomes the dominating mechanism over heating by conduction from the platform [206]. Because of the very small "thermal mass" of the sample material, the

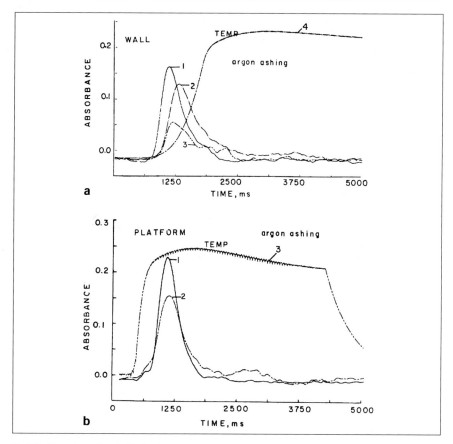

Fig. 3.3. Transient signals for lead atomization (ashing temperature 880 °C); *trace 1*: 0.5 ng Pb (as nitrate) in a reference solution; *trace 2*: approx. 1.1 mg Bovine Liver (from: [183], with permission).
a Test samples deposited on the *tube wall*; wall heating rate 0.75 °C/ms (*trace 4*).
b Test samples deposited on a *platform*, wall heating rate 3.0 °C/ms (*trace 3*)

temperature delay is reduced and its gradient is steeper than that of a dried reference solution, out of which the analyte can only be atomized when the platform has reached a sufficient temperature [236].

Trace c in Fig. 3.2 gives the temperature for a hypothetical sample mass of only 2 mg heated by radiation that is not thermally coupled to the graphite platform, e.g. as a model for spongy carbon residues (charcoal) from a biological sample after the pyrolysis step. It shows a significantly smaller delay than the temperature of the platform. At low temperature gradients of the wall, the peak shapes are also significantly different to those from aqueous reference solutions due to this effect, however, these become more and more similar and the delay between the signals is small if the temperature gradient of the tube (i.e. wall) is increased (compare Fig. 3.3 b and Fig. 3.4 d). In [237] these observations are confirmed with the atomisation out of the residues of polymer sorbent particles.

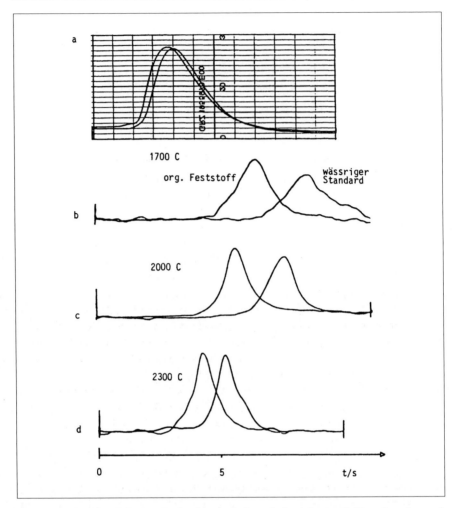

Fig. 3.4. Transient signals for atomization from a *platform*; the "early" traces (*left*) are these from reference solutions; the "late" traces (*right*) are these from solid test samples of "Orchard Leaves":
a Atomization of Cu, wall end temperature 2600 °C.
b–d Atomization of Cd at different wall end temperatures

It should be noted that a double and irregular signal from a *slurry* may have originated from this effect, due to fractions of analyte that behave thermally differently, i.e. partly extracted and thermally coupled to the platform and partly enclosed in the solid material and heated mainly by radiation (e.g. see [238]). A reduced delay for Pb atomization from the platform is also observed for sediment slurry samples [239].

So, to a certain extent it must be reckoned with reduced isothermal conditions for analyte atomization, e.g. if solid materials (in first order organic samples) are deposited directly onto the platform.

Fig. 3.5. Transient signals for lead atomization from reference solutions; without modifier (*trace 1*); with 100 mg $Nh_4H_2Po_4$ + 10 mg $Mg(NO_3)_2$ (*trace 2*); with 32.5 mg $Pd(NO_3)_2$ + 10 mg $Mg(NO_3)_2$ (*trace 3*, ×5 abs. scale expansion) (from [14], with permission).
Trace a shows the temperature of the tube wall (HGA600), *trace b* the vapor phase temperature if the tube was equipped with a platform, and *trace c* the vapor phase temperature if the tube was equipped with a cup-in-tube

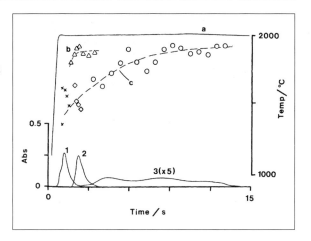

Matrix modification may be an appropriate measure to delay analyte atomization, if interferences due to unsufficient isothermal atomization out of a solid material are presumed. Figure 3.5 shows the influence of two modifiers to the peak shape for lead atomization from reference solutions. It can be assumed that a modifier can shift the signals from biological materials into the more isothermal region as well. For example, in [240], it was shown that the atomization of cadmium out of a tabacco leaves slurry is significantly delayed by a modification by 15 µg of palladium (as nitrate).

Analyte atomization out of *refractory materials* (e.g geological and metallurgical samples) is determined by further thermo-physical properties to a notable extent. The thermal contact to the deposition area and heat conduction through "compact" materials cannot be neglected. Furthermore, the heat capacity of the samples may be significant larger. Hence, the vaporization/atomization behavior for these materials cannot be described adequately by the model characterised by Eqs. 3.1 and 3.2. Additionally chemical processes of the analyte/matrix/graphite complex in solid-liquid-gasious phases may have a strong influence. Because these effects hamper the vaporization/atomization process, the signals for these materials are delayed so that improved isothermal conditions can be supposed.

Nevertheless, atomization from a platform has proven to reduce vapor phase interferences for solid sample to a degree that true results can be achieved. The advanced technique of side-heated tubes promises a substantial improvement also for the direct analysis of solid materials.

In Sect. 2.1.3, the necessity of a suitable sample introduction system for solid samples – preferably realized as a carrier – is discussed. The atomization process can be crucially effected by a sample carrier if it remains in the furnace during the measurement step. Considerations (or investigations) must be carried out thoroughly if a new system is used to ensure that isothermal atomization conditions are fullfilled. A general prediction cannot be made for the behavior of a specific solid material in a given furnace/sample carrier configuration.

Following the description of the *different sample introduction systems* in Chapter 2 (see Table 2.2 and Fig. 2.5) these will be discussed below with respect to their influence to the atomization condition.

Obviously, the *platform boat* [206], the *solid sampling platform* [204], and the *sampling platform* [202] are acting as a L'vov platform if these carriers are placed in the furnace as the sample holder, thus providing temporal isothermal atomization conditions as far as it is feasible with this technique (compare discussion above). Heat conduction is not prevented with the *microboat* [199] because of the rectangular cross section of the accessory tube. No experiments have been published from which a potential reduction of the vaporization/atomization delay can be recognized. Simple modifications of the boat or the tube in order to minimize the contact area can lead to improved isothermal conditions.

The *miniature cup* [207] is in close thermal contact with the tube wall. However, because of the enlarged cross section of the tube due to the integrated cup in the bottom, the temperature gradient of the support area is flatter or even delayed. Although no data are available from which a delay in the atomization can be read, it can be supposed that the vaporized sample is entering a gas atmosphere which is already in a quasi-isothermal state [122]. The observation of a distinct increase in sensitivity if the sample is placed on the miniature cup [9] is an indication that radiation is the dominating heating mechanism for the vaporization/atomization process. Similar considerations are valid for the *Mini-Massmann graphite cup* [137]. It was shown that for the inner cup, isothermal conditions are achieved if an adequate design is chosen.

At first glance, it seems that the *cup-in-tube* may act as a L´vov platform as well. Although the graphite body of the cup-in-tube is heated mainly by radiation from the tube wall, it must be noted that the cup shields the inner space of the absoption volume from radiation from the heated tube wall. As a consequence, the heating characteristic for the gas and the sample is not dominated by radiation but by heat conduction from the inner surface of the cup-in-tube [236]. The earlier signal appearence from reference solutions compared to these from solid samples confirms this assumption [220]. Because of its large mass, a significant delay in atomization can be observed in comparison to direct wall or even to platform atomization [155]. This has led to the erroneous impression that, with this technique, enhanced isothermal conditions can be achieved [151]. However, a distinct delay *between gas and sample temperatures* (inside the cup i.e. in the atomization volume), which is responsible for reduction of gas phase interferences – cannot arise to the same extent as with the platform. This was experimentally shown by Baxter and Frech [14], whose results are shown in Fig. 3.5 (trace c). It must be suspected that the reduced isothermal conditions inside the cup-in-tube are partly responsible for difficulties which have been reported with solid sample analysis [151, 150, 241, 242, 243].

The probe introduction systems (the *autoprobe* [211] and the *solid sampling probe* [203]) are acting as a L´vov platform if these are introduced into the tube before starting the atomization cycle. Enhanced isothermal conditions are provided if the probe is introduced into the preheated tube. As a further advantage, it has been mentioned that such a system allows thermal pretreatment prior to the introduction, thus reducing the "potential vapor phase interferences and that part of the background absorption which would have been otherwise caused by the smoke condensed at the cooler end of the graphite tube" [183] (in principle this is possible with all removable sample carrier systems as well). Because of the extreme increase of the temperature gradient that the sample is exposed to, the preheating technique may lead to a significant increase in sensitivity [55]. In particular, the furnace design of Pye Unicam instruments allows a simple introduction for solid test samples by a sample carrier. Lücker et al. [203] have modified this system for a better application in direct solid sample analysis. However, attention must be paid to smooth movement with the transportation and introduction of the sample carrier [183] (this is a general requirement for powdered samples).

Kitagawa et al. [244] used a constantly heated *glassy carbon tube atomizer* (power >7 kW) for the analysis of biological solid samples, which where introduced by a small cup probe system. The absorption was observed radially through holes in the tube wall. The sample vapor and the carrier gas passes the "package" of activated charcoal, leading to isothermal atomization conditions in the absorption volume. It was found that measurements were "free from background absorption".

The *dual-chamber furnace* [245] and the *ring-chamber tube* [205] have been developed in order to eliminate interferences in the optical light path by the sample body and the sample residues. Additionally, a delay in the vaporization occurs so that the atomization takes place under more isothermal conditions than with (inner) wall deposition. The enlarged cross section of the graphite as the electric current conductor produces a lower temperature in the tube center and thus the axial temperature gradient is "flattened" [205]. In combination with the sample placement outside the absorption volume, a delay in the sample vaporization may be postulated. The observed decrease in sensitivity is not necessarily a disadvantage for the direct analysis of solids, however, for the more refractory elements the temperature in the outer chamber may not be sufficient for complete volatilization.

3.2
Sensitivity and Range of Analyte Determination

In atomic absorption spectrometry, the general absorption law (Bouguer-Lambert-Beer) is only valid in a relatively small range of analyte density in the absorption volume because the emitted lines cannot be achieved strictly monochromatically (line broadening e.g. caused by Doppler effect, hyperfine structure, also e.g. stray light, incompletely separated emission lines). This leads to deviation from linearity with increasing absorption level. If the Zeeman effect for background measurement is created, the spectroscopic situation is even more complicated by incomplete separated split lines or/and an anomalous Zeeman pattern [213], which may lead to enlarged unlinearity.

Since modern AAS-intruments are equipped with software which permit determination to be made in the region of curviture, the usable "working range" in terms of analyte mass will be between two and three orders of magnitude [14].

If the analyte amount in the samples is too large with respect to the usable range of the calibration curve, in conventional solutions analysis, these can be easily diluted (even automatically performed by the autosample system). For solid samples, this treatment is usually inadequate for reaching a suitable analyte amount. The equivalent step for solids will be the reduction of the test sample mass. However, this measure may be limited e.g. by sample heterogeneity (see Sect. 2.3) and the problematic handling of very small microgram amounts. For the extension of the working range, two alternative measures may be considered, reduction of line sensitivity or setting of an insensitive analytical line and these are discussed below.

3.2.1
Selection of a Suitable Analytical Line

With fixed setting of the instrumental conditions (e.g. analytical line, temperature program, gas flow), the range of analyte detection is given, in GF-AAS expressed in terms of analyte mass a_L-a_U for the lower and the upper limit. The determination of the (lower) *detection limit* a_L and examples are discussed in Sect. 2.5.2. Generally one can make use of a range not exceeding two to three decades, even if curviture of the calibration curve is accepted (see Table 3.2).

The range for element determination for solid materials is given by the *lower and upper limits of analysis c_L and c_U, respectively,* according to the equations

$$c_L = \frac{a_L}{m_{min}}$$
Eq. 3.3

$$c_U = \frac{a_U}{m_{min}}$$
Eq. 3.4

The range of the *test sample mass m_{min}-m_{max}* is mostly restricted to two orders of magnitude, e.g. 0.05–5 mg.

Generally the lower limit of analysis c_L for the direct analysis of most solid materials with GF-AAS is sufficient even in the case of very low contents, it is at least one order of magnitude better than with slurry or decomposition analysis.

Example

Docekal and Krivan [180] compared the limit of analysis with GF-AAS of direct solid sampling and after prior sample decomposition. For the analysis of molybdenium metal they found the limit of analysis for the determination of Mn by

Table 3.2 Working range of the most sensitive and some alternative (insensitive) lines with the ZAA spectrometer SM20/30 using the platform boat. For comparison some attainable values for the lower limit of detection of other graphite furnace/sample carrier systems are listed: The miniature cup show comparable values, while the cup-in-tube represents a more sensitive system (for the latter the characteristic mass values are only found in the literature).

Element	Line (nm)	Working range/limit of detection/characteristic mass(*) (ng)		
		GA SM 20/30 platform boat	PE cup-in-tube [155, 150, 241]	Hitachi miniature cup [86, 312]
Ag	328.1	0.005-1		
	338.3	0.5-50		
Al	309.3	0.05-10	0.01-2	0.07
	256.8		0.2-40	0.5
As	193.7	0.05-20		
	197.2	1-80		
Au	242.8	0.01-5		
	267.6	1-100		
Cd	228.8	0.001-0.2	<0.001*	0.002
	326.1	1-100	0.41*	2.5
Cr	357.9		0.004*	
	359.3	0.05-5		
	427.5	1-40	0.01*	
	301.7		0.22*	
Co	240.7	0.1-40		0.02
	241.2			0.1
	352.7			1.0
Cu	324.8	0.03-10	0.011*	0.02
	327.4		0.015*	0.06
	249.2	10-1000	0.42*	
	244.2	100-6000	2.2*	
Fe	248.3	0.02-10		
	344.1	10-150		
Mn	279.6	0.01-2		0.007
	403.1	0.1-10		0.06
Ni	232.0	0.2-15		0.2
	305.1	5-200		
Pb	383.3	0.02-4	0.01*	0.04
	261.4	5-150	1*	15
	368.3	25-1000	2.4*	20
Si	251.6	0.5-100		
	288.2	50-4000		
Te	214.3	0.1-15		
	225.9	50-1000		
Tl	276.8	0.01-2		
	377.6	0.5-15		
	535.1	50-5000		
Zn	213.8	0.002-0.1		0.004
	307.6	0.5-200		1.0

direct solid sampling to be 0.09 ng/g (m_{max}=80 mg) and after decompositon only 20 ng/g (m_s=200 mg). Because with solid sampling of molybdenium silicide, the molten material crept from the carrier if the test sample was larger than m_{max}=10 mg, the limit of analysis was 0.7 ng/g. Similar differences in the limits of analysis between the methods were found for Cu, K, Mg, Na, Zn. The authors concluded that "only highly intrumental methods such as GDMS and RNAA, which are seldom applicable to routine analysis, provide comparable limits". (Compare also examples in Sect. 2.5.2.)

In AAS, the upper limit of analysis c_U is determined primarily by the unlinearity of the calibration curve, i.e. decreasing sensitivity, if the calibration curve is bent to the analyte amount axis. Accordingly, at a certain value of the analyte amount a, an unacceptable imprecision occurs due to the declining sensitivity that is represented by the value of a_U for the given application. For the sensitive analytical lines, which are commonly used in conventional AAS, the upper limit of analysis often causes problems in direct solid sampling. For some elements which are present in the environment at a relatively high level, these lines are often unsuitable for the analysis of solid materials.

A reason for the frequent "misadaption" of sample content and the upper limit of detection a_U for the sensitive line is caused by the development of modern AAS-tools which gives extreme detection power. Extreme low sensitivity is often required in order to compensate for the dilution of analyte content that the original solid sample undergoes due to the chemical decomposition process. In addition to improved spectral sources, sensitivity was increased by smaller absorption volumes (graphite tubes) and faster heating gradients. Thus a large graphite tube in direct solid sampling is not only advantageous for the handling of solid test samples (discussed in Sect. 2.1.3) but also because this is relatively insensitive, showing generally a better adaption to the content of real (solid) materials. In Table 3.3 a comparison of the range of analysis for AAS-instrument SM20/30 given, from which it can be recognized that the relatively large (insensitive) graphite tube of these instruments both ranges mostly fit to high degree.

Fortunately for most elements *insensitive analytical lines* exist and can alternatively be used for analysis as well (see Table 3.2). Commonly in solid sample analysis, a rough calculation considering the approximate content, suitable sample mass and given detection range is performed before setting the instrument in order to choose the most suitable line (Eqs. 3.4 and 3.5 and see Sect. 2.5). Using solid sampling in routine analysis for a variety of elements, it is worthwhile building up tables of the limits of detection, test sample masses, and ranges of analysis for the different analytical lines, the sample materials to be analyzed, and different instrumental conditions.

Example

For the determination of cadmium, the resonance line 228.8 nm is generally used, whose range of detection may be 1 pg–0.2 ng (GA SM 20). If the sample

Table 3.3 Comparison of the range of analysis for direct solid sample analysis with the graphite tube (SM20/30, Grün Analysentechnik) and typical element contents in biological materials

Element	Analytical line (nm)	Working range[a] ($\mu g/g$)	Range of content[b] in plant materials
As	193.7	0.1-200	<0.05 - 20
Cd	228.8	0.005-2	0.05 - 2
Cr	359.3	0.1-50	0.02 - 5
Cu	324.8	0.01-100	1 - 100
Fe	344.1	0.04-100	20 - 1000
Mn	403.1	0.02-20	0.2 - 300
Ni	232.0	0.5-100	0.03 - 2
Pb	283.3	0.05-40	0.03 - 50
Zn	307.6	1-2000	10 - 200

a) Range of analysis estimated from 10 x lower limit of detection to upper limit of detection (see Table 3.2) and a range of test sample masses 0.1-5.0 mg
b) Ref.: Saari E., Paaso A. (1980) Acta. Agric. Scand. Suppl. 22, 23; list of biological CRM/SRM Table 2.2)

mass can be varied between 0.1 mg–10 mg, the range of analysis (c_L–c_U) for the sample content can be estimated as being 0.1 ng/g–2 mg/g. Thus materials highly contaminated with cadmium such as sediments, sludges, and fly ashes cannot be analyzed on this line. Because the analytical line 326.1 nm is 2 to 3 order of magnitude less sensitive, these materials can be analyzed easily on this line.

This example shows that for cadmium using the two lines, the contents ranging from low ng/g (ppb) up to several thousand $\mu g/g$ (ppm) in solid materials can be analyzed, although there might be a small "gap" in between the two lines. Unfortunately not for all elements do lines of close-fitting detection ranges exist (compare Table 3.2).

The very insensitive lines are mostly non-resonant lines, i.e. the lower atomic state is not the ground state. In this case the electron population of the lower state is temperature dependent, hence the sensitivity is temperature dependent as well. For the non-resonance line of lead at 261.4 nm, Baxter and Frech calculated that "should the mean temperature prevailing during the standard singal be 100 K higher than that for the sample concentration … the systematic error is… -22%" [14]. As discussed above, the appearence of signals of reference solutions and solid samples may differ, thus "although good results have been reported, much more care is required to ensure that either the temperature is strictly constant or the analyte signals from both standards and samples appear in the same interval" [14].

3.2.2
Extension of the Analytical Range

The lower limit of detection is a fixed value for a given atomizer/spectrometer system and if all instrumental parameters are optimized (e.g. largest possible slit width, maximum temperature gradient, gas stop). A further improvement in the

limit of analysis is possible by increasing the test sample mass. However, this
parameter is also limited by the given instrumentation and the analytical behavior of the material (see discussion in Sect. 2.5).

In order to enlarge the mass that can be introduced into the sample carrier and
the graphite tube respectively, Sonntag [110, 198] used the *pellet technique*. Powdered biological materials were compressed into a pellet of 5 mm diameter using
a hydraulic press (SPECAC, UK). Up to 30 mg of test samples could be loaded into
the platform boat. Thus, the limit of analysis for Tl and Pb was lowered, so that
low-contaminated materials could be analysed successfully (e.g. for Tl sample
contents down to 0.03 mg/kg). This similar technique was described earlier from
Lord et al. [46] for the direct analysis of mussels with GF-AAS. Homogenized
materials were mixed with pure solid naphthalene and pressed to a pellet
between nylon pistons in a uniform bore, thick walled, glass tube.

In practice, it is often the upper limit of analysis, rather than the lower limit of
analysis that causes problems in direct solid sampling. For a more detailed discussion on the dilution of powderd solid samples see Sect. 2.5.4.2. If a minimum test
sample mass yields absorption values out of the usable range, two instrumental
parameters – the gas flow and temperature program – can be easily varied in order
to shift the working range of a chosen analytical line towards a larger value of the
upper limit of detection by "dilution" of the atom density in the absorption volume.

A higher internal *gas flow* through the furnace during the atomization step
lowers the residence time of the atoms in the tube, and thus the senisivity can be
reduced. The upper detection limit can be raised by the factor of about 3, if the
internal gas flow is increased to 50–100 ml/min where the atom transport is dominated and forced by the flowing gas (instead of diffusion in the case of "gas stop"
conditions) [14, 80, 243, 246]. However, an extreme gas flow should not be used
[150] because the vapor phase temperature may be significantly lowered which
might increase the occurrence of interferences [14]. Additionally sample particles may be blown out by an internal gas "storm" [247].

Larger analyte masses can also be accommodated by a *reduction in the temperature gradient* during the atomization step. This results at a lower atomization rate and diffusion loss from the furnace can often maintain the atom density within the furnace within a measurable range.

Reduction of senitivity was described when the *deposition of the solid sample*
in the tube was changed [206]. For Cd, a reduction of 50% occurs if the sample
is unidirectionally moved 8 mm out of the center of a side-heated tube [49].

These procedures may be appropriate for setting a suitable working range for
a given analyte content of the solid sample which is to be analyzed. However,
since each measurement may be connected with a deterioration of the isothermal condition, a combination of these instrumental settings or the adjustment
of one parameter to an extreme extent should be avoided.

This problem can be overcome to a certain extent if the *two-step atomizer* is
used because the atomization-tube temperature can be selected independently
of the volatilization-cup temperature (see discussions above). Baxter and Frech

[14] pointed out that "by selecting a slow cup heating rate it should therefore be possible to gradually volatilise large analyte amounts without causing the maximum absorbance to reach the level where the signal is no longer linear ...". They showed that, with this technique, the linear range of the calibration curve can be extended, e.g. for cadmium (228.8 nm), by more than three orders of magnitude.

If the pressure in the absorption volume is decreased, the rate of diffusive loss increases and the absorption profile of the analyte atoms becomes more narrow as a consequence of reduced "collision broadening". Accordingly absorption is reduced leading to a lower sensitivity in AAS. Using a *reduced pressure furnace,* it was found [248] that a reduction from atmospheric pressure (~1 bar) down to 0.0013 mbar (0.001 Torr) reduces the sensitivity of resonance lines by more than two orders of magnitude. At a pressure of 13 mbar – which is more easily achieved – the reduction is still more than one order of magnitude (Cd, Cu, Mn, V). For the analysis of solid samples, Holcombe and Wang [49] used a side-heated tube (integrated contact cuvette, ICC), which is enclosed in a stainless steel chamber equipped with quartz windows through which light from the hollow cathode lamp is passed. The solid test sample is transferred and delivered into the furnace with the help of a sample holder and a stainless steel crucible through the side window. A small amount of water was added on top of the sample before drying to prevent expulsion of very fine particles during evacuation. For the determination of Cd in a coal fly ash sample the test sample mass was limited to 0.1 mg (sensitive resonance line), while at a pressure of 13 mbar it could be increased to 2 mg. Consequently the precision of the measurement was significantly better (~10% instead of ~20%). For the determination of Cu it was shown that the sensitivity can be controlled by an appropriate adjustment of the pressure, so that the optimum test sample mass could be used (here: 0.5–2 mg), e.g. for a sample containing ~20 µg/g, the analysis was carried out at a pressure of 13 mbar, while for ~100 µg/g, 0.026 mbar was more suitable.

The authors reported further effects, which may be connected with *atomization at reduced pressure*:

- The absorption peaks are found to be sharper at lower pressure for Cu, Mn and V. "The generally observed reduction of tailing at reduced pressures may be a reflection of the reduced interaction of the element and the graphite surface for those elements which exhibit stronger metal/graphite interaction". For Cd down to 13 mbar no significant change of signal shape was found, but at 0.007 mbar the signal was significantly broadened – "this undoubtedly reflects a change in the vaporization characteristic" [248].
- The background, like the analytical signal, may be reduced and even can often be neglected under low pressure atomization [49].
- The well known signal depression for Cd in the presence of NaCl was observed to a greater extent at a 13 mbar pressure than that at atmospheric pressure. A slight similar effect was observed also for Mn (both effects were seen using atomization from the wall) [248].

For making use of the reduction in sensitivity, the space in and around the graphite tube must be evacuated to constant pressure. For this purpose, either the furnace design has to be modified so that it is vacuum-tight [249] or the complete furnace must be placed in a vacuum chamber [248]. Additionally, a vacuum system is required, including a pump, vacuum lines, gauges, and valves. Further, a control system for pressure, atomization program and sample handling (vacuum tight introduction port) may be useful. This expenditure may only be justified if solid sampling is used as a routine method when it will give notable advantages. In this case, however, the proposed technique may offer flexibility for direct solid sampling under optimum analytical conditions.

In addition to the instrumental measures, *signal evaluation* may also give us the opportunity to extend the working range.

Koizumi and Sawakabu [250] proposed the evaluation of the AAS transient signal by the *width of the peak* at a fixed absorbance level e.g. 0.1 (similar to the "half-width" concept). Correspondingly, the dependent variable is not the absorbance but the time (-interval of the peak). Irregular courses at high absorption values, e.g. "overflow" due to extremely high analyte amounts in the furnace, do not influence this measurement. They showed that – despite reduced sensitivity (i.e. slope of the calibration line) – with this technique, extremely small amounts of analyte contents are quantitatively detectable on the sensitive lines. This approach was made to overcome the disadvantage of the "roll-over" effect in ZAAS (see below), however, there is no reason why this technique may not be generally applicable.

Sommer and Ohls [107] applied *multible atomization cycles* on each test sample for the determination of Mn and Cr in alloys or oxides. Because of the extremely high content (compared to traces), the first signal will be outside the dynamic range and the total amount of the analyte is not released in a practicable atomization interval. Further atomization cycles gave successive decreasing signals, which were evaluated. For this procedure the vaporization rate method (see Sect. 2.2.2.4) would be an appropriate calibration.

L´vov et al. proposed a model of *linearizing AAS calibration curves* [251, 252] that was further extended for Zeeman-AAS [253]. The calibration curves in Zeeman-AAS are characterized by stronger unlinearity and the "roll-over" effect [213]. The latter means that at high analyte densities in the absorption volume, the background corrected absorbance is smaller than that at lower densities, leading to a "dip" in the center of transient signals from large analyte amounts. For the proposed calculations of linearization and straighten up calibration curves, three parameters are needed: characteristic mass m_0, roll-over absorbance A_r and sensitivity ratio R (Zeeman – to conventional AAS). The authors pointed out that the values for these parameters are easily determinable by the use of only two reference solutions. With the evaluation of calibration curves based on the proposed model, the upper limit of the (linear) analytical range was extended by 1.6–1.8 fold. Of similar importance for solid sampling might be that "the method provides constancy of the analytical signal irrespective of possible variations in the shape of peaks for different matrices". However, "among the drawbacks of the method are the transition in the dip region from the Zeeman-effect background correction to the less efficient contimuum background correction" [253]. Nevertheless, the positive features of the "straighten-method" – if verified in future experiments – promise particular advances for solid sampling analysis.

3.3
Experience with Direct Solid Sampling

This section will discuss concrete experience with the analytical behavior of particular matrices and elements as far as these are not apparent from the general discussion in other sections of this book. This will concern various aspects. However, as generally in this book, these considerations are not given as "cookbook"-instructions, but as information for developing one's own method.

3.3.1
Particulars of Sample Materials

The analysis of **organic materials**, particularly of biological origin, have been successful in particular since the introduction of Zeeman background correction and the techniques of isothermal conditions. The atomization conditions for these materials with platform atomization have already been discussed above in Sect. 3.1.2.2. With these kinds of material, the advantages of the graphite furnace show to full advantage. Chakrabarti and coworkers [10, 11, 183] performed some basic investigations into the analytical effects and conditions with the determination of numerous elements in plant and animal tissues and these show that accurate results can be achieved without sophisticated analytical procedures. Applying the advanced techniques in GF-AAS Rosopulo et al. [8] gave a first comprehensive methodological description for direct sampling of biological materials.

Even though the matrix composition from organic materials can vary considerably, matrix differences generally have a minor influence to the signal shape. This can be understood by different recurrent observations:

- Calibration lines from different biological materials show identical slopes (in the limits of the measurements uncertainty) [205, 9, 254, 8] (compare Fig. 2.8),
- the evaluation from peak height and peak area often yields nearly identical results [254, 8],
- accurate results can be obtained with a calibration based on reference solutions (see Table 3.5).

However, these positive feature cannot be taken for granted if unknown materials are to be analyzed. Distinct interfering matrix effects may occur if the particular biological material has an unusually large salt content (e.g it is of aquatic origin), silicates (e.g surface contamination by dust or soil particles), or a large content of signal depressing species (e.g sodium). Hence, it follows that even for samples of this "easy" matrix a specific method developement and analytical quality control should not be neglected. For this purpose a limited but adequately large number of CRMs should be available in the laboratory.

The analysis of **geological materials** such as *stones, minerals, soils, sediments* has been a favorable attempt to use direct sampling with GF-AAS since its invention. The thermal decompositon in the graphite furnace seems to be as effective (complete) as the methods of chemical sample decomposition. As long ago as 1977, Langmyhr [4] gave a review of the applications on this field that deal with all analytical essentials.

The geological/mineralogical matrix has a strong influence on the atomization process. Distict differences in peak shape (apperance time, slope, width, structure, tailing) between different geological materials and from a reference solution are generally observed. Figure 3.6 a and b shows examples for analyte signals for these materials. The variance in signal shape reflects the complex physical and chemical processes which happened nearly simultaneously during the atomization step, so that thermodynamic processes are extremely difficult to understand, e.g. the rates of vaporization, dissociation, atomization and diffu-

Fig. 3.6. Transient signals of atomic absorption signals from geological samples in comparison of these from standard solutions:
a Signals from the determination of Cd in standard solution (*a*), Calcareous Loam Soil (*b*), Light Sandy Soil (*c*) (from [178], with permission).
b Signals from the determination of Pb in standard solution, and three rock reference materials, for matrix modification 20 ml of a 1% H_2PO_4 and 5% $(NH_4)_2HPO_4$ mixture was added on top of the test sample (from: [256], with permission)

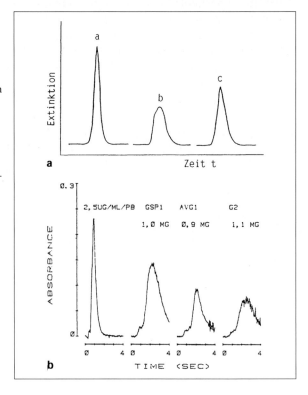

sion in the presence of numerious matrix species in solid, liquid and/or gaseous states. Additionally the properties of the involved species are often not known and even not determinable (e.g. the place of the analyte element in relation to the surrounding matrix inside the mineral or crystallinity of these minerals).

So the different attempts at thermo-chemical models at the present stage are only concerned with isolated effects and are generally not satisfying and may not be helpful yet for method development in solid sampling with geological samples (compared to this situation, those of the volatile components of biological materials or even to the comparable "pure" metallurgical substances are simpler). It is to be hoped that the efforts and the advances in this scientific field will spread to solid sample analysis and lead to further progress in the future.

However, measurements consequently evaluated by the peak area of the analytical signals yielded true analytical results even based on reference solutions [255, 190, 256, 151, 257]. The application of AAM may be successful in overcoming matrix effects (e.g. [96], compare the example in Sect. 2.2.2.3), but numerous workers recommend CRMs as appropriate for calibration [154, 96, 192, 178, 258]. Kersabiec and Benedetti [122] recommend selecting a CRM of the same "petrological family". Fortunately, today in this field, a very large number of reliable reference materials are available [70]. The matrix composition of numerious geological/mineralogical CRMs is quite well specified, providing an adequate selection.

However, in an extensive study Schrön et al. [258, 259, 260] found no general strong matrix effects between different materials. In Figs. 3.7 a-d, their results for some elements in different geological reference materials are visualized. From these data, the authors conclude that for the elements investigated, a matching of a calibrant reference material with the unknown sample seems generally *not* to be required even some deviations are found when comparing different reference materials. However, it should be stressed again that if a discrepancy to a reference value of a CRM is observed, this does not necessarily point to a matrix effect with this particular material, instead it may be caused by the uncertainty (i.e. bias) in the certified value (compare Sect. 2.4.2.4).

Example

The statement above should be discussed for the determination of Pb in "Gabbro" MRG-1 (see Fig. 3.7 c). This shows a distinct bias compared to the other materials under investigation (different slope of signal vs analyte mass, the latter calculated from the reference value, for MRG-1 of 10 µg/g). Based on the calibration curve established by the other reference materials a content of 6.6 µg/g is calculated for MRG-1. This value is quite compatible with the data on which the reference values are based (range from 4.1 µg/g up to 29 µg/g!). If the standard deviation from the reference data is estimated as >4 µg/g, the determined content of 6.6 µg/g must be regarded as being "true" (compare Sect. 2.4.2.1, Eq. 2.29). If, vice versa, such an "uncertain" reference material is used for the

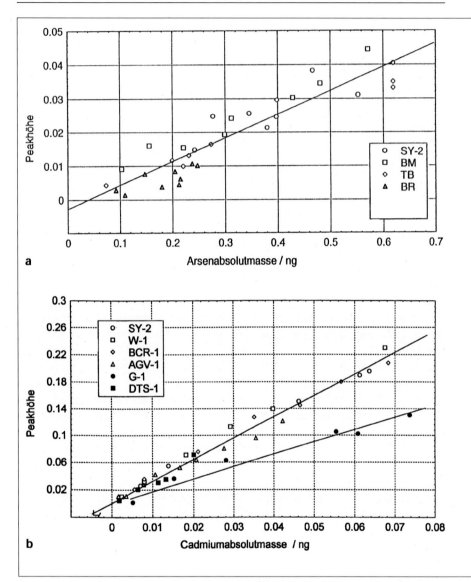

Fig. 3.7. Calibration curves established by variation of the test sample mass of different geological reference materials for
a Arsenic (193.7 nm); **b** Cadmium (228.8 nm); **c** Lead (283.3 nm); **d** Sb (235.4 nm); the diagrams show "absorbance" (peak height) vs. analyte mass (ng) (from [258], with permission).
The reference materials are (released from USGS:) Granite G-1, Diabase W-1, Granodiorite GSP-1, Andesite AGV-1, Basalt BCR-1, Peridotite PCC-1, Dunite DTS-1, (released from CCRMP:) Syenite SY-1, Syenite SY-2, Syenite SY-3, Gabbro MRG-1, (released from ZGI:) Granite GM, Basalt BM, Clay shale TB, Limestone KH, (released from CRPC:) Granite GA, Granite GH, Granite GR, Basalt BR

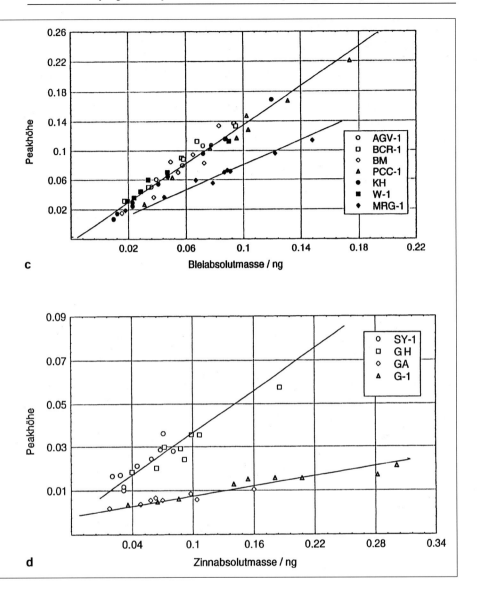

analysis of unknown samples, the expanded uncertainty in the reference value has to be considered with the estimated uncertainty in the unknown sample results. For Cd analogous problems of the reference values must be considered, when judging the deviation of the slope for "Granite" G-1 from that of a variety of geological samples from different "families" (Fig. 3.7 b).

The interpretation of the data for the determiantion of Sn is not clear where two significant different regression slopes are found from two reference materials in each case (Fig. 3.7 d, see below Sect. 3.3.2). Similar discrepancies are documented in [122] for the determination of Sb and Cd.

In an early methodological investigation of the direct solid sample analysis of geochemical reference samples Luecke et al. [261] found significant differences of the calibration curve slopes for Cr, Co, and Ni, however, these were calculated "on the basis of the reference contents, that are assumed to be true" (translated from the author), thus the uncertainty in the reference contents (i.e. bias to the true value) may be (mis-)interpreted as being caused by "matrix effects" occurring in the solid sample analysis. Gong and Suhr [255] have bypassed this "trap" by visualizing the range of the values that are used for the estimation of the reference content, in the comparison with their own results. This can be regarded as a simple approach to an uncertainty estimation.

The reference materials discussed above are from the "first generation" of geological reference materials, which were prepared in the 1950s and re-investigated over a period of two decades. The required corrections of the reference values posed what Roelands called "the salutary shock caused to the geochemical and geoanalytical community by the granite G-1 and diabase W-1 standards" [262]. The "third generation" of certified reference materials (CRM) available today can be regarded as much more reliable [70]. However, the principal problem of using a reference value for the detection of "matrix effects" still remains.

The *mixture with graphite powder* method has proven to be a suitable or required method particularly for the modification of geological matrices (see Sect. 2.5.3.2).

Analyte heterogeneity may influence the precision strongly, however, by the common grinding techniques sufficiently small particle sizes are achieved [259] so that, even in the case of pronouced mineralogical heterogeneities, the RSD can mostly be held below 10%.

The composition of the environmental materials e.g. *sludge, waste, compost, ash, fly ash, coal, dust, particulate matter* can varying substantially also within a specified type of material. Often these samples are composed of organic (biological) and inorganic (geological) components, hence the matrix may show the characteristics of both materials. Also in this field, CRMs with different compositions and varying contents are becoming more and more readily available today (see Table 2.2). The differences in the matrix affect solid sampling results to a less degree than often expected. Frequently it was shown that different materials can be analyzed based on the same calibration material [153, 152, 24, 257, 241, 263, 117] (compare Sect. 2.2.1.2, Fig. 2.8).

However, a study of matrix effects is recommended for the analysis of unknown materials because unexpected effects may occur if the sample material contains a component that has a strong influence on the evaporation/atomization process.

Example

In a method developement study for "coal" the determination of Mn based on reference solutions was reported to be not possible, in contrast to Pb and Zn [51]. The influence of **metals** on the atomization of the analyte element is generally strong, while background absorption plays a minor role. Much more than with other sample types, there is controversy in the literature about the applicability of reference solutions as calibrant material for the analysis of metals. Although successful analyses based on reference solutions are reported, some investigations point out the requirement of matrix-matching calibrant material for obtaining unbiased results. For this reason, some results and models explaining interfering mechanisms should be discussed in detail:

L´vov [3] documented analyses of tungsten samples (Al and Zn). He shows that the traces of signals for Al from a reference solution ("pure aluminium") and from solid tungsten are very similar in appearence, width and height. He conclude: "In the fractional vaporization of elements from materials with low volatilities, it is possible to use calibration graphs plotted for the pure element." The graphite furnace used gave good isothermal conditions. As Lundberg et al.[264] pointed out, isothermal atomization procedures give better precision as well as trueness and must therefore be recommended for metallurgical materials.

However, it is well established [265] that condensed-phase interference due to incomplete extraction occurs if the analyte element is soluble in a matrix component, thus resulting in pronounced tailing of the signal because of a large "destillation time". Frech et al. [235] pointed out that "Pb and Bi are almost insoluble in iron, chromium, nickel and manganese, and bismuth in iron and chromium. ... , which explains why lead and bismuth can be easily extacted from steels and determined only with minor problems". However, a negative bias of about 10% are found with the determination of Bi in steels based on reference solutions at low atomization temperature (1800 °C) which increases if the chosen temperature is higher.

A supplementary model for the vaporization/atomization mechanisms of analytes from a metallurgical matrix was given by Hinds et al. [54], a model which might be promising to improve understanding. They observed slim signals for wall atomization of Si out of gold [266] (see Fig. 3.8, similar signals are found for the atomization of Pb and Bi out of steel [235]) and explained this effect as follows: "The rapid exhaustion of Si, ... , results from transport of the analyte in the convective cells of the matrix. These cells, induced by surface tension inhomogeneities and the temperature gradients in the molten sample, move the analyte between the bulk and surface regions. This mixing, combined with vaporization

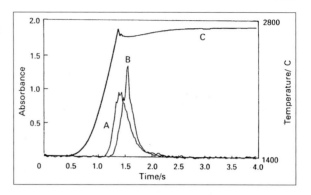

Fig. 3.8. Peak profiles of atomic absorption signals for 10 ng Si from *A*, a standard solution and *B*, 0.4 mg solid gold sample (wall deposition); *C*, tube wall temperature (from: [54], with permission)

of the species near the matrix surface, results in a depletion of the analyte before much of the gold matrix is vaporized. ... It is shown that the liquidus lines of the binary phase diagrams associated with a gold solvent can be used to predict the existence of analyte carryover in the gold. Consequently, these diagrams offer a means of identifying, a priori, candidate elements that may be determined by solid sampling ETAAS in gold, via aqueous solution calibration". This model was verified with the determination of Pd and Fe in gold. For the latter the phase diagram show that Fe bonding is independent of temperature for low concentrations near the gold melting point, so that a slow release might be expected. In fact, a marked tailing of the Fe-signal was observed. In a further study Brown et al. [63] monitored the gaseous species that evolved from aqueous and solid gold samples by mass spectrometry in vacuo and in real-time. They summarized: "Free Si and SiO vaporized from both sample types, but $SiC_2(g)$ and $Si_2(g)$ were unique to the aqueous and to the solid-gold samples, respectively. These observations imply that atomisation occurs via dissociative adsorption processes during heating of the aqueous samples. For the solid-gold samples gas-phase collisions can enhance dissociation of the $Si_2(g)$."

Takada and Hirokawa [108] studied the observed appearance of *double peaks* from metallurgical samples (Fig. 3.9 a): "Atomic-absorption signals of trace lead, bismuth, silver and zinc in steel, obtained by directly atomizing one sample particle, were found to consist of a small first peak and a large second peak. ... The first peak was caused by the analyte element existing around the grain boundaries of the steel and near the sample surface and the second by the fraction of the analyte element existing within the crystal grains of the steel". They found that the first peak was removed by annealing the test samples for 24 hours at reduced pressure and temperatures at 600 °C (Fig. 3.9 d).

Wang and Holcombe [267] used the reduced pressure furnace (see Section 3.2.2) for a study of the distribution of Pb in metals. Because at low pressure atomisation the residence time is reduced, the absorption profile approaches the generation function of free atoms, accordingly a better temporal resolution is achieved. The consistent appearence of three peaks for Pb out of a copper alloy sample was interpreted as follows: "It is proposed that the first peak belongs to the release process

Fig. 3.9. Peak profiles of atomic absorption signals for Zn in steel: **a** no annealing, **d** after annealing, conditions 600 °C, 1.3 x 10^{-5} Pa (from: [108], with permission)

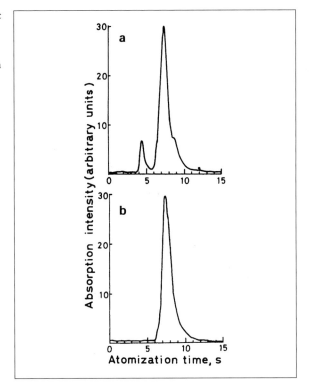

of Pb from the surface, the second peak may originate from grains until the Cu-Pb binary phase transformation to a homogenious solution occurs, and the third peaks arises from Pb in the bulk of the sample." On the basis of their observations, they speculated that the expulsion of analyte during the vaporisation of the copper matrix may reduce the third (bulk-) peak (approx. 77% of total Pb), because these atoms are generated coincident with copper vaporisation, so that the quantitative determiantion with the use of a reference solutions may not be possible.

Bäckman and Karlson [22] analyzed nickel-based alloys for various elements (Pb, Bi, Zn, Ag, Sb). They reported a reduction in the signal for Pb when the steel sample contained more than 0.1% sulphur. They explain this effect by the high melting point of sulfides of lead (1114 °C). Zn has not been observed to be prone to oxide interference, even its oxids are very stable (1975 °C). They used reference materials for calibration in routine analysis, proving that no significant matrix effect occurs between those materials. Generally they concluded that "the results agree, within the range of measuring accuracy, with the results reported in the literature and with those obtained by other methods". The influence of manganese as a main component of alloys was studied, however, no interference was found [268].

In their series of investigations, Headridge and coworkers [268, 187, 269, 270, 271, 272, 268] successfully used solid reference materials for the determination of a large variety of analytes in nickel-based alloys and copper. In one study, the analysis based on reference solutions was investigated, however, only "for certain elements, such as bismuth and lead, can semiquantitative results be obtained…" [272]. Also for the analysis of steel [48, 268, 273, 107] and the determination of Cu in aluminum [195] reference materials were used for quantification.

Marks et al. [48] found acceptable agreement of the results for Pb, Bi, Se, Te, and Tl of direct atomization of nickel based alloys using synthetic reference metals. The preparation and use of synthetic reference materials for metallurgical samples are described in [39, 82, 84, 54] (see Sect. 2.2.2.3).

Irwin et al. [243] reported results from the analysis of nickel-based alloys reference materials using the "STPF technology" based on reference solutions. The results were found to be in very good aggreement for Pb, Tl, Te, Bi and Se. Fig. 3.10 a, b shows results in comparison with reference values. The authors deplore the disagreement of their results for the high contents of Se (9–13 µg/g), however, in the light of the above discussion concerning comparative measurements related to reference values of solid materials, these may be judged as satisfactory (particularly for an element such as Se that often gives rise to analytical problems). The authors investigated the use of Ni-modifier in the reference solution and conclude (note added in proof): "The residue acted as a matrix modifer in the same fashion as the 1 mg reduced nickel that was added to the standards. Aqueous standards needed only to be pipetted onto the nickel residue in the cup in order to obtain accurate calibration".

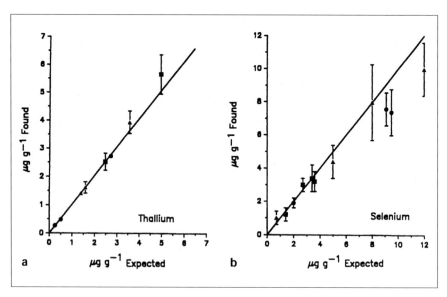

Fig. 3.10. Correlation plots of analyte contents for **a** Tl, **b** Se; expected (reference value) versus those found by direct solid sampling and calibration with standard solutions modified with 1 mg of reduced Ni (from: [243], with permission)

Takata and Hirokawa found, for tin samples for Pb and Cd [274] and Zn and Bi [275], respectively – although very strong differences occurred in the peak shape compared to these of reference solutions – that the analysis of unknown tin samples yields reliable results if solid "pure tin" is added to the reference solution as a matrix modifier when running calibration measurements. A similar method was applied by Hinds [80] for determinations of Au, Pd, Pt in silver. He added 0.4 mg dissolved silver (as nitrate) to the reference solutions. The determinations compared favorably with results from other methods.

Docekal et al. [180] determined trace impurities of Cu, K, Mg, Mn, Na, Zn in powders of molybdenum and molybdenum silicide by direct solid sampling based on reference solutions. The results were compared to those obtained by differend methods (e.g. GF-AAS after sample decomposition, RNAA, GDMS, IDMS, TMS-AES within a joint project). Despite some discrepancies, the results were found to be satisfactory, in several cases even in excellent agreement, so that the authors concluded that "these comparisons elucidate the importance of the two developed direct solid sampling ETAAS techniques". Friese et al. [181] analyzed tantalum powders with an improved solid sampling system. In Fig. 3.11, the peak profiles for this material for several elements in comparison to these of reference solutions are documentated. The authors state: "For Cu, Mn, and Zn, nearly identical transient signals were obtained in processing solid samples and standard solutions, and they were in acceptable agreement for Fe and Na. The agreement of the analytical results obtained by calibration using aqueous standards, based on peak height and peak area evaluation, also supports the assumption that the chemical or physical state of the analyte does not significantly affect the time dependence of the absorption signal".

The integrated absorption signals are generally not or only slightly affected by the form of the metallurgical test samples (e.g. balls, flacs, drills, needles) or the test sample size [3, 48]. Although researchers are generally aware of this potential problem, such an effect is only described for lead, where the difference between ball and thin flaces is 15%, however this difference was caused by imperfection in the peak area measurement system [22]. These authors also tested to see if the mass of the test samples affect the measurement by using several reference materials of nickel-base alloy and steel samples with various contents fo trace elements. They found "that absorbance is a linear function of the absolute amount of the element measured and that it is independent of the mass of the sample in the area examined". Furthermore they investigated the influence of the sample content using a test sample mass of 2 mg in each case, the results for Pb are given in Fig. 3.12. The authors stated: "The points coincide well along a straight line, which means that wide variations in the concentration of the main elements of the matrix have little or no effect on the result of the analysis".

So the conclusion drawn from the results of a series of basic investigations by Lundberg and Frech [264, 53, 235] can be followed: "The applicability of the solid-sampling technique with respect to the number of elements as well as the type of metallurgical materials should be related to the solubility of the impurity in

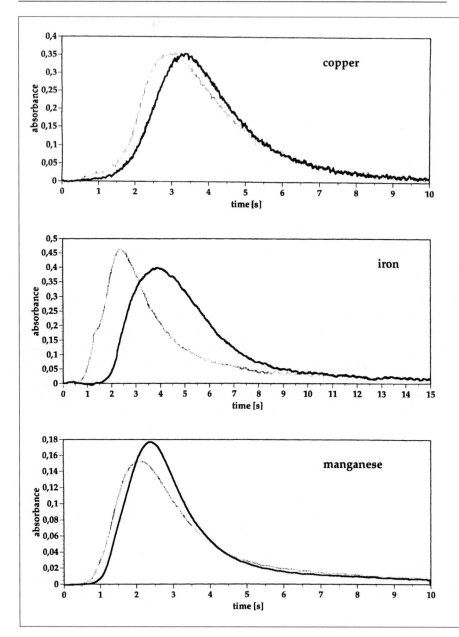

Fig. 3.11. Absorption signals from the analysis of tantalum powders (*black lines*) and reference solutions (*grey lines*) for different elements (Cu, Fe, Mn, Na, Zn) (from: [181], with permission)

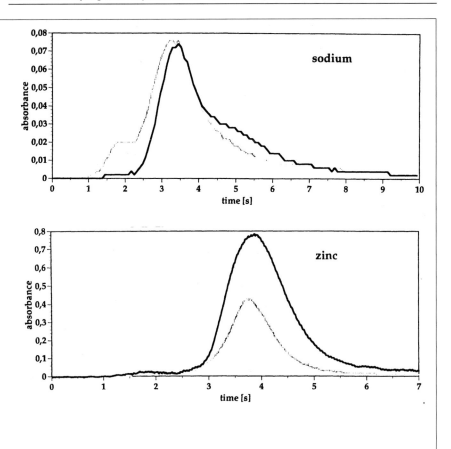

Fig. 3.12. Calibration graph for various types of alloys with different contents of Pb, using a constant test sample size of 2 mg (from: [22], with permission)

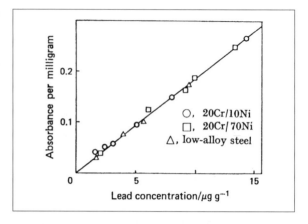

matrix components and, in addition, to the volatility of the impurity in matrix components and to the volatility of the impurity as well as that of the matrix. If favorable conditions exist for the determination of an element in a certain matrix, a suitable standard alloy should be used for standardization" [235].

Because a large number of metallurgical CRMs and reliable reference materials are available, it will be mostly not a problem to find an suitable material for routine analysis of metals (see [72, 73, 74, 75, 76, 77, 78]).

3.3.2
Particulars of Element Determination

In this section only some remarks about the "popular" elements are made and singular observations for rarely determined elements are reported. The general scientific knowlegde about the behavior and the properties of analyte elements should be considered also for solid sample analysis (e.g.[23, 276, 277, 278]).

Al

Generally a significant problem in the analysis of aluminium is the high risk of contamination because of its ubiquitous occurrence, e.g. if the low content in body speciments are to be analyzed [155]. Although this danger is drastically lower in direct solid sampling, the sample handling instruments must carefully checked for potential contamination sources (e.g. class capillaries even after acid washing [156]).

As

Characteristic for arsenic is the large difference in the volatility of its compounds. While the oxides are highly volatile, other compounds are very stable. These properties may lead to analyte loss – particularily in biological samples –

during pyrolysis and even in the first stage of atomization, which the temperature must be relatively high (e.g.>2400 °C). It is well known that this problem can be overcome by the "stabilization" of arsenic during pyrolysis by the addition of nickel in different forms [208, 151] (see Table 2.14).

Cd

1. With the determination of cadmium special attention must be paid to the exact adjustment of the pyrolysis temperature (see Table 2.14), in order to avoid analyte losses. Specially when biological samples are to be analyzed, where the background must be reduced to a minimum, particularily if only a D_2-correction is used, a recovery study is mandatory.
2. The working range of both analytical lines show a gap (compare Table 3.2). This often requires compromise conditions, e.g. if the amount of analyte introduced at the optimum sample mass is too large or too low respectively (Eq. 3.3). The criterion may be e.g. supposed sample heterogeneity (using the sensitive line with a very large test sample mass) or high background level (using the insensitive line with a very small test sample mass) (compare Sect. 2.5.1). Additionally the measures discussed in Sect. 3.2.2 may be considered.

Cr

1. The use of the insensitive alternative lines of chromium is seriously hampered by the low intensity of these lines, thus the determination of chromium in environmental sample with a high content may be hampered, particularly if a small (sensitive) tube is used [241].
2. The 357.9 nm line should not be used because, on this line, spectral interference may occur caused by the CN molecule, see Table 3.2 (which may probably occur with the direct vaporization of solid materials). Furthermore this line shows an unfavorable line spitting pattern in Zeeman-AAS leading to strong curvature of the calibration line.
3. If solid sampling results are compared with those based on sample decomposition, it should be considered that chromium may not be soluble with certain digestion techniques, leading to lower contents (e.g. compare [279]) . Rosopulo [263] has shown this effect for a sewage sludge reference sample using direct solid sampling also for the decompositon residues. The results are shown in Table 3.4.

Ge

A serious difference in atomization efficiency between solid samples and the reference solution has been observed. The first was found to be about 500 times more intense [280]. The authors explained this effect by the considerable higher volatili-

Table 3.4 Determination of chromium in Sewage Sludge (BCR CRM 145), comparision of results based on the digested sample and direct solid sampling (both with GF-AAS) (from: [263], with permission)

Method	Content of Cr (mg/kg)		
	soluble[a]	non-soluble[b]	total
Extraction			
HNO_3 + 3 HCl	70	54	124
2M HCl	50	69	119
Pressure decomposition			
HNO_3	80	41	121
$HClO_4$ + HF	125	–	125
Direct solid sampling			120
BCR report (not cert.)			
total			105.4
aqua regia	85.2		

a) analysis of the solution
b) analyis of the solid residue with direct solid sampling

ty of Ge compounds in the solid sample (e.g. nitride or sulfide), whereas the vapor pressure of elemental Ge should be very low (boiling point 2830 °C). The observation that no signal could be obtained when the lignite samples was mixed with graphite powder(!)is inconsistent with the outlined theory. The amount of Ge determined in the lignite sample, achieved with AAM (portions of the solid sample are spiked by a reference solution in a preparation step), are assumed to be accurate.

Hg

For the special properties of mercury and appropriate instrumentation for solid sampling see below (Sect. 3.3.3).

Mn

1. Although for manganese, two triplets of analytical lines exist, the limits of detection for the two groups are partly overlapping (see Table 3.2) and the upper limit of detection may be too low compared to the content of some materials of about several thousend ppm (e.g. in aquatic plants). Thus the direct determination of manganese in these samples can be impracticable if the reduction of the line sensitivity is not sufficient [8].
2. With the analysis of coal "aqueous standards resulted in a much steeper calibration plot and gave extremely low values for the SRMs" [51]. An explanation for this effect was not given.

Pb

1. Although lead is one of the most frequently determined analyte, it seems that the determination of Pb is still connected with serious, partly unknown inter-

ferences. As an indicator for this, the comparablly large confidence intervals of CRMs may be taken, which are an expression of unknown systematic effects (compare discussion in Sect. 2.4.2.5). This has been documented repeatedly, e.g. in a recent case study [281] for the certification of trace element content in a spinach material. The results for Pb show the largest scatter of all 34 determined elements, the standard deviation between means from 25 laboratories was 42%, although the results of 22 laboratories were not accepted (the solid sampling results from the author's laboratory was in very good agreement with the certified content).
2. Especially for Pb, problems with heterogeneity on the mg- and g-level of biological and environmental samples have been reported (e.g. see [124, 127, 13] or examples in Sects. 2.3.2 and 6.6.3).

Se

1. It is well known that the determination of selenium in the graphite tube may be connected with serious problems because of the high volatility of some of its compounds (e.g. boiling point of SeO_2 315 °C and Se_2Cl_2 127 °C). It seems that for this element matrix modification is mandatory in order to get a sufficient recovery of the analyte. For anorganic samples the mixture with nickel and graphite powder was succesful (e.g. portions 1+2+4) [282] (compare Table 2.14). However, this modifier failed with organic samples [283, 242]. Lindberg et al. [284] showed that for biological materials the addition of palladium leads to a stabilizing effect yielding a sufficient recovery even after a thermal pretreatment up to 1000 °C. Accurate results are found applying AAM (see example in sect. 2.2.2.3). For metallurgical samples no modifier seems to be required [271].
2. The analytical lines for selenium may be interfered with by phosphorus species originating from the thermal decomposition of the sample. This "structured" background is not corrected accurately with the D_2-technique, for the extremely high background not even completely with the Zeeman technique [308].

Sn

Figure 3.7 d shows that for the determination of Sn in four geological reference materials, two slopes with a significant difference are yielded, each established by two materials. It is noticeable that the materials that show an enhanced sensitivity both contain approx. four times more Sn then the two that show reduced sensitivity (~11 µg/g to ~3 µg/g). This may indicate a signal suppression with increasing test sample mass. However, no conclusions can be drawn because no further experience with Sn in solid sampling has been reported.

Tl

With the analysis of samples with a high salt (e.g. chloride) content, a strong signal depression (volatile species) may occur even if the sample is loaded on a L´vov platform e.g. incinerator ash [648]. Hence, a recovery experiment should be performed, if a high salt content is suspected. If necessary, matrix modification should be considered (see Sect. 2.5.3.1).

Zn

The sensitive line 213.9 nm, usually used in flame AAS is unsuitable for solid sample analysis, because the zinc content is generally too high compared to the working range of this line. Fortunately, the insensitive line shows high intensity, favorable linearity, and a satisfactory working range.

3.3.3
Mercury Determination with Direct Solid Sampling

Two properties of mercury are responsible for the special role of this element for its determination with atomic spectrometric methods. First, its high volatilization leads to difficult decomposition procedures and losses during the storage of the digest solutions and allows absolutely no thermal pretreatment prior to the atomization. On the other hand, a high contamination risk exists from the glassware used and the surrounding environment in the laboratory. Second, the usable spectral line at 253.7 nm is comparatively insensitive (the most sensitive resonance line at 185.0 nm lies in the "vacuum-UV" region, so that this line is not available using normal spectrometers).

   These characteristics make it infeasible to apply the usual analytical techniqes because the required limit of analysis often cannot be met e.g. with the analysis of pollution and pathways of Hg in air, water, soils, and its storage in plants, animals and residues as well as food and clinical specimens, which play an important role in environmental research.

   Consequently, special methods and techniques have been developed for the trace determination of mercury. In atomic spectrometry the cold-vapor technique offers excellent "detection power" because the analyte amount introduced into the absorption volume (mostly a quartz tube) is considerably increased by a large test sample mass. However, this improvement can only be reached by considerabely increased chemical and technical expenditure. A long term aim of analytical chemists should be to find out a rapid and secure but more simple method for analyzing the environmentally hazardous element mercury.

   In this section, no comparison between direct solid sampling and the "cold vapor" technique should be made, but rather the development and application of direct measurements of Hg in solid materials should be discussed. The determination of Hg with AAS and direct solid sampling may be an advantageous alternative for many applications.

3.3.3.1
Direct Determination of Hg in the Graphite Furnace

For solutions from chemically decomposed solids, the determination of Hg by the conventional graphite furnace technique is not applicable, because the chemical procedure yields a highly diluted Hg content in the solution. With the limited volume of test samples (10–20 µl) the required limit of analysis usually cannot be met.

Kurfürst and Rues [152] used the graphite furnace in a Zeeman-AAS (SM 1) for analyzing sewage sludge reference samples directly. In comparison with the reference values, reasonable results for Hg were obtained based on reference solutions. No thermal pretreatment was applied, but the atomization step with a temperature up to 1000 °C was started after introduction of the (liquid and solid) test samples into the tube. The authors point out that due to the moderate temperature rise of the platform below 200 °C (compare Fig. 3.2), no spray effect occurs from a solution aliquote, but the droplets are evaporated within the first second. The test sample mass was restricted to about 1 mg because of high background absorption (no thermal pretreatment). Since a detection limit of 0.2 ng was found, the analysis limit is approximately 0.2 µg/g.

The direct determination of Hg with the graphite furnace in biological materials was tested by Fleckenstein [285, 286]. He found that the Zeeman correction technique allows a sufficient separation of the Hg absorption from the high background (smoke) absorption. Conditions are specified for direct determination in samples with a Hg content of more than 0.2 µg/g in the dry matter. The procedure described was applied to biological materials (waste composts, sewage sludge, mushrooms, and earthworms). Samples up to 1–2 mg are directly atomized in the graphite boat in the graphite furnace at 1700 °C without a prior drying or ashing step.

DeKersabiec and Benedetti [122] used the graphite cup cuvette (Hitachi Z7000) (see Fig. 2.5) for mercury determination in soils. The limit of analysis was reported as being 0.1 µg/g for these materials. A good correlation between different soil reference materials was achieved, from which a regression curve was calculated. The amounts of Hg so determined in various other reference samples are in excellent aggreement with the reference values. Hiltenkamp and Jackwerth [83] analysed gallium metal for Hg (PE Z3030) and found a similar limit of analysis.

It can be concluded that for the determination of Hg in the graphite tube some prerequisites must be given: The atomization has to be carried out without drying and ashing steps (A short drying step at an adjusted temperature of 150 °C may be possible for fluids without losses [122], because the sample on the platform or the cup may not reach this temperature.) A decisive factor is the behavior of the matrix destroying process – sometimes the material may blow out into the beam, producing an irregular signal. This may be avoided by covering or mixing the sample with graphite powder. Even with Zeeman correction, the background absorbance is not allowed to be more than 1.5–2, hence the sample amounts may be limited depending mainly on the organic mass fraction. Thus, only relatively high Hg content >0.5 µg/g is determinable with this technique.

3.3.3.2
The Hadeishi Furnace

Hadeishi [287] reported in 1972 on a new type of mercury atomizer for flameless atomic absorption spectrometry which allows a rapid determination of trace amounts of Hg without prior separation from the host material. The required limit of analysis is obtained by using a large absorption tube (200 mm) and the introduction of a large test sample mass (up to 100 mg).

Church et al. [288] tested this *two-chamber furnace* with a platinium sample cup. The absorption tube was 30 cm long, had a diameter of 1.25 cm, consisted of stainless steel, and was *constantly heated* at 850–1000 °C. An approximate theoretical description of the time dependance signal was given, considering the volumes of both chambers and the carrier gas flow. Liquid samples in the range of 20–50 µl and solid samples up to 30 mg were combusted with oxygen as the carrier gas when organic samples were analyzed (flow rate 400 ml/min). A strict linearity for standard solutions from 0.1–50 ng Hg was found. Several organic and inorganic samples were analyzed under various conditions using oxygen or argon as the carrier gas. The excellent performance for the determination of Hg was proven by comparative measurements of reference materials and all the measurements showed an excellent agreement with the reference values. The authors concluded: "The two-chamber furnace offers a useful and accurate means for reproducibly vaporizing, combusting, and dissociating samples in solid, liquid, or gaseous form containing volatile elements. An oxidizing carrier gas serves both as an aid to combustion of the sample and as a means for reproducibly carrying the vaporized sample from the combustion chamber to the absorbtion tube. Consequently, no prior chemical sample preparation is required for solid samples and drying and ashing cycles are not necessary. Sample measurement times are reduced to about one minute" [288].

A precondition for using this furnace is a very powerful background correction system, because the combustion products are transported into the atomization tube simultaneously with the Hg compounds, a thermal separation prior to atomization or in the atomization process is not practicable. Consequently the development of this furnace was connected with the development of the Zeeman-technique for background measurement in AAS [289] (see Chapter 1). However, because of the large diameter of the outer combustion tube, only the direct Zeeman-technique (magnetic field at the "lamp") can be applied (alternatively the S/H-correction technique may be used).

With the development of the commercial Zeeman atomic absorption spectrometer (SM 1, Grün Optik, Germany 1979), where the direct Zeeman-technique is applied, an impulse to direct solid sampling was given. As a "solid sampling" tool the Hadeishi furnace was commercially produced and has proven its worth for the determination of Hg [290].

Figures 3.13 a and b shows this atomization unit in principle and in the real configuration. The *furnace tubes consist of nickel*, which seems to influence the combustion of the sample and the release of Hg by a catalytic effect. A temperature of 900 °C has proved to be the threshold temperature for a rapid release of Hg from the host material. The use of quartz tubes (of identical geometry), as they were inserted in hydride systems, gives reduced sensitivity and increased background.

Fig. 3.13. The design of the Hadeishi furnace (two chamber nickel tube furnace) for the direct determination of Hg in solid materials.
a The principle of combustion and absorption chamber and sample introduction.
b The real configuration, upper shell opened (containing constant heating elements and thermal isolation).

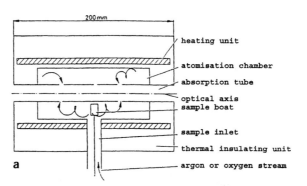

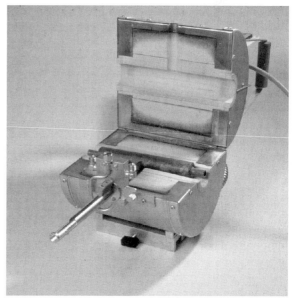

For taking advantage of the Zeeman effect for background correction the "pen ray" lamp can be used, which remains stable in a high magnetic field. However, attention must be payed to the temperature drift of the lamp bulb, which can be increased by the lamp current or due to room temperature. This may lead to severe self-absorption (increasing intensity), thus reduced sensitivity and linearity.

It is possible to introduce up to 130 mg of inorganic sample material into the furnace by using a platinum crucible with a volume of 250 mm^3. The limit of detection with this tool is reported as being 0.1 ng. The applicable sample amounts are between 1–10 mg using argon as support gas and up to 50 mg using oxygen, thus the limit of analysis can be calculated to 0.1–0.002 µg/g, however a realistic value for biological samples is approximately 0.02 µg/g. Although this is one to two orders of magnitude worse than that reachable with the cold vapor technique, it is sufficient for analysis for a variety of problems, even for checking food samples (e.g. the legal limit for Hg in fish in Germany is 1.0 µg/g).

A similar atomisation technique was developed by Berndt [209]. Solid samples on a sample probe are burnt inside a quartz tube in a stream of air with the aid of focussed infrared lamps – the analyte is brought into the gas phase. Campos et al. [291] coupled this flushed burning chamber with an electrically heated T-shaped quartz tube as an absorption cell for AAS determination. For mercury, they achieved a characteristic mass of 0.65 ng. Because of the excessively large background, the mass for biological samples with this configuration is restricted to 2 mg. As a consequence, the limit of analysis is comparitatively high (approx. 0.5 µg/g), thus this technique can be used only for the analysis of contaminated materials. This investigation comfirms the importance of a long absorption pathway, the use of oxygen and the positive influence of nickel for a sensitive determiantion of Hg. (The described system was successfully used for the direct determination of other easy volatilizable elements, e.g. Cd, Cu, Pb, Zn in biological samples for these the limits of detection and the used test sample mass were suitable.)

In the last few years, results with the Heideishi furnace in various fields of application from several working groups have been documented. The values from CRMs determined with the system described above show generally good agreement with the certified values [287, 179, 292, 290, 293, 41, 294, 295].

For extended analytical experiments, Tobies and Großmann [292] used Zeeman-AAS for the determination of Hg in soils, coals, and ashes without a decomposition step by means of direct input of the samples into the Hadeishi furnace, which was heated to a constant temperature of 1000 °C. As support gas Argon or air was used. The calibration was carried out with reference materials. For test measurements annealed seasand containing no detectable Hg, was used after being spiked with different amounts of reference solutions. Peak areas correlate to the Hg amounts and the time integrated signals were used. In routine analysis, a detection limit of 0.2 ng was found, the test sample mass was 0.3–130 mg. The total time period was 5–10 min for 3 replicates. Figure 3.14 shows the results of the determination of BCR CRM-142 "Light Sandy Soil" compared with the results from different laboratories on which the certification is based. The repeatability (long-term stability with the method) is evident from the 4 determinations on different days.

Bachmann and Rechenberg [295] compared the Hg determination with the Hadeishi furnace and the cold vapor technique after sample decomposition for different materials processed in cement plants. The different reference solutions of identical volume in the platinium cup were covered with solid $CaCO_3$. They found excellent agreement for all analyzed materials (e.g. coals, filter dusts). Figure 3.15 show the calibration curve obtained from the measurement of different

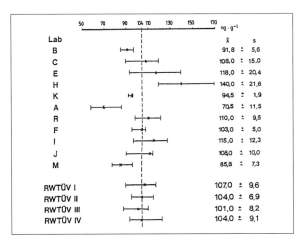

Fig. 3.14. Results of the determination of Hg in "Light Sandy Soil" (BCR CRM 142, certified 104 ± 12.3 ng/g) by direct solid sampling with the Hadeishi furnace (RWTÜV I-IV). For comparison the authors took the plot of mean values and standard deviations from different laboratories (B-M) on which the certified value is based from the certification report (from: [292], with permission)

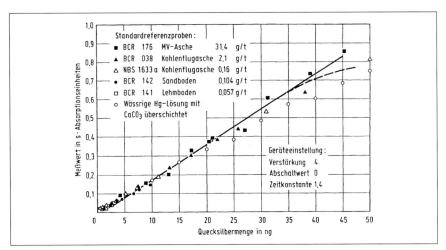

Fig. 3.15. Calibration curve for Hg obtained with the Hadeishi furnace from the measurement of different CRM/SRM and of a reference solution (o); the diagrams show peak area (s) vs. analyte mass (ng) (from: [295], with permission)

CRM/SRM in comparison to that of a reference solution. Although the content of the different reference materials varies by a factor of 10^3, only in the upper region can a deviation be recognized.

A special instrument based on direct Zeeman-AAS and the Hadeishi furnace is available for the determination of Hg only (AMZ 010, Grün Analytische Meßgeräte, D-Ehringshausen). A simple Hg-tool can be assembled in the laboratory by constructing the two chamber furnace of Hadeishi´s design (see [288]) and applying a (relatively) small permanent magnet on the pen-ray lamp. The optical components and the casing may be used from an old AAS tool that otherwise can no longer be used. (The nickel tube assembly, as well as technical and eletronic support are available from Ing.-Büro O. Schuierer, D-85737 Ismaning).

A Zeeman-AAS equipped with a Hadeishi furnace may be the method of choice for a majority of samples and the advantages of the technique are obvious. The purchase of such a tool for a laboratory may be justified if the determination of Hg is part of the research program, necessary for continuous process control or for rapid measuring of environmental pollution.[3] The initial idea from Hadeishi has proved itself, significant improvements in the analytical performance have not been achieved although more than 20 years have passed (but there have been, of course, developments in instrumentation, signal aquisition, and data evaluation).

3.4
Analytical Applications

In Table 3.5 those methodological applications of solid sampling with graphite furnace AAS are listed which have made a contribution to the development of the method or which show the analytical conditions for the analysis of solid materials. (Applications, which are not of analytical character, but present the solid sampling method as an adventageous tool for other sciences, are discussed in Chapter 6).

The table is arranged first in order of the "Element" and second in order of the "Matrix", the latter is differentiated to characterisic groups of materials (biological, geological/mineral, metallurgical, environmental, industrial) additionally characterized more specifically.

Papers mentioned in the column "Ref." in which a number of elements are determined are described in detail only for one element, the other elements can be found in the column "Further References". In this column, additional papers are also listed.

---

3) As a consequence of the fire accident in a chemical factory in Basel 1986, the river Rhine was heavily polluted with mercury compounds. The described tool made fast and reliable control analysis of a huge number of samples possible (at comparatively low cost) over a period of more than one year.

The column "Calibration" gives information about the reference material and the calibration method, which are applied. It should be mentioned at this point that the table may give reason for misinterpretation about the role of reference solutions for calibration in solid sampling. These are used predominantly in the methodological papers, in order to investigate the potential of the method. However, in routine analysis and if solid sampling is used for obtaining data for investigations in other scientific fields, calibration based on CRMs has become a prevailing alternative because some problems which may limit the trueness of solid sample results based on reference solutions are easier to solve (compare Chapter 2).

In the column "Instrumentation", the applied background measurement technique and the sample introduction system (see Table 2.1) are the main entries.

The column "Special analytical features" gives details of the specific methodological development which are discussed in the various chapters of this book (the RSD of the respective investigations are stated, only if this value gives information about a special quality of the analysis or the material).

Table 3.5 Compilation of methodological investigations of direct solid sample analysis using GF-AAST

Element	Matrix /Samples	Ref.	Instrument/Furnace Sample Introduction	Calibration Material & Method	Special Analytical Features	Remarks	Further References
Ag							
	environmental coal fly ash (real samples)	[49]	IC-cuvette steel crucible,	ref. solution introduction sampler	method. comp.	reduced pressure furnace	[40] [197] see Au
	geolog./mineral. quarzite rock (ref. samples)	[96]	AA5/90CRA	ref. solution, AAM	effect of particle size	graphite mixture	[81] see Bi, [154] see Tl [205] see Cu
	metallurgical steel (CRMs)	[273]	H170-50/GA-2 nippers	ref. solution	atomisation model		[205] see Cu [187]
	steel (CRMs)	[53]	AA-6/CRA63 autom.solid dispensor	ref. solution	comp. of peak high/c With the Hadeishi furnace (see Chapter 3.3.3.2) area	background recorded	[313] [22] see Bi
	industrial semiconductur	[84]	GA SM20 platform boat	synth. ref.mat. cal. curve	method. comp. (certif. campaign)	study of sampling error	[24] see Cu [55] see Mn [97] see Sb [98] see In [315] see Bi
Al							
	biological biopsy/necropsy samples & CRMs	[155]	PE Z3030/HGA600 cup-in-tube	ref. solution	control of contamination	oxygen ashing,	[156], [316] [46] see Cr [86] see Zn [26] see Fe
	industrial						[184] see Mg

As	*biological* CRMs & marine organisms	[208]	H Z8000 miniature cup	ref. solution	Ni as matixmod. comp. of methods	pyrolysis curves	[151] see Cu [254] [317]
	geolog./mineral.						[122] see Sb
	metallurgical nickel-based alloy	[271]	VT AA6/IL 555	ref. material	Limit of analysis method comp.		
	industrial						[179] see Zn [55] see Mn
Au	*environmental* red mud (real samples/CRM)	[197]	PE 503/HGA2100 platform	ref. solution	comp. of methods recovery study	urea dilution	[141]
Bi	*geol./mineral.* sulfide ores (real & ref.mat.)	[81]	PE 303/induction heating solid sampling spoon	ref. solution syn. solid ref. mat.	methode comp.	graphite mixture sulph. acid added to ref. solution	[131] see Rb
	metallurgical steel/alloy (ref. samples)	[22]	IL 251/IL 555 microboat	ref. solution ref. material	various mat. drillings, matrix effects	routine analysis	[268], [269] [53] see Ag
	gallium	[83]	PE Z3030 cup-in-tube	syn. solid ref. mat.	better limit of analysis by partial dissolution	electrolytically spiked ref. material	[235] see Pb [48] see Te
	industrial glas	[315]	VT AA6/IL 455 microboat	CRM	method. comp.		[98] see In

Table 3.5 Continued

Element	Matrix/Samples	Ref.	Instrument/Furnace Sample Introduction	Calibration Material & Method	Special Analytical Features	Remarks	Further References
Be							[131] see Rb
Ca	geolog./mineral. feldspar, lime (real samples)	[196]	PE 503/HGA2100 solid sampling spoon	ref. solution cal. curve	method. comp. (flame)	graphite mixture (1+999)	[131] see Rb
Cd	biological bovine liver (CRM)	[207]	H 170-70 miniature cup	ref. solution, AAM		interference from residues	[10,11] see Co [318] see Cs [19] see Cr [86] see Zn
	foodstuff	[132]	GBC carbon rod system 1000	ref. solution	method. study, sampling error, matrix modification	deuterium background corr.	
	CRMs	[104]	GA SM20/30 platform boat	ref. solution "zero matrix"		extrapolation to zero matrix cal.	[240] see Co, [127] see Fe [203] see Zn
	CRMs	[14]	VT AA6/two step atom. PE Z3030/HGA600 cup-in-tube	ref. solution, CRM cal. curve, AAM	spectral interference, matrix mod.	verification of calibr. methods,	[9], [319], [254], [320], [283], [321]
	environmental sewage sludge (CRMs)	[152]	GA SM1 platform boat	ref. solution, cal. curve		transient signals	[197] see Au [24,151] see Cu [43,241] see Ni [42] see Pb [257,322] see Zn, [323] see Cu [202] see Sb [122] see Bi,
	geolog./mineral. rocks (ref. mat.)	[255]	PE 305A/HGA2000	ref. solution		transient signal	[81] see Bi,

Element	Material	Ref.	Instrument	Calibration	Study	Remarks	References
			solid sampling spoon				[324], see Pb [190]
	metallurgical tin (real samples, CRM)	[274]	H 180-80/cup cuvette nippers	ref. solution	method comp. matrix effect	effect of residues transient signals	[53] see Ag [270] see In [83] see Bi [325]
		[210]	graphite lid to the cup				
	industrial plastic (CRM)	[123, 326]	GA SM30	CRM cal. curve	homogeneity study	certification measurements	[97] see Sb [327] see Te [328] see Cd [329, 330] see Rb [717]
Co	biological bovine liver oyster tissue (CRM)	[10] [11]	PE603/HGA 76B spoon	ref. solution	comp. of wall and platform atomis	pyrolysis curves	[101] see Zn [101] see Ni
	tobacco (ref. mat.)	[240]	CZ AAS3/EA3 ring camber tube	ref. solution, cal. curve, AAM	methode comp. matrix mod. (Cd)	background reduction	[87]
	geolog./mineral. rocks (ref. mat.)	[261]	PE 305A/HGA70 solid sampling spoon	cal. curve from different ref. mats.	strong "matrix effects"	(see discussion in Sect. 3.3.1)	[131] see Rb
	metallurgical steel (real samples)						[107] see Mn
	industrial pharma. B12 prep.	[312]	H Z8000, miniature cup	ref. solution, synthetic ref. mat.	method comp.	very low analysis limit: 0.15ng/mg, transient signals	[87]

Table 3.5 Continued

Element	Matrix/Samples	Ref.	Instrument/Furnace Sample Introduction	Calibration Material & Method	Special Analytical Features	Remarks	Further References
Cr	*biological* mussel, hair (CRM)	[19]	H Z9000 miniature cup	CRM cal. curve	various CRMs for calibration	simultanious multielement	[254], [331], [101], [102]
	mussel (real samples)	[46]	JA/HGA2000 solid sampling spoon	ref. solution lab. sample spiked	method comp., effects of pyroysis, peak height/area, sample mass	dilution with naphthlene, pellet sampling, homogeneity test	[151] see Cu [318] see Cs
	environmental sewadge sludge (CRM)	[263]	GA SM1 platform boat	ref. solution, cal. curve	method comp.	too low contents with digestion	[197] see Au [241] see Ni [43] see Ni [197] see Au
	sediment (CRM)	[202]	JA AAS5 EA, sampling platform	ref. solution, CRM, cal. curve	method comp. effect of grinding	transient signals	
	metallurgical						[107] see Mn
	geolog./mineral.						[261] see Co
	industrial plastic (real samples)	[150]	PE Z5000/HGA500 cup-in-tube	ref. solution cal. curve	characteristic cup-in-tube	transient signals	[182] see Fe [328] see Cd [85] see Ni
	semicond. (GaAs, real samples)	[332]	IL Video11/IL-555 microboat	ref. solution	background study	transient signals	[333] [98] see In [184] see Mg
Cs	*biological* milk powder (CRM & real samples)	[318]	PE-Z5000/HGA500 cup-in-tube	ref. solution AAM		matrix mod.	
	geolog./mineral. silicate rocks (ref.)	[94]	PE303/induction heating solid sampling spoon	ref. material	method comp. interference study		[131] see Rb

Cu	*biological* CRMs	[88]	H Z8000 miniature cup	CRM, syn. ref. mat. cal. curve	comp. ref. materials, pyrolysis curves	transient signals, effect of concommitants	[46] see Cr, [320],[254],[320], [50], [322] [10,11] see Co
	CRMs	[151]	PE Z3030/HGA600 cup-in-tube	ref. solution	precision < 10% bias > 10%	transient signals	[203] see Zn [19] see Cr [101] see Ni
	fingernails	[25]	PE 403/HGA2000 solid sampling spoon	ref. solution	method comp. (flame) good agreement	inhom. distr. in nail pieces	[86] see Zn [318] see Cs [56]
	CRMs	[205]	JO AAS3/EA3 ring chamber tube	ref. solution cal. curve, AAM			[247] see Mn
	environmental coal (real samples)	[153]	PE400S/PU SP01	ref. solution,CRM "solid injector"	cal. curve, AAM	[43] see Ni	[49] see Ag
	coal fly ash) (CRM	[24]	VT AA375/GTA95 graphite cup	ref. solution		second surface atomizer	[88] see Pb
	sediments, soils (real samples)	[323]	GA SM20 platform boat	ref. solution	method comp.	slight differences to decomposition	
	groundwater colloids	[334]	PE5100/HGA600 micro-piston	ref. solution	method comp.	results of field measurements	
	geolog./mineral. carbonte scales & rocks	[130]	H Z8000 miniature cup	ref. solution study of interferences, and particle size	mixture of graphite powder		[194] [261], see Co [131] see Rb
	metallurgical aluminum	[195]	H 180-80 nippers	ref. mat.	method. comp.	effect of graphite powder	[273] see Ag [181] see Na

Table 3.5 Continued

Element	Matrix/Samples	Ref.	Instrument/Furnace Sample Introduction	Calibration Material & Method	Special Analytical Features	Remarks	Further References
	industrial paper (real samples)	[335]	PE 303, /induction heating solid sampling spoon	ref. solution, AAM	precision 10-40%	samping by cutting paper discs	[328] see Cd [182] see Fe [85] see Ni
	Calcium fluoride	[260]	GA SM20 platform boat	CRM, cal. curve	confidence interval cal. curve	mixture of graphite powder	[180] see K [329,330] see Rb
Fe	*biological* cod muscel	[127]	GO SM20 platform boat	CRM, cal curve	homogeneity study	distribution of results (n=85)	[10,11] see Co [19] see Cr
	geolog./mineral.						[196] see Ca
	metallurgical						[54] see Si [181] see Na
	industrial polymers (real samples)	[182] [26]	PE305B/HGA2000 solid sampling spoon	ref. solution cal. curve	method comp.	discussion of sampling error	[97] see Sb [184] see Mg
Ge	*environmental* lignite (real sample)	[280]	VT AA6/HGA72 steel spoon	ref. solution AAM	effective atomisation with solid sample	propose atom. mechanism	
Hg	(see section 3.3.3)						
K	*industrial* molybdenium metal and silicide	[180]	PE 5000/ HGA 72 and CRA 90 cup and boat/platform	ref. solution	limits of analyis, precision	study of blank control	[184] see Mg [336]

Element	Sample	Ref.	Instrument / sampling	Calibration	Remarks	Remarks	References
Li							[131] see Rb
In	*metallurgical* nickel-based alloys	[270]	VT AA6/IL555 PE 300S/induction heating	ref. materials (RM)	method comp. limits of analysis		
	industrial semicond. (GaAs, real samples)	[98]	AA6/IL-555 microboat	ref. solution	method comp. limit of analysis	matrix separation in gas stream	[89] [327] see Te
Mg	*biological*						[244]
	geolog./mineral.						[196] see Ca
	industrial graphite	[184]	PE Z5100 pipette tip/microspatula	ref. solution, AAM	method comp.	oxygen ashing, transient signals	[180] see K
Mn	*biological* fish meal (real samples)	[320]	PE 303 solid sampling spoon	ref. solution AAM	method comp.		[50] see Rb, [254] see Pb, [19] see Cr, [101] see Ni
	CRM	[843]	PE HGA74 solid sampling spoon	ref. solution AAM		ref. solution added after pyrolysis	[151] see Cu, [203] see Zn
	environmental coal (CRMs)	[51]	H 180-80 pipette injection	CRM cal. curve		bias with ref. solution graphite mixture	[43] see Ni, [257,322] see Zn
	air particulate matter	[337, 338]	TJA IL457/655 tungsten rod	ref. solution	electrostatic precipitation	nearly real-time analysis	[338]
	metallurgical steel (ref. samples)	[107]	Two-channel-AAS/IL555 microboat	rel. to main component (Fe)		transient signals, memory effect	[273] see Ag, [181] see Na

Table 3.5 Continued

Element	Matrix/Samples	Ref.	Instrument/Furnace Sample Introduction	Calibration Material & Method	Special Analytical Features	Remarks	Further References
	industrial semicond.(GaAs) real samples	[55]	lab. build ZAAS inject. funnel	ref. solution			[47] see Sb [98] see In [180] see K [329,330] see Rb
	paper (ref. mat.)	[26]	PE-403/HGA2000 solid sampling spoon	ref. solution	method comp. (flame)	discussion of solid sampling features	
Mo	*biological*						[318] see Cs
	industrial						[339] see Rh
Na	*geolog./mineral.* magnesite (CRM)	[47]	PU SP2900/SP9-01 graphite cup (platform)	ref. solution cal. curve, AAM		content 0.05%	[180] see K
	metallurgical tantalum powders	[181]	GA SM1(modif.: D_2-BC) platform boat	ref. solution	method comp. for limits of analysis	transient signals	[184] see Mg
Ni	*biological* plants (ref.mat.)	[340]	PE 5000 cup-in-tube	ref. solution, cal. curve	method comp	matrix mod.	[86] see Zn [240] see Co [205] see Cu
	margarine (real samples)	[341]	GA SM1 platform boat	ref. solution cal. curve	limit of analysis 0,2 mg/kg		[316]
	plants (ref. mat.)	[101]	H Z8000 miniature cup	ref. solution, AAM	application of the "three-point" method	preashing outside graphite furnace	[87]

environmental particulate matter (real samples)	[43]	PE 5000/HGA500 sample probe	ref. solution bracketing	recovery for a CRM	probe collection for air particles	[153] see Cu [241] see V
geolog./mineral.						[261] see Co [131] see Rb
industrial polyethylen (synth. samples)	[85]	GA SM1 platform boat	ref. solution, synth. ref. mat. cal. curve	matrix effect using ref. solution	recovery study	[184] see Mg

Pb

biological CRMs	[254]	H 8000 miniature cup	ref. solution, CRM cal. curve	effect of pyrolysis time	transient signal	[46], [320], [283] [185], [9] [14] see Cd [132], see Cd [10, 11] see Co [86] see Zn
CRMs, ref. materials	[8]	GA SM1 platform boat	ref. solution CRM cal. curve	method comp.		[151] see Cu [19] see Cr [104] see Cd
CRMs	[183]	PE503/HGA2100 direct through injection hole, graphite probe	ref. solution cal. curve	effect of pyrolysis comp. of wall, probe platform atomistion	oxygen ashing transient signals	[240] see Co [127] see Fe [318] see Cs [203] see Zn [247] see Mn
environmental susp. matter (real samples)	[44] [42]	GA SM1 platform boat	ref. solution cal. curve	sampling error	analysis of filter, sampling of particulate matter	[257,342] see Zn [43] see Ni [152] see Cd [51] see Mn [197] see Au [343] see Se [323] see Cu [202] see Cr

Table 3.5 Continued

Element	Matrix/Samples	Ref.	Instrument/Furnace Sample Introduction	Calibration Material & Method	Special Analytical Features	Remarks	Further References
	PVC	[20]	PE 3030/HGA direct through injection hole	ref. solution, cal. curve	effect of pyrolysis method comp.	no background correction	[40], [246]
	cellusose filter	[344]	Il 551/Il655 micoboat	ref. solution	detection limit, precision, accuracy	filter cuts 2mm diameter	[337, 338] see Mn
	geolog./mineral. rocks ref. materials	[256]	PE Z5000/HGA500 pipette injection	ref. solution cal. curve	matrix mod.	transient signals	[189] see Pb, [122] see Sb, [154] see T1, [261] see Co, [131] see Rb
	teeth (real samples)	[324]	PE 303,/induction heating solid sampling spoon	solid ref. material cal. curve	hydroxyapaite as reference material		
	steels/alloys (ref. materials)	[264] [235]	VT AA6/CRA90 pincer	ref. solution, ref. material	study of isothermal conditions, interference study	effect of test sample mass	[274] see Cd, [243] see T1, [22] see Bi
	metals	[106] [249]	VT AA375/GTA95 graphite cup	ref. solution	study of vaporization kinetics, effects of sample mass	second surface atomizer	[267], [269], [48] see Te, [83] see Bi
	industrial paper (real samples)	[328]	GA SM1 platform boat	ref. solution cal. curve	method comp.	matrix mod.	[150] see Cr, [98] see In, [97] see Sb
	graphite (ref. and real materials)	[345]	PE503/HGA74 solid sampling spoon	ref. solution	method comp. (emission spec.)	low limit of analysis required	[85] see Ni, [315] see Bi
	glasses (SRM and real samples)	[189]	VT 63CRA cup atomizer	ref. solution	heating rate study, peak area vs. hight	graphite mixture, transient signals	[194], [717]

Element	Matrix	Ref.	Instrument	Calibration	Study	Notes	Ref.
Pd	*metallurgical*						[54], see Si
	industrial						[339] see Rh
Rb	*biological* CRM	[50]	PE4000/HGA500 pipette injection	ref. solution, AAM		oxygen ashing	[346]
	geolog./mineral. ref. materials	[131]	H Z8000 miniature cup	ref. solution, cal. curve	particle size and grinding time	graphite mixture, transient signals	[94] see Cs
	industrial polymer	[329, 330]	PE Z3030/HGA600 cup-in-tube	ref. solution cal. curve	method comp.	transient signals	
Rh	*industrial* radioactive waste	[339]	PE Z5000/HGA500 cup-in-tube	ref. solution cal. curve	comp. of transient signals liquid/solid	graphite mixture for dilution, memory effect	
Ru	*industrial*						[339] see Rh [131] see Rb
Sb	*geolog./mineral.* ref. materials	[122]	H Z7000 miniature cup	ref. solution	matrix effects matrix mod.	graphite mixture	[22] see Bi
	metallurgical						[315] see As

Table 3.5 Continued

Element	Matrix/Samples	Ref.	Instrument/Furnace Sample Introduction	Calibration Material & Method	Special Analytical Features	Remarks	Further References
	industrial silicon, (real samples)	[97]	V AA6/IL-555 microboat	ref. solution	method. comp. low limit of analysis	peak hight	[98] see In
	polyester (real samples)	[47]	PU SP2900/SP9-01 graphite cup (platform)	ref. solution cal. curve	method comp.		
Se	*biological* liver biopsy	[242]	PE 5100/HGA600 cup-in-tube	ref.solution, CRM, cal curve, AAM	matrix modifier	transient signals	[318] see Cs [347]
	CRMs	[284]	PE Z3030/HGA-600 cup-in-tube	ref. solution, CRM cal curve, AAM	matrix modifier, interference effects	transient signals	
	environmental NI-APDC	[343]	H Z9000 miniature cup	ref. solution cal. curve, internal standard	effects of pyrolysis pH, coprecipitation	enrichement of traces in tap water	
	CRMs	[282] [283]	GA SM1 platform boat		Ni, ZnS, graphite as matrix mod.	interference study	
	geolog./mineral.						[122] see Sb
	metallurgical						[271] see As [48] see Te
Si	*metallurgical* gold	[266]	PE 5000/3100/HGA500, 4100ZL, direct through injection hole, funnel	ref. solution	atomisation study, transient signals, method comp.	atomisation model based on binary phase properties	[26] see Mn

Element	Sample	Ref	Device	Calibration	Method comp.	Notes	Cross ref.
Sn	*industrial*						[327] see Te
Te	*metallurgical* copper	[269]	PE 300S/indution heating	ref. material			[269] see As
	nickel alloy	[48]	PE 305B/HGA2100 VT AA5/CRA63 loading funnel	ref. solution	methode comp. comp. of tube and cup	influence of residues (CRA63)	[243] see T1 [271] see As
	industrial semiconductor	[327]	PE3030B/HGA600 graphite platform	ref. solution		sputter sampling	
Tl	*geolog./mineral.* silicate rock (ref. & real samples)	[154]	PE303/induction heating solid sampling spoon	ref. solution two point cal.	methode comp.	sulph. acid added	[179] see Zn [24] see Cu
	metallurgical nickel alloys (CRMs ref. mat.)	[243]	PE Z5000/HGA500 cup-in-tube, funnel tip	ref. solution cal. curve	nickel matrix mod. to ref. solution	transient signals	[48] see Te [187]
	nickel alloys	[39] [82]	ZPMQ II, massmann furnace, steel spoon	ref. solution, syn. ref. mat. cal. curve	validiation of syn. ref. materials	syn. ref. materials based on urea, impantation	[83] see Bi
V	*environmental* CRMs	[241]	PE Z3030/HGA600 cup-in-tube	ref. solution, cal. curve	study of matrix effects	transient signals	[43] see Ni [49] see Ag [153] see Cu

Table 3.5 Continued

Element	Matrix/Samples	Ref.	Instrument/Furnace Sample Introduction	Calibration Material & Method	Special Analytical Features	Remarks	Further References
Zn							
	biological						
	CRMs, ref. mat	[203]	PU 9200/9390 autoprobe	ref. solution, CRM cal. curve	method comp., validiation of calibr. methods	transient signals	[46], [254] [10, 11] see Co [19] see Cr [104] see Cd
	CRMs	[86]	H Z-8000 miniature cup	synthetic ref. materials		thermal pretreatment	[127] see Fe [318] see Cs
	environmental						
	soil (CRM)	[257, 322]	PSP9 & PU9000 sample probe	ref. solution, cal. curve	method comp.		[43] see Ni [51] see Mn
	geolog./mineral.						[154] see T1
	metallurgical						
	tin	[275]	H 180-80 nippers	ref. solution cal. curve	matrix effect	transient signals	[53] see Ag [22] see Bi [270] see In [181] see Na
	industrial						
	cement & CRMs	[179]	GA SM1 platform boat	ref. solution	transient siganls	graphite mixture interlab. studies	[97] see Sb [180] see K

Solid Sampling by Electrothermal Vaporization-Inductively Coupled Plasma-Atomic Emission and -Mass Spectrometry (ETV-ICP-AES/-MS)

PETER VERREPT, SYLVIE BOONEN, LUC MOENS AND RICHARD DAMS

4.1
Introduction

Electrothermal Vaporization (ETV) as a sample introduction technique for inductively coupled plasma (ICP) atomic emission spectrometry (AES) and mass spectrometry (MS) has been studied intensively for the last 10 years. Although most of the investigations were limited to the analysis of liquid samples (especially in the area of ETV-ICP-MS), some dealt with the direct determination of traces in solid material.

In this chapter solid sampling (SS) with ETV-ICP-AES/-MS will be discussed. Some points of solution analysis with ETV-ICP-AES/-MS will also be reported since most of the items such as the coupling of the ETV device to the ICP, adaptation of the different gas flows, modification of the furnace, etc are the same for both types of samples. However, during the discussion, attention will be drawn to the possibilities of the described system for the analysis of solid samples.

Many research groups have investigated the ETV technique and have coupled it to several plasma sources, such as microwave induced plasma (MIP), direct current plasma (DCP) and capacitively coupled microwave plasma (CCMP). However only very few applications of solid sampling with these plasmas have been reported in the literature; therefore the discussion will only cover the coupling of the ETV-device to an inductively coupled plasma. For completeness, however, some references on coupling ETV to other plasma sources, are included in the text.

In the early sections, some instrumental requirements are discussed. The discussion involves those requirements which are needed to couple an ETV-device to an ICP. Different solutions are reviewed for some of the difficulties encountered. Since solid sampling with the aid of a graphite furnace is most often used with atomic absorption detection, a comparison of some analytical features is given in subsequent sections. This part also deals with the advantages of ETV as a sample introduction technique over the conventional pneumatic nebulizer. In the last sections, some typical interferences and the ways of overcoming them are discussed and some examples of "real" solid sample analyses with ETV-ICP-AES/-MS are given.

4.2
Characteristics of the ICP Technique

4.2.1
The Inductively Coupled Plasma

The *inductively coupled plasma*, as reported in the pioneering articles [348, 349, 350], has become one of the most widely used sources for excitation in the case of *atomic emission spectrometry* and as an ion source in the case of *mass spectrometry*. A detailed discussion of this ICP is beyond the scope of this book, however, for better understanding of some of the consequences of coupling an ETV-device to an ICP, it is necessary to give a brief description.

The ICP can be visualized as a very hot electrical flame, with excitation temperatures going up to 7000 K [351], generated at the end of a torch (Fig. 4.1).

Because of the high temperatures, the ICP is a superior excitation and ion source. The torch consists of three concentric quartz tubes mounted in a two or three turn load coil. Through each quartz tube, a gas flow, generally Ar, is led as shown in Fig. 4.1. The energy transfer to the ICP is as follows. An RF-generator, mostly operating at 27.1 MHz, supplies an alternating current to the load coil. This forms an oscillating magnetic field co-axial with the torch. In this magnetic field, electrons, generated by a tesla discharge, are captured and accelerated. These high energy electrons collide with incoming Ar atoms which are ionized. This causes an increase in the number of electrons and ions, and of the gas temperature leading to a cascade effect forming the plasma. When coolant gas is continuously supplied the plasma will stabilize and sustain.

Each of the three gases mentioned satisfies a specific requirement, and is of major importance when coupling an ETV device to an ICP. The coolant gas flow

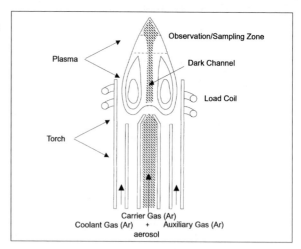

Fig. 4.1. The inductively coupled plasma (*ICP*) generated at the end of the Quartz torch

rate (Fig. 4.1) is around 14–18 l/min and is the main supporting gas to sustain the plasma. The auxiliary gas flow (flow rate of about 0–1 l/min) is not so important when using a pneumatic nebulizer, but when coupling ETV to an ICP, this gas can be used to keep the plasma in its normal position when the furnace is open (see further). The third gas, the *carrier* or *transport gas* (0.3–1 l/min), delivers the *aerosol* formed into the plasma. This flow is, of course, the most important gas flow of the ETV-ICP combination since it is this gas which sweeps the aerosol out the furnace and transports it to the plasma. The use and importance of the different gas flows will be clarified during the discussion of the coupling and optimization of the ETV-ICP system. However at this point it is necessary to describe the different processes the sample goes through before emission signals or ions are detected.

In a conventional ICP configuration with a *pneumatic nebulizer*, the sample solution is aspirated by a peristaltic pump and turned into a suitable aerosol by means of a nebulizer and spray-chamber combination. To make this possible, the samples must be liquids, which means that every solid material to be analyzed must be dissolved (e.g. acid digestion, fusion) or extracted. The aerosol formed, when entering the plasma, consists of small droplets resulting in a wet plasma. In the plasma, the aerosol is desolvated, resulting in dry particles which are then volatilized, atomized, ionized and finally excited. Finally *emission light* is detected in AES and ions are counted in MS.

ICP-AES and ICP-MS plasmas are comparable as far as set-up, start-up, and fundamental plasma processes are concerned. However, there are some differences. The ICP, in the case of AES, is usually placed vertically so that light is viewed side-ways (see Sect. 4.2). The distance between the spot in the plasma were the highest emission intensity is observed and the top of the load coil is called the *observation height*. In the case of MS detection, the ICP is placed horizontally so that ion extraction by means of a sampling interface is possible (see Sect. 4.2.3). In this case, the distance between the orifice in the sampler cone and the load coil is called the *sampling depth*.

4.2.2
Atomic Emission Spectrometry

As mentioned above, in ICP-AES the emission radiation is viewed side-ways from the plasma. Figure 4.2 illustrates schematically a typical ICP-AES instrument, in this case a Czerny-Turner configuration.

Light emitted by an analyte element is viewed at the optimum observation height for this element. The light beam is focused on the entrance slit by the focusing lens. The grating diffracts the polychromatic light into its different wavelength components (e.g. $\lambda_1 \neq \lambda_2$), so that for light falling on the grating at a certain angle of incidence, only one wavelength λ_2, actually a narrow wavelength range, is detected by the photomultiplier. Scanning a spectrum is done in this configuration by changing the angle of incidence by rotating the grating. This

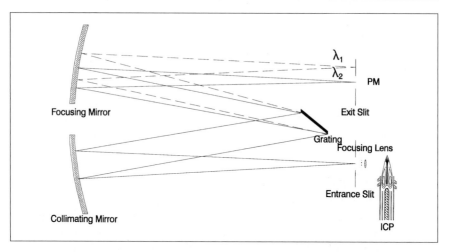

Fig. 4.2. Schematic overview of a typical ICP-AES instrument using a sequential monochromator Czerny-Turner configuration. The emission light (*ICP*) is viewed side-ways, focused on the entrance slit, diffracted by the grating and detected with a photo-multiplier (*PM*)

rotation of the grating is a rather slow movement, and is too slow for the detection of rapidly changing *transient signal* as occurring when using ETV for *sample introduction*. Different solutions for this problem will be discussed further on.

4.2.3
Mass Spectrometry

ICP-MS has, since its introduction in the early 1980s [352], gained popularity and has proven its detection power and capability for determining ultra-traces in different kinds of matrices. Coupling an ETV-device to an ICP-MS instrument has also been tried out intensively for the analysis of solutions. During the last years few researchers have used the ETV-ICP-MS combination for the direct analysis of solid materials.

Since in ICP-MS ions are detected by means of a mass spectrometer, these ions must be extracted from the plasma. Therefore the plasma is placed co-axially (or at least in the same plane) with the path way of the mass analyzer. A schematic diagram is given in Fig. 4.3 for the conventional ICP-MS configuration with a pneumatic nebulizer.

The aspiration and importing of the aerosol into the plasma is identical to the one described above (see Sect. 4.2.2). The plasma works at atmospheric pressure which is in contrast to the high vacuum of the mass analyzer. The large difference in operating pressure is possible by using an interface consisting of a series of chambers at steadily decreasing pressure. Because of this pressure drop at the

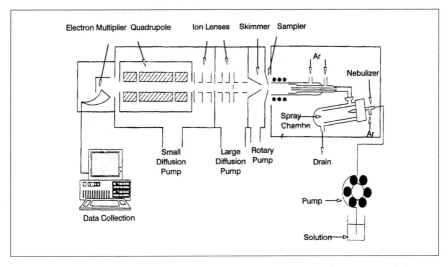

Fig. 4.3. Schematic diagram of an ICP-MS instrument with a conventional pneumatic nebuliser for sample introduction

interface, a part of the plasma is sampled (sampler cone) via extraction. The ion beam is focused with the aid of a set of electrostatic lenses and a quadrupole mass analyzer is used for separation on the basis of mass/charge. The ions are detected by an electron multiplier tube, and the resulting signal is sent to a multi-channel analyzer for data collection and interpretation.

4.3
Electrothermal Vaporization

4.3.1
The Need for an Alternative Sample Introduction System

From the instrumentation described in the former sections, it is clear that, despite the many advantages of plasma spectrometry, one of the major drawbacks is the lack of an "ideal" sample introduction method [353, 354]. Such an sample introduction system could be described as a system through which small volumes of any sample (solid, liquid or gaseous) without any pretreatment can be fully turned into an aerosol and transported to the ICP with an efficiency of 100%.

In the literature, many alternative sample introduction techniques are described, however, most of them were specially developed for the analysis of a specific kind of sample e.g. the V-grove nebulizer for the analysis of samples with a high salt content, laser and spark ablation for the analysis of solid samples, flow injection for the analysis of small volumes of liquid samples, and direct sample insertion (DSI) for the determination of mainly volatile elements in both liquid and solid samples.

By virtue of the design of an ETV system, this device almost matches the ideal situation. Small volumes of both liquid and solid samples can be analyzed. Under optimal conditions, the conversion and transportation processes are efficient. Perhaps the most important advantage of ETV is the separation of the desolvatation and vaporization step from the atomization and excitation step. This implies that the different steps can be optimized separately and independently so that matrix effects can be drastically reduced. The many advantages of such an introduction system will become clear during the discussion of the coupling of an ETV to an ICP. Many different designs of ETV systems exist and analytical performances can be different for different designs. Therefore, it is necessary at this point to describe in more detail the operation of an ETV device.

4.3.2
The Principle of ETV

A sample (liquid or solid) is introduced onto a *resistive filament*. The heating filament can be made of graphite or a highly refractory metal such as W, Ta, or Re. Computer-controlled *ohmic heating* of the filament is used to warm up the sample in different steps. The first step can be used as a *desolvatation* or *drying step*, when solutions are to be analyzed. In a second step, an attempt is made to *ash* or destroy the matrix, so that during the determination of the analyte element, less *matrix interferences* will occur (see further). In this step, some *matrix modifiers* can be used to increase ashing, transport or measurement efficiencies. Up to this point the furnace acts identically to a furnace used in GF-AAS.

In the next step, the sample is *vaporized* or *volatilized*. Whereas in graphite furnace AAS, the atomization occurring in the furnace forms the light absorbing atom cloud, the furnace or filament in ICP is merely used to produce an aerosol and *atomization* and *excitation* occurs in the plasma source. The aerosol is formed and *transported* with the carrier gas to the ICP (or other plasma source, MIP, DCP, CCMP). The design and dimensions of this filament are of major importance.

When utilizing metal filaments, a metal coil or tip (W [355, 356, 357, 358, 359, 360,], Ta [361]) can be used as the sample holder. In this case, the ETV device can only be used for the analysis of solutions. Micro-liter amounts of a sample adhere to the turns of the coil. Other research groups, use a metal strip or ribbon (W [362, 363, 364, 365], Re [366], Ta [367]) as the sample holder in the ETV device. The latter devices were coupled to different kinds of plasmas (*MIP* [368, 361, 367], *DCP* [358], *CCMP* [356, 357] and ICP). Marawi et al. [369] utilized Pd-coated metal strips (Ta, Mo, W and Re) as sample holders for the determination of As. The main purpose of this investigation was to trap As-hydrides in order to extend the linear dynamic range.

Since it is the purpose of this book to give an overview of solid sampling using a graphite furnace, an in-depth discussion on the design and operation of metallic coil or strip devices is beyond the scope of this chapter. These devices can, however, overcome some problems occurring with carbide forming elements (V, Cr and Ti) when using graphite furnaces. Impurities of the metals, on the other hand, can cause a serious increase in the blank signal [370, 371]. Park and Hall [372, 373, 374] have designed an ETV device which combines a metal strip with a graphite rod so that the filament of choice can be used depending on the volatility of the analyte element, the nature of the solution, or the presence of interfering impurities in the filament material.

The use of graphite allows high purity ovens to be produced in different shapes and dimensions. An important advantage of using graphite includes the transfer of knowledge of GF-AAS (e.g. pyrolytic coating, matrix modifiers) to ETV-ICP with a graphite furnace. Even though there is a wide variety of shapes and dimensions of the furnaces two main groups can be distinguished.

First there is the graphite rod or cup consisting of a graphite bar with a notch which is used as the *sampler holder* [375, 376, 377, 378, 379, 380, 381, 372, 382, 383, 384, 385, 386, 367, 387, 373, 388, 374, 389, 390]. Most of these devices were specially designed for plasma sample introduction. The characteristic of this group is that the sample is "*top loaded*" on the graphite rod or in the cup. The heating part of the ETV device is shielded from the air using a *bell-jar* [391] as shown in Fig. 4.4.

Fig. 4.4. Electrothermal Vaporization (ETV) Device: (*A*) sample and holder gas inlet to the plasma torch; (*B*) sample injector port; (*C*) ground glass stopper; (*D*) glass dome; (*E*) vaporization cell; (*F*) support electrodes; (*G*) holder gas inlet; (*H*) fixed support stainless steel blocks ; (*I*) water cooling system; (*J*) Teflon base; (*K*) aluminium base; (*L*) power cable (from [379], with permission)

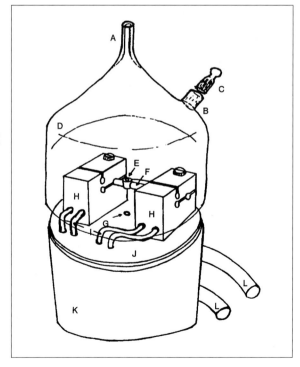

The apparatus is usually coupled to the ICP with tygon tubing. Since these ETV-systems use a shielding chamber (Bell-jar), which cannot be removed while the plasma is turned on, the introduction of a sample can only be done via inlet B (Fig. 4.4). Therefore, these systems are limited to the introduction of liquids using a micro-pipet. The knowledge of the connection of such a system to an ICP can however be transferred almost completely to the other types of furnaces usable for solid sample introduction. Therefore cross-references will be made to some of the articles dealing with this type of furnace but for the in-depth description the reader is referred to the original references.

A few of these furnaces (top loading) have, however, been modified especially for introducing solid samples. In these cases the furnace was brought almost directly under the plasma torch to improve sample transport. A more detailed description will be given in the next section.

The second group of furnaces are the *"flow through"*-type of ovens. Most of these devices are modifications of existing furnaces used for GF-AAS, e.g. HGA 74 [392, 393, 394, 395, 396], HGA 300 [397], HGA 400 [398, 399, 400, 401, 402, 314, 403, 404], HGA 500 [405, 314], HGA 600 [400, 401, 406, 404], HGA 2000 [407, 408], HGA 2200 [409], WF-1 [92, 410], FLA-100 [411], CRA-63 [412, 413, 414], CRA-90 [415], CRA-99 [416], GA-3 [417, 418], IL 455 [419], IL 555 [420, 421], IL 655 [422], SM-20/30 [423, 171, 103]. Some were specially designed for the introduction of solid/liquid samples into the ICP e.g. Halonarc [398, 424], Microtherm [425], HGA 300 [369], HGA 2200 [427, 459], SM 20/30 [428, 132, 429]. Perkin Elmer recently developed a graphite furnace (the HGA 600 MS) which allows vapors generated during the drying, ashing and cleaning steps to be removed with an Ar-beam (coming from both sides of the furnace) through a central hole in the graphite tube [707, 430, 431, 432]. During the vaporization stage, this hole is shut by means of a graphite pin so that the vapor is swept to the ICP. This type of furnace can however only be used for the analysis of liquid samples.

Finally some ovens have been designed specifically for the introduction of solid samples and have a completely different design to those described above [415, 420, 433, 434, 435, 436, 421].

4.3.3
Different Designs of Furnaces to be Coupled with the ICP

When designing an ETV-device for the direct analysis of solid materials, one should be aware of the *system requirements* on a *furnace design* in connection with solid sample introduction as these are discussed in Sect. 2.1.3. However, for ETV-ICP systems the demands may be different in some aspects and these will be discussed below.

As discussed in Chapter 2, the possibility must exist to *weigh* a small amount of (powdered) sample and transfer it quantitatively to the sample holder. This means especially that a separate sample holder which can be introduced and heated in a graphite tube is much more useful for solid sampling than sample loading in a notch in the heating filament.

However, in the case of ETV-ICP-AES there is no necessity for the sample holder to act as a L'vov platform in order to achieve isothermal conditions. Only vaporization or particles are needed, not a complete atomization as in AAS [399]. Some researchers established that vaporization from the graphite tube rather than from a surface platform gave better sensitivity [437].

The sample must be vaporized completely after its introduction into the oven, and the aerosol formed must be transported efficiently to the ICP. This requires a *transport without condensation or impaction* in the furnace or in the transport tubing and further implies that dilution by the carrier gas should be kept as low as possible.

The furnace should be designed or equipped with special tools to optimize *sample introduction without disturbing the plasma*, ensuring a high sample throughput. For this purpose different gas flow systems may be necessary (see the discussion below).

Because of these requirements (easy sample handling and automatization) some of the top-loading ETV devices can be considered as less suitable for analyzing solid powders. Bearing these requirements in mind, different approaches have been tried out and reported in the literature.

Several approaches have been proposed to optimize the overall performance. This has resulted in sometimes totally different ETV systems depending on the philosophy used by the designer.

One design is the top loading graphite cup or crucible placed directly under the torch, as illustrated in Fig. 4.5. The modified furnace block can be moved to one side away from the torch for loading [420]. The plasma however, must be kept stable during the loading of the sample into the cup, therefore an auxiliary gas flow (argon) is used to replace the carrier gas flow. This has, at a later stage been replaced by a Meinhard nebulizer through which the carrier gas can flow continuously [421].

The main advantage of these systems is a good transport efficiency since the furnace is connected almost directly to the plasma (see Fig. 4.5). Some difficulties were, however, encountered because solid particles can be swept out of the crucible causing clogging in the capillary tubes in the nebulizer [415]. A modification was introduced by Reisch et al. [416] by replacing the nebulizer with a new perpendicular gas feed device. In this way the rise in pressure during the analysis could be reduced. Similar graphite cup concepts were reported elsewhere [435, 438].

The second furnace design for the analysis of solid samples is the flow through type. Most of these furnaces are equipped with a sample boat and therefore they are also often referred to as "boat in tube" type furnaces. A schematic diagram of such a system is presented in Fig. 4.6.

Some of these furnaces were originally designed for the analysis of solids in AAS. For coupling them to an ICP, some modifications were necessary such as

Fig. 4.5. Graphite crucible ETV device, coupled directly to the plasma torch (from [420], with permission)

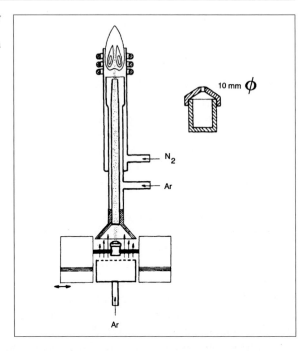

10 mm ϕ

N$_2$

Ar

Ar

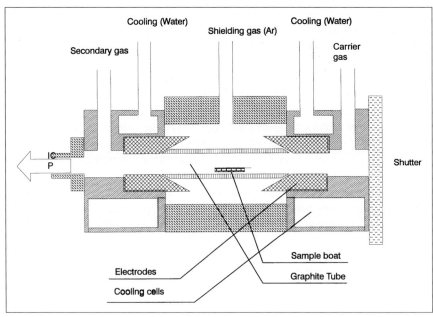

Cooling (Water)

Shielding gas (Ar)

Cooling (Water)

Secondary gas

Carrier gas

Shutter

P

Sample boat

Electrodes

Graphite Tube

Cooling cells

Fig. 4.6. Platform boat ETV device after modification for the adaption to the ICP (see Sect. 2.1.3.2)

changing the gas flows and closing one side of the furnace with the aid of a shutter. Those modifications will be discussed now and it will become clear that the ideal ETV device will be a mixture of all the existing devices.

The sample is usually weighed on a tared sample boat which is then inserted into the graphite tube. In this way, an effective solid powder handling procedure is accomplished. It is probably one of the main advantageous features of this type of furnaces that there is no longer a need to transfer the sample from a weighing flask to the sample holder. The transport of the sample into the furnace has also become much easier. Most of the furnaces are equipped with a special tool such as a pair of tweezers or a small pincer to pick up the sample boat for transport. This enables a reproducible positioning of the sample holder in the furnace which is important in view of temperature gradients existing in the furnace tube. The use of a pair of tweezers keeps the ETV introduction system compatible for later automation which, without any doubt, will be of major importance in the future.

The sample can be transferred from the boat to the furnace walls [392, 395] but normally it is left in the boat, which serves as the sample holder during the vaporization [398, 423, 399, 419, 407, 424, 402, 171, 418]. The use of the sample boat as the sample holder during vaporization overcomes the difficult quantitative transfer of the sample from the boat onto the bottom of the furnace. Only when analyzing compact solid samples such as resins as analyzed by van Berkel et al. [392, 395], can the use of a sample holder be omitted.

Modifications on this concept aiming at improved analytical performance, are possible and will be described in Sect. 4.6.

4.3.4
Coupling the ETV Device to an ICP

One of the advantages of ETV over the conventional pneumatic nebulizer is the better transport efficiency of the aerosol to the ICP. Therefore coupling an ETV device to an ICP should be done in such a way that diffusion and condensation (on the tubing walls) of the analyte are kept to an absolute minimum. Different concepts have been described in the literature.

Mitchell et al. [419] coupled an ETV system to a DCP using an interface of 20 cm length (1/4 in. bore tygon tubing). To overcome problems with melting of the glass or tygon tubing on the ETV side, boron nitride inserts were used.

Kántor et al. [398] modified the furnace so that a more stable aerosol was formed. Their furnace is a combination of the flow through type, for the analysis of solid, and the top loading device, for the analysis of aqueous solutions. By replacing the graphite tube with a hole (top loading) by a graphite tube without a hole and consequently exchanging some cocks, they can turn the top loading device into a side loading device (boat in tube). To improve the transport of the flow through type, they introduce cold argon in a counter flow at one side of the furnace. This mixes with the sample vapor at the relatively cold ends of the sample tube. In this way the aerosol vapor is condensed into particles which are kept from sticking to the tubing walls by the gas flow. A more detailed discussion on various modifications to promote aerosol formation is given in Sect. 4.6.1.2.

During desolvatation and especially during ashing of the sample a large amount of material is introduced into the ICP. When the ETV is coupled to an ICP-AES and, even more, when it is coupled to an ICP-MS instrument, it is advisable to avoid the introduction of bulk material (solvent vapor and matrix material). Even when the matrix elements are introduced before the analyte element some memory effect may occur. Therefore a valve system is placed after the furnace (in between the furnace and plasma), allowing the vapors produced during drying and ashing to be vented [405, 439, 394, 396, 409, 428, 132, 429]. Carey et al. [397] modified the furnace by introducing a drying vent, which consists of a two-way valve at the opposite side of the furnace (loading side of the furnace), which is opened during desolvatation and ashing of the samples. These modifications overcome most of the interferences in the ICP. However, when the furnace is opened for a longer time or when the carrier gas is not entering the ICP, another gas should be introduced to stabilize the plasma and to avoid the plasma shifting towards the load coil, what could result in the torch melting[405, 171]. This is also the reason for introducing the nebulizer in the case of the top loading furnaces (see above). This problem can be overcome by replacing the carrier gas by a secondary gas [171] or combining it with a slightly higher auxiliary flow rate in the plasma [405]. Verrept et al. [171] modified the furnace gas flows in such a way that, while the furnace is open (e.g. for introducing a powdered sample), a second argon gas flow is introduced in counter flow to the carrier gas, entering the furnace from the opposite side (valve at connection between the furnace and the transfer line) so that the graphite tube is permanently flushed with argon. In this way, no air can enter the furnace, and the plasma is kept from being blown out after closing the furnace and transferring entered air to the plasma.

Gunn et al. [391] and Crabi et al. [314] investigated the influence of the *length of the transport tubing* on the signal. They found little influence of this length on the total integrated time resolved signal. This implies that, when using long transport tubes, the analyte pulse will only be broadened but that all the analyte will finally reach the plasma if a fine aerosol is obtained. On the other hand, it is obvious that the tubing should not be longer than needed. However, since the furnace must be easily accessible, a length of 30–60 cm is recommended.

The physical coupling is mostly done using Tygon, glass, silicon rubber, or Teflon tubes with ball and socket joints. Daniels et al. [440] investigated the influence of different materials and the internal diameter on the amount of absorption of atomic mercury vapor and found that the analyte loss is kept to a minimum if the smallest tube diameter, the shortest length, and the highest possible flow rate, are used. This highest possible flow rate is, of course also determined by other factors such as the stability of the plasma or the dilution of the sample vapor, and in practice an intermediary flow rate must be used. Even though these results are only valid for atomic mercury vapor, they are indicative for the transport of aerosols. For ETV it seems that for typical flows of 0.3–0.8 l/min, an inner tubing diameter of 3–4 mm is appropriate.

Lamoureux et al. modified a graphite furnace (Model HGA 76B, Perkin Elmer) to allow simultaneous measurement of the atomic absorption and mass spectrometric signals [437, 441]. This system is based on the extraction of sample vapour through the dosing hole rather than one end of the graphite tube.

4.4
Transient Signals in the ICP Technique

When an ETV-system is coupled to an ICP with atomic emission detection (AES), some instrumental compatibility typical for this hyphenated technique is needed. As a consequence of using ETV, for sample introduction in ICP, transient signals are obtained. It is evident that the instrument must be able to handle such signals in order to get reliable response values. Real time *background correction* should also be carried out in the case of transient signals and an appropriate *data-acquisition* system should be available. In this section, different methods for background correction of transient signals and data-acquisition for obtaining response values of such signals will be discussed.

4.4.1
Background Problems and Correction Techniques in ICP-AES

As in AAS, in ICP-AES the importance of an accurate background correction method is obvious. Background correction, however, is one of the most difficult problems in emission spectrometry. The reason for this is twofold. First, it is not possible to measure the emission intensity under the analyte line (analyte signal + background + interferences) directly. Secondly this background depends strongly on the matrix of the sample.

When the sample aerosol is introduced into the plasma with an ETV system, background correction becomes even more complex because transient signals are obtained. Also, the background level is changing in time during one measurement.

In his book on methodology, instrumentation and performance of the ICP – AES, Boumans summarized the required performance of the spectrometer in order to ensure an accurate correction; "The essential information for background correction must be derived from wavelength scans centered about the analysis line and extending over a wavelength interval of at least several times the spectral bandwidth of the apparatus" [442].

In general two different approaches for background correction can be used, the *on-peak* or the *off-peak* correction. In the case of the on-peak correction, a blank measurement (a material with the same matrix composition as the sample but without the analyte element) is used to determine the background, which is then subtracted from the total analyte peak. For transient signals and certainly for solid samples, this type of background correction is almost impossible to obtain, since it cannot be supposed that the background enhancement during the measurement of the blank, is exactly the same as during the sample measurement. Even more annoying is that unless the matrix is known completely and a procedure to produce a synthetic blank is available, the real blank level for solid sampling cannot be measured. The background enhancement during a measurement can therefore not be determined in this manner. Consequently the off-peak method must be used instead, even though, in this way, no overlapping spectral interferences can be detected and corrected for.

To use the off-peak method, however, information is needed on the emission intensity of the analyte line and the intensity of a range of wavelengths close to the analysis line. This is illustrated in Fig. 4.7.

When using ETV (and other sample introduction systems yielding transient signals, e.g. laser ablation, flow injection, thermospray nebulization), off-peak background correction requires a spectrometer which is able to measure transient emission intensities at the peak and at neighboring wavelengths. In the literature different possibilities are mentioned, of which only a limited selection will be discussed in detail.

Fig. 4.7. Off-peak background correction method. The integrated net peak area (I) is found by subtracting the integrated background from the total peak area (I_A). The integrated background intensity ($I_{B, \text{Calculated}}$) is calculated by an interpolation algorithm using the emission intensities of some neighboring wavelengths ($I_{B1, B2}$)

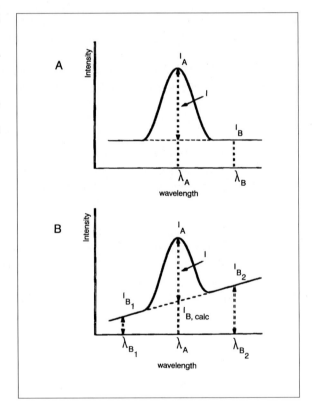

If the configuration of the sequential spectrometer (Fig. 4.2) is considered, again a wavelength scan can only be done by rotating the grating. This is true for a slew scan spectrometer, where the grating is rapidly 'slewed' from one analytical line to the other, and a wavelength scan around the peak is done by moving slowly (stepping) over the peak. Also for a direct reading spectrometer (polychromator), a movement of the grating is needed.

As mentioned previously, this movement is rather slow and, being the limiting factor, it can hamper the detection of short-lived transient signals. To correct for alternations of the background during such a measurement, an alternative method for measuring a small wavelength interval must be introduced. Different solutions do exist.

The use of a faster scanning drive to speed up the rotation of the grating is the most obvious method. Alternating measurements of the analytical line and background intensities should be done at the highest practicable frequency (>300 Hz) to reach almost simultaneous measurements [443]. At present the fastest commercially available spectrometers are those equipped with magnetically or galvanometrically driven gratings. Some can slew over a wavelength range of 600 nm in less then 20 ms. This high speed will be somewhat lower when actually measuring but must stay high enough to ensure the registration of several measuring points at least at a rate of several Hz. Speeding up the system even more is a critical solution since the grating is the most valuable part of the spectrometer.

Another more popular method is the use of an oscillating refractor plate placed behind the entrance slit of the spectrometer (Fig. 4.8), which serves as a wavelength modulator.

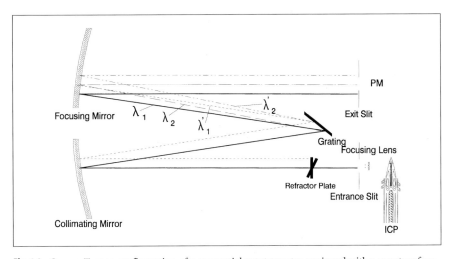

Fig. 4.8. Czerny-Turner configuration of a sequential spectrometer equipped with a quartz refractor plan-parallel plate, which serves as a wavelength modulator. In this way information of the peak and the background is obtained and can be used to calculate the real time background and correct for it.

This technique was introduced in the early 1970s in FAES [444] and has been upgraded for use in different fields [445, 446, 447, 448, 449, 450, 451, 452, 348, 453, 454, 455, 456, 457, 458, 459]. The method is based on the property that light is laterally displaced over a distance D when going through a plane-parallel plate [460]. The distance (displacement D) is given by

$$D \cong \frac{t(n-1)\alpha}{n} \qquad\qquad\qquad\qquad \text{Eq. 4.1}$$

where
t= thickness of the plate (typical 6 mm),
n= index of diffraction of the used material and
α= angle of entrance

Most commonly, a quartz plate is used because quartz absorbs significantly only below 185 nm. For quartz at 198 nm n=1.65. Rotating this quartz plate (changing α) displaces the light so that a higher or lower wavelength is detected (Fig. 4.8, λ_1 to λ_2). The use of high speed galvanometer optical scanners makes it possible to scan small wavelength intervals ($\cong$0.2 nm) at a rate of up to 80 Hz [445]. The maximum scan speed over this interval depends on the number of measuring points in this interval which is usually set at $\cong$80points/0.2 nm.

In this way information about the peak (around 40 measuring channels) and about the background (about 20 channels on each side of the peak) is found in a fast way, so that when transient signals are detected (residence time in the plasma is about 5–10 s depending on the gas flow rate [423]) the wavelength interval is scanned many times during one measurement. A calculation of the background can then be done for every different scan ($\cong$time slice). How data-acquisition is further carried out will be discussed in the next section (see Sect. 4.4.2). The use of refractory plate is particularly interesting and attractive because the plate can be installed at low cost in almost every spectrometer. This technique is also used to scan wavelengths with fixed gratings in polychromators.

A third possibility of gaining information about the peak and the background can be derived by moving the detector over a set of exit slits using a fixed grating. When the 3 point background correction (measurement points before, on and after the peak) is used, three 2 s integrations are typically needed for 1 line [442]. Therefore this technique is not fast enough to handle fast transient signals.

A more attractive method was introduced some years ago, and is based on the use of *array detectors*. Nowadays some manufacturers of ICP-AES equipment offer commercial systems equipped with this kind of solid state detector. Different types of array detectors were used, beginning with the electron beam scanning array detector (e.g. silicon vidicon and *SIT vidicon* (silicon intensified target vidicon detector)). This type has now been replaced by the lighter more compact and more rugged photodiode arrays (self scanning) and charge transfer devices (*Charged Coupled Device* (CCD) and *Charged Injection Device* (CID)).

Solid state detectors can be used in a linear configuration or as a 2-dimensional array (e.g. with an echelle grating to have wavelength scans in different spectral orders). The number of active cells can be as high as 2048x2048 with a wavelength interval range over which the quantum efficiency exceeds 10% from ±200 nm to ±950 nm [461]. With these self scanning detectors, a continuous readout at 30 frames per second can be achieved which is more than sufficient for data-detection and response acquisition of transient signals. The wavelength interval that is scanned at fixed wavelength depends on the linear dispersion of the spectrometer used. The main disadvantage of these systems is their higher cost.

4.4.2
Data Acquisition and Interpretation

As described previously, ICP-AES determinations are usually made by scanning a wavelength interval around the peak for a certain time. The net integrated signal is obtained by subtracting the calculated background from the peak area. Since scanning is involved, the peak and background are measured at different moments. This is justified when a continuous signal is obtained. However, in the case of transient signals, wavelength scans are made as a function of time. In the previous section, different detection techniques were described which have the particular option of detecting a small (or larger) wavelength interval so rapidly that peak and background can be recorded (quasi-) simultaneously. This gives us the possibility of measuring this interval as "*time-slices*" at a relatively high frequency.

Sometimes the software which comes along with the new instruments does not have the option of handling these different time-slices, so that in many cases new software must be introduced (mostly self-written). This software may differ in many aspects, however the algorithm used must meet some common requirements. Each measurement must be stored in a three-dimensional array (y, λ, t). The element y of the array represents the signal intensity for a certain wavelength at a certain moment. λ and t are the corresponding wavelength and time of the detected signal intensity. For example $(y, \lambda, 0)$ represents the first wavelength scan at the initial starting time while $(y, \lambda, 1)$ corresponds with the next wavelength scan.

For each wavelength scan, the background can be calculated using cursors to indicate which areas in the spectrum will be used as peak or background signals (see Fig. 4.9). Depending on the matrix, different options are possible. If a flat background is obtained a simple linear fit will be satisfactory in most cases. However if a sloping background is encountered, a higher order fit should be used instead. As an illustration, Fig. 4.9 shows a wavelength scan obtained during the determination of Pb in River Sediment (BCR CRM-320) using 2.2 mg powder with SS-ETV-ICP-AES [171].

River Sediment contains ±82 mg/g Al which is the interfering matrix element. In this case a quadratic fit is obviously a better choice than a linear one [171].

Every obtained net peak area (per time slice) can then be integrated as a function of time, yielding the time resolved signal as illustrated in Fig. 4.10.

In this figure, emission signals corrected for the background are plotted against time for Cu, Cd and Pb in different environmental materials. At this stage,

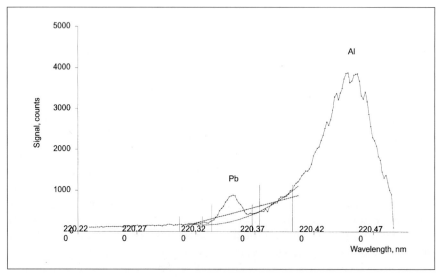

Fig. 4.9. Determination of Pb in River Sediment (BCR CRM-320); sample mass 2.2 mg. Spectrum of Pb (220.353 nm), interfered by Al (220.462 and 220.467 nm), obtained 15 s after the start of the vaporization step. With a linear approach for the background correction a net peak area of 3280 counts was obtained with this cursor set-up. A quadratic fit resulted in a net peak of 7047 counts

Fig. 4.10. Background corrected time resolved signals for solid sample analysis using SS-ETV-ICP-AES. Time resolved signals (normalized to 50 ng) for Cu in River Sediment BCR 320 (324.754 nm, observation height (OH) 9mm, holder gas flow rate (GFR) 0.61 l/min), Cd in Incineration Ash BCR 176 (226.502 nm, OH 7 mm, GFR 0.18 l/min) and Pb in Fly Ash BCR 038 (283.306 nm, OH 11 mm, GFR 0.38 l/min).

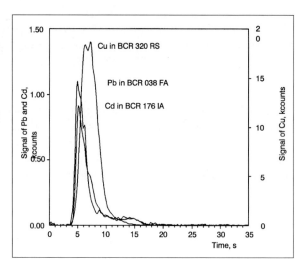

another pair of cursors can be used to settle the time integration borders (not included in the figure). In this way, it should also be possible to determine the concentrations of different species volatilizing at different temperatures from the furnace when gradual heating is used. An application by Záray et al. [90] on this *speciation* is given in the last section of this chapter.

4.4.3
Coupling the Furnace to the ICP-MS

When coupling an ETV-system to an ICP mass spectrometer (MS), few instrument modifications or software developments are needed. Most ICP-MS manufacturers have, to a certain extent, made the optional software commercially available. Generally the software allows a detection of several elements quasi-simultaneously. This is because the speed of sweeping of the quadruple mass analyzer over the mass region of interest (in peak hopping or peak jumping mode) is in fact usually high enough to detect the transient signals of different elements. There are some articles on data-acquisition describing how to optimize the time setting of the quadrupole for transient signals [462]. Although coupling of the ETV to the ICP is more or less the same as in the case of ICP-AES, very little work has been reported in the literature (at the time of writing this chapter) on powdered solid sampling ETV-ICP-MS as reflected in the application in Sect. 4.7. This will undoubtedly change in the near future.

4.5
Special Methodical Features

4.5.1
ETV-ICP Versus GF-AAS

Since most of the graphite furnaces used for coupling to an ICP are modified furnaces taken from GFAAS instruments, a wealth of knowledge on furnace processes has been inherited from the latter technique. However, having it coupled to an ICP involves transport of the aerosol formed and excitation of the atoms or ions in the plasma. As a result of these different processes, the analytical features connected with this technique (SS)-ETV-ICP cannot always be extrapolated from GF-AAS. Some differences will be discussed here.

4.5.1.1
Atomization Versus Vaporization

One of the explicit differences between GF-AAS and ETV-ICP-AES with a graphite furnace is undoubtedly the different *function of the furnace*. Since in AAS, atomic absorption is measured, the furnace must fulfill all the needs that are necessary to produce a dense atom cloud in the path way of the hollow cathode light. Hence atomization temperatures must be reached to produce the atom cloud and during measurement, a gas stop is in some cases introduced to enlarge the mean residence time of the atoms in the path way.

When ETV is operated with a graphite furnace, the function of the furnace is to produce a dry *aerosol* which can be transported to the ICP. This involves a twofold difference to AAS. Other temperatures (often lower than atomization temperatures) will be used and a continuous gas flow is applied to transport the aerosol. In some cases, however, the aerosol is sucked out of the furnace by low pressure at the end of the furnace [438].

The description of the different processes occurring during vaporization and aerosol generation is beyond the scope of this book, however, a short explanation is necessary to understand the different effects (e.g. of carrier gas flow rate, matrix composition, sample mass), for instance, on the transport efficiency which will be discussed later. To understand the aerosol formation processes fully, the reader is referred to the literature concerning this topic. The following description is based on an excellent overview article by Kántor [399] on this topic in relation to ETV-ICP and two monographs, one by W. Hinds [463] on aerosol technology and the other by Twomey on atmospheric aerosols [464].

One of the advantages of ETV over the conventional pneumatic nebulizer is that a dry aerosol is obtained. This means that during the heating of the sample the solvent is vaporized separately from the analyte. The aerosol formation thus starts from a dry sample present in the graphite sample holder. The formation of an aerosol in principle is based on the condensation of supersaturated vapor on small particles which function as sites or nuclei. A supersaturated vapor is obtained when a saturated vapor is cooled down. In ETV, the sample is vaporized in the sample holder and with the carrier gas flow is swept towards the colder ends of the tube resulting in a cooling of the vapor. Since aerosol formation is preferred in ETV a "*cooling gas*" can be introduced into the furnace to enhance the supersaturation and thus to increase condensation of the vapor on the sites [398, 370, 424, 397, 402, 465]. The degree of supersaturation can be expressed by the *saturation ratio S*, defined as

$$S = \frac{P_{vap}}{P_s} \qquad\qquad\qquad \text{Eq. 4.2}$$

where
p_{vap} =*partial pressure of the vapor*
p_s =*the saturation vapor pressure of the component*

Thus, if at a given temperature, the partial pressure of, for instance, Cd (g) exceeds p_s of Cd (g) the vapor is supersaturated and condensation on germs will take place. Since the vapor pressure remains almost constant (at constant total carrier gas flow pressure) with decreasing temperature the saturation ratio increases as mentioned earlier because of the fast drop of the saturation pressure.

A formed particle is stable for a given saturation ratio S, if its diameter corresponds to the *Kelvin diameter* given by [399]:

$$d_p^* = \frac{4\sigma V_m}{k\,T\ln S}$$ Eq. 4.3

with
σ =*the surface tension and*
V_m =*the molecular volume of the vapor species.*

Thus for a given saturation ratio at a certain *temperature T*, only those particles of which the diameter equals the Kelvin diameter are stable. Particles with a diameter smaller than d_p^* will evaporate, while particles with a diameter greater than the Kelvin diameter will grow.

As initially stated, condensation happens on small particles which act as sites or nuclei. The process of formation of these sites is called *nucleation*.

Homogeneous nucleation also called *self-nucleation* is the process where particles are formed by colliding clusters (agglomerates), without the aid of condensation nuclei. Clusters are always present in the vapor as a result of collisions of vapor species. The higher the saturation ratio the higher the number of clusters resulting in a higher frequency of collisions. This affects the number of agglomerates which have a diameter greater than the Kelvin diameter and thus more stable nuclei are formed.

Heterogeneous nucleation or *nucleated condensation* is the process of particle formation and growth by condensation of vapor onto foreign condensation nuclei or ions. In ETV with solid samples this is the more important process since often molecular matrix species [399], and/or graphite particles of the furnace [399, 466] can be released on which the saturated vapor can condense.

Once a stable nucleus is formed (diameter d corresponds to d_p^*, under the given conditions, Eq. 4.3) the particle will keep on growing by condensation of the vapor onto the formed nucleus (heterogeneous condensation). This is expected to be the major process in ETV-ICP methods, and is also referred to as *the carrier effect of foreign particles* [423, 399, 466]. A second way of particle growth is *coagulation*. Coagulation is the process where two particles collide with each other to form a new larger particle. When this collision is the result of the Brownian motion, it is called *Brownian* or *thermal coagulation*, and in the other cases, where other forces increase the kinetic energy, it is called *kinematic coagulation*. In any case, this results in a decrease in nuclei and fine particles and in an increase in the mean particle size. The rate of growth (the rate of these condensation and coagulation processes) depends on the saturation ratio, particle size, and *particle size* relative to the gas mean free path and can be formulated in the following simplified equation [399]:

$$d_p = k_s\, c_v^{0.4}$$ Eq. 4.4

where
d_p =particle size,
c_v =the initial vapor concentration at the nucleation temperature (g/cm³),
k_s =constant, characteristic for the vapor species and the operating conditions.

When particles grow too large, sedimentation in the transport tubes becomes important. An optimum particle size is thus required. An important implication of this equation is that, with increasing sample mass (solid sampling ETV), only the number of particles in the vapor (c_v) will increase. The mean particle size d_p will only slightly increase. This implies that, vaporizing solid samples does not necessarily lead to great sedimentation problems. This is important when discussing dynamic ranges and calibration lines since the transport efficiency is not influenced significantly.

In general, for ETV-ICP, the principles mentioned above can be understood and summarized in the following way. The sample is heated in the furnace and cooled down on its way to the ICP. The supersaturation ratio therefore increases towards the ICP so that nucleation will start for a given analyte at its corresponding temperature of supersaturation (this temperature can be found from the calculated supersaturation curves [399]).

For volatile elements such as Cd, As, Se, and Zn the cooling process will take longer so that nucleation will start further along in the tubing. This process may cause severe condensation of atomic vapor on the cold transport tubes, resulting in a low transport efficiency (see below). In these cases, a carrier effect (e.g. supplying additives or cooling down the vapor in the furnace) could increase the transport significantly. Crabi et al. [314], e.g. found that for cadmium the atomic vapor can resist at a flow-rate of 1.2 l/min up to 3 m of tubing.

For less volatile and refractory elements, condensation on existing particles is more important since these analytes enter the vapor only at higher temperatures. Sometimes it is found that, for less-volatile elements e.g. Cu, high losses are found during transport through the cold ends of the graphite tube. This is because condensation for non-volatile elements occurs very rapidly and thus not only on particles. This condensation on cold parts of the furnace is even more pronounced when low carrier gas flows are used [423, 467]. To increase the transport of such less-volatile analytes different groups have introduced gaseous reagents e.g. CCl_4 [398, 424, 281], CCl_2F_2 [91, 427, 459] or solid reagents e.g. Teflon powder [420, 468, 469], AgCl [415, 416], sea water [470], or complex modifiers e.g. CoF_2 + BaO [435, 471, 438, 472, 148], to improve the vaporization of refractory elements and possibly to form molecular species that increase the transport efficiency. The use of coolant gases can also reduce the condensation drastically.

4.5.1.2
Calibration Line

One of the more important analytical features of a technique is the *dynamic range*. In the ICP excitation temperatures over 7000 K are reached (see Sect. 4.1.1). This is the fundamental reason for the large dynamic range which is one of the main advantages of the ICP. This wide linear relationship (4 to 6 orders of magnitude [442]) between the emission signal and the concentration of the analyte element is very convenient, since calibrations can, in practice, be done with only one reference solution and without the need of diluting the sample. Moreover, it results in a longer analytical range which implies that the technique can be used in a larger concentration range with resulting precision below a pre-defined value (often set at $3\sigma_B$ of experimental set-up).

This large dynamic range of 4 to 6 orders of magnitude was originally established for the analysis of aqueous solutions with a pneumatic nebulizer as the sample introduction system. In view of above considerations on the different vaporization and condensation processes, which have an effect on the transport efficiency, it is obviously also necessary to check the linearity of the calibration graphs when coupling an ETV to an ICP for the analysis of solid samples. Since the introduction of solid samples also involves the introduction of large amounts of matrix elements in the plasma, the stability of the plasma can be influenced and some matrix effects can occur.

The influence of the volatility of certain elements and their transport efficiency on the linearity of the analytical curve can be explained in the following way. As mentioned in the former section, volatile elements e.g. As, Cd, Se, and Zn, can be partly transported in atomic vapor, partly as particles. The transport of particles is the more efficient way. Several researchers have investigated the Cd transport (for analysis of Cd – mono compound solutions) in detail and by measuring the atomic absorbance they could determine the amount of Cd transported as atomic vapor [314, 390]. Further investigation (multi-component solutions) showed that the Cd transport can be increased by supplying additives [399, 473]. Kántor [399] found that, in these cases, the Cd is condensed on molecular species formed by the additives. Therefore, increasing the amount of transported mass will decrease the ratio between the Cd in atomic vapor form and the Cd in particulate form. From this, it is obvious that the signal to analyte mass ratio will not be constant. In solid sampling, however, the amount of foreign particles and additives is, in comparison to the amount of analyte mass (trace-elements), enormously high so that, in this case, the amount of condensation will be very high (increased carrier effect). In fact, the analytical curves reported for solid sampling ETV-ICP-AES were found to be linear.

Several elements were investigated by different research groups. Linearity was found for volatile (e.g. Cd – Fig. 4.11), less-volatile (e.g. Pb – Fig. 4.12) and refractory elements (e.g. Ti – Fig. 4.13).

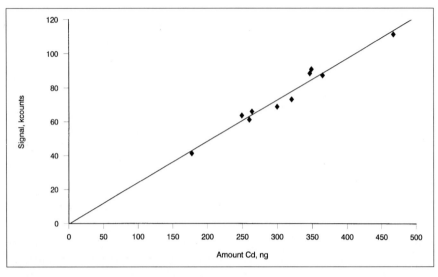

Fig. **Fig. 4.11.** Calibration line of Cd in City Waste Incineration Ash (BCR CRM 176). The calibration lines where obtained using SS-ETV-ICP-AES by varying the sample mass

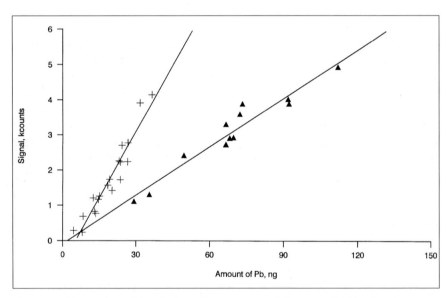

Fig. 4.12. Calibration line of Pb in biological (BCR CRM 062 "Olive Leaves" (+)) and environmental (BCR CRM 320 "River Sediment" (Δ)) samples. The calibration lines where obtained using SS-ETV-ICP-AES by varying the sample mass

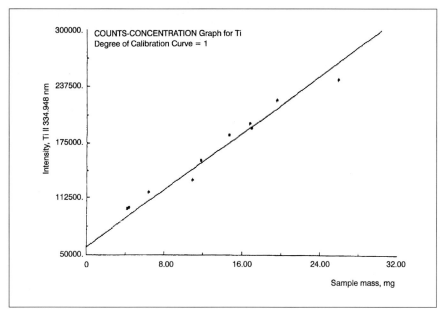

Fig. 4.13. Line intensity versus sample mass relationship for titanium, applying chlorination at 2100 °C for 30 s. The concentration in the original sample was 280 µg/g (from [424], with permission)

Figure 4.11 shows the signal intensity of Cd as a function of the absolute amount of Cd in an environmental material "Incineration Ash" (BCR CRM-176) [S. Boonen, unpublished].

As in AAS, this type of curve is obtained by varying the sample mass. The linearity obtained is partly explained by the large number of particles (solid sampling, carrier effect) but also Eq. 4.4 predicts that, with increasing sample mass (SS-ETV), mainly the particle concentration will increase which favors the condensation process.

For Cd (Fig. 4.11)and also for Pb (Fig. 4.12), a good linearity is obtained. It is at this point interesting to draw attention to the difference in slope of both curves shown in Fig. 4.12. This, of course, is a result of a difference in vaporization and transport processes due to different matrix composition. This subject will be discussed in a separate section on interferences and overcoming them.

Záray et al. [424] investigated the linearity of the calibration curve for the more refractory elements e.g. V, Ti, Cr after halogenation with carbon tetrachloride vapor (Fig. 4.13).

From all three figures, it can be seen, that for ETV-ICP-AES with solid samples, a good linearity is found and that the dynamic range is, in these cases, about 2 to 3 orders of magnitude. This is in sharp contrast to GF-AAS (liquids and solids) and makes the calibration in SS-ETV-ICP-AES much easier and more reliable.

The reason for this linearity is two-fold, namely, better transport because of the carrier effect and the high temperatures of the ICP as stated in the introduction to this section.

In the case of solid sampling, this dynamic range is limited to the lower side of the mass region by the detection limit or by the smallest analyzable mass. On the high amount side, the line is limited by the maximum sample mass which fits in the sample boat or, by the maximum sample mass for which the plasma is still stable (see below).

Vanhaecke et al. [428] investigated the linearity of calibration graphs for As in standard reference materials of biological origin using solid sampling ETV-ICP-MS. On plotting the ETV-ICP-MS signal intensity as a function of the sample mass, an increase of signal suppression with increasing sample mass was observed. This behavior is ascribed to matrix effects in the plasma and/or during the ion beam formation process, caused by matrix components which are not removed during ashing and are vaporized and transported into the central channel of the ICP together with As. The plasma zone with maximum M+ density is thereby shifted and not efficiently extracted by the interface, resulting in a decrease of the signal intensity observed. Since, however, for Sb (added as an internal standard) the signal intensity was found to be suppressed to the same extent as for As (same shift of the zone with maximum M+ density), plotting the 75As+/121Sb+ signal ratio as a function of the sample mass, a linear calibration graph was observed. For larger sample masses (e.g. >1.4 mg for BCR CRM 279 Sea Lettuce), the curvature observed was even more pronounced, so that the curve became almost horizontal. For these larger sample masses, Sb could no longer be accurately corrected for the signal suppression effects. In contrast to solid sampling ETV-ICP-AES, the dynamic range is limited on the high amount side by the effects of matrix-induced signal suppression and therefore strongly depends on the sample under investigation. Similar results were found for Se [429] in standard reference materials of both biological and environmental origin.

These observations clearly indicate the power of the concept internal standardization using EV-ICP-MS. In general, however, selection of a suitable internal standard is not self-evident. Firstly, an appropriate internal standard should show a (furnace) chemistry analogous to the analyte to be determined. This means that they have to behave identically in vaporization and transportation processes, i.e. they give similar signal profiles (signal intensity as a function of time). Secondly, the element chosen as an internal standard can only be present at negligible levels in the samples under consideration. Therefore, attempts were made to use the Ar^{2+} signal as an indication of plasma loading effects and non-spectral interferences. This concept was introduced for pneumatic nebulization ICP-MS by Beauchemin and successfully used by Grégoire et al. [430] and Vanhaecke et al. [706] to correct for matrix-induced signal suppression effects.

4.5.1.3
Test Sample Mass

To improve the precision of the technique, coincidental errors should be overcome as much as possible. One of the uncertainties introduced with solid sampling is the weighing error. As discussed in former chapters, this uncertainty is very low when a new generation micro-balance is used. However when the weighed sample mass of a powder is further reduced to very small masses (10–500 μg), an increase in the RSD values can be introduced, since the probability of taking a representative sample aliquot becomes smaller. In previous chapters it was shown that inhomogeneity of the sample can cause serious problems (see Sect. 2.3.2 "nuggets"). Therefore in SS-ETV-ICP-AES, masses of material which are as large as possible are used to improve the representability of the sub-samples. The use of large masses is justified through the linearity of the calibration lines. Typical masses are 1–10 mg depending on the matrix composition. A check on the variation in transport efficiency for highly varying sample masses can be obtained by calculating the RSD on the signal to mass ratio.

The low mass side is limited by the *limit of detection* of a particular element and the corresponding *limit of analysis* for a certain matrix (see discussion in the section below). When a high concentration of the element in the sample is found, the low mass side is limited by the smallest possible weighable sample (physical weighing limitations, some μg). However, this is mostly not the case in ETV-ICP-AES. Here the more common problem is the large mass side. Again there are two possibilities. When a dense aerosol is introduced (high sample mass) in the plasma, the plasma can become instable and in the worst case it can extinguish. The amount of sample which can be analyzed depends very much on the matrix composition. First, when a "light" matrix (e.g. coal, ash,…) is analyzed the plasma remains stable up to 20 mg [171]. The sample mass is then limited by the dimensions of the sample holder. In the more common case, when "heavy" matrices are analyzed (e.g. biological samples), the plasma will extinguish at a certain sample mass.

Verrept et al. [171] investigated some parameters that influences the maximum amount of sample mass that can be vaporized without destabilizing the plasma. They found that the R.F. power has a significant influence: A slight increase in the R.F. power ensures a stable plasma even when higher amounts are introduced [395, 171]. They also found that the maximum sample quantity strongly depends on the type of the solid material. For instance the plasma will tolerate larger amounts when environmental materials (4–15 mg) are vaporized than in the case of biological materials (2 mg). Similar observations have been reported in [475] when hair samples are analysed. A tentative explanation might be, that in the case of biological materials, larger amounts of gases are formed causing extra loading of the plasma. Also the influence of the carrier gas flow rate on the maximum sample amount was studied. It seems that at lower carrier gas flow rates the plasma can tolerate more solid material (e.g. for river sediment 14 mg at 0.28 l/min, but only 6 mg at 0.86 l/min). The feeding of the aerosol into the plasma is in fact extended over a longer period of time due to the diffusion of the solid in the transport tubing.

4.5.1.4
Limits of Detection – Limits of Analysis

According to the IUPAC definition, the *limit of detection*, in the ET-technique expressed as an analyte amount a_L, is derived by the analytical function from the smallest response R_L, that can be detected with reasonable certainty for a given analytical procedure. The estimation of the smallest response is as follows

$$R_L = \overline{R}_B + k\sigma_B \qquad\qquad\qquad\qquad \text{Eq. 4.5}$$

where
R_B =*response of blank measurements*
σ_B =*standard deviation of the blank response population*
k =*coverage factor chosen in accordance with the confidence level.*

σ_B is usually estimated by the experimantal standard deviation s_B of a sufficient large number of blank measurements. When the factor k is chosen to equal 3 ("3σ-criterium") there is a 0.13% chance that a response at R_L is the result of random fluctuations of the blank signal.

Obviously, this definition requires a good knowledge of the response of the blank signal. As a blank reading a response is usually understood to be *yielded* from test samples consisting of the *same matrix compounds* as the unknown sample but *not containing the analyte*. Thus during the measurement of the blank, conditions and matrix influences must be obtained identical to those during the unknown sample measurement.

When analyzing liquid samples, the application of this definition does not involve serious problems since synthetic blanks (solvents + major matrix elements) can be prepared in the same way as the sample.

However, in solid sampling the so defined blank reading can generally not be obtained since the required blank material will mostly not exist and can hardly be prepared synthetically (see discussion below). Furthermore, it is almost impossible to transform a smallest response R_L obtained into the detection limit a_L by an analytical function established from only one reference material (e.g. a CRM), because, with the reduction of the introduced analyte amount, the matrix amount is reduced simultaniously (compare Eq. 2.4). So, in order to define a real blank level, different solid reference materials with an identical matrix but different analyte contents would also be required (which seems impossible to achieve).

So the classical definition of the detection limit (Eq. 4.5) for a certain element no longer holds. This definition does not reflect the smallest determinable amount of element in real solid samples.

Being aware of these questions, different approaches were reported in the literature to determine a figure of merit usable in interlaboratory comparisons by different groups, e.g. the $3\sigma_B$ minimum quantity (MQ) – this is only possible when a blank is available, see applications for further information [392, 395], background equivalent concentration (BEC) [424, 476], background equivalent mass (BEM) [398], limit of determination [171]. Some of these will be discussed below.

In most papers [e.g. 419, 420, 407, 418] the limit of detection is found in the classical way, i.e. by using a reference solution of a certain element and the corresponding blank solution (solvent). The value of this limit of detection can only be a reflection of the conditions used (optimal or not). Such a limit of detection reflects the possibilities of the technique for the analysis of small liquid samples. This value can, however, hardly be used as a guide for a real analysis of solid sampling.

Another approach is the measurement of the background equivalent mass or background equivalent concentration [398, 424, 476]. In these cases, the background is determined by measuring peak heights while running an empty boat or while running a single element oxide standard [434].

Verrept et al. [171] used an empty boat as a blank, which they refer to as the *"zero-mass signal"*. From these measurements they determined the smallest response R_L, that can be detected according to the $3\sigma_B$ criterium (Eq. 4.5). The transformation to the limits of detection in terms of analyte mass a_L, they also used single element reference solutions.

Additionally, they determined the ratio of the analyte mass specific signals (i.e. response value divided by the analyte mass) obtained by the analysis of liquids to those obtained for solids (which may be significantly different in ETV-ICP-AES due to different transport efficiency, see below Sect. 4.6.3). In this way, they were able to determine a correction factor (for each measured element in the different materials used) which includes matrix influences on the signal in the cases of solid sampling. By multiplying the estimated limit of detection with this correction factor a value for the limit of detection in the solid material can be obtained.

Finally, after determining the maximum possible sample load at stable plasma conditions, the limit of determination is determined by dividing the calculated limit of detection by the maximum sample mass following Eq. 2.53.

Example

With SS-ETV-ICP-AES for Cu a limit of detection of $a_L=0.32$ ng was found [171] by comparing "zero-mass signals" with response values from reference solutions. For "Light Sandy Soil" (BCR CRM-142) a maximum sample loading for a stable plasma was determined as approximately 10 mg under optimum analytical conditions. Thus, the limit of analysis for Cu in this sample can be calculated as $c_L=0.032$ mg/kg. (For copper the transport efficiency for reference solutions and the solid samples were found to be not significantly different [423], hence no correction was required.

As discussed in the former section on the sample mass, the useable mass of a sample depends very much on the kind of sample and its matrix composition. Therefore the limit of analysis (or, in practice, the smallest detectable amount) for a certain element, in e.g. a biological material, is not equal to the one in an environmental material.

For a correct limit of analysis, a real blank can only be obtained with synthetic samples which are identical to the unknown sample but contain no analyte. Such synthetic samples can of course only rarely be prepared.

Van Berkel et al. investigated the possibilities of loading poly(dithiocarbamate) (PDTC) resins by sequestration analyte from different types of samples [392, 395]. In this way, it should be possible to produce blank resins acting as real blanks. Unfortunately the resins were not available in sufficiently pure form, so that contamination of the samples was unavoidable [392]. However minimum loading quantities of the resins (MQ) in order to obtain significant analytical signals were determined using the $3\sigma_B$ criterium. The limit of determination of the method then depends on the MQ values as well as on the attainable concentration (loading) factor. Another synthetic blank was made for the analysis of SiC powders by heating the SiC at 2200 °C to release the impurities [90, 91]. The residue may in this case possibly be used as a blank. (Compare further attempts at obtaining synthetic reference materials in Sect. 2.2.1.3.)

Because such procedures will mostly be not feasible, one of the former alternatives should be used. However, to make interlaboratory comparisons possible a *consensus should be found*. In all cases the figure of merit that is to be compared in solid sampling can only be the limit of analysis for a certain element in a specific matrix, so that boat dimensions and plasma conditions should also be included in the comparison since in conventional methods the operating conditions also have to be reported.

4.5.2
ETV-ICP Versus Pneumatic Nebulization ICP

Electrothermal vaporization was initially introduced as a sample introduction system for the ICP, to avoid the obvious drawbacks of nebulization ICP. Minimizing the sample volume, increasing detection power and an increase in transport efficiency were major hopes of the first investigators. However, it soon turned out that the use of an ETV not only changes some features related to the sample introduction but that the plasma processes are also influenced.

One of the advantages of ETV is the desolvatation prior to the vaporization of the analyte. This, as described in detail before, results in a dry aerosol. Therefore, no water is introduced into the plasma resulting in a so called *dry plasma*. When using a nebulizer for sample introduction, a wet plasma is obtained.

Many researchers have investigated the differences in plasma conditions for the wet and dry plasma and came to the conclusion that the observed differences might be explained in terms of thermal conductivity [395]. In *wet plasmas* (conventional nebulizer), the introduction of water is found to cool down the plasma in the lower regions [395, 477]. Higher in the plasma, water loading appears to promote energy transfer between electrons and Ar atoms, leading to a smaller difference in gas and electron temperature [477]. *Dry plasmas* on the other hand are characterized by a steep excitation temperature gradient towards the central channel [395].

This might explain why in dry plasmas, lower optimal observation heights are found [423, 395] as illustrated in Fig. 4.14 a and b for Pb and Cu. These figures were obtained by measuring a few mg of solid material (CRM). The signals were

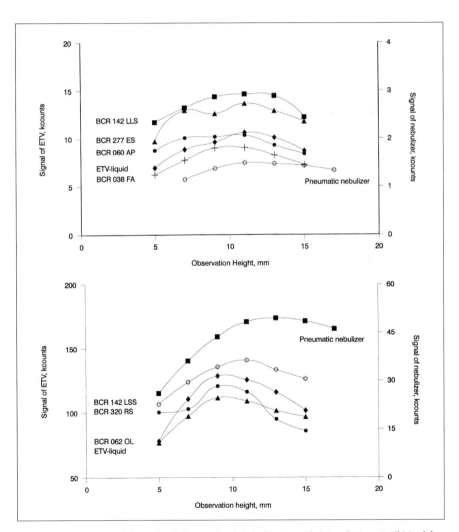

Fig. 4.14. Comparison of the optimal observation height for 50 ng Pb (**a**) and 50 ng Cu (**b**) in different matrices, at a holder gas flow rate of 0.61 l/min

normalized to 50 ng of each element so that a comparison could be made with a pneumatic nebulizer (Meinhard TR-50-C3).

Since water loading promotes the energy transfer in wet plasmas, it might be that the addition of a molecular gas (e.g. H_2) to the carrier gas will in the case of dry plasmas, improve excitation conditions. The electron density in dry plasma is lower than in wet plasmas [478]. Matousek and Mermet found that this deficiency can be compensated for by the addition of hydrogen gas, without consuming energy in contrast to what is the case for water loading [479, 478]. As a result, they obtained an improved performance and an enhancement of the emission intensity of ionic lines and consequently improved the RSD.

4.6
Interferences and Overcoming them in AES and MS

Two distinct types of interferences can hamper the analysis of solids using ETV-ICP-AES/-MS: spectral and non-spectral (or matrix effects) interferences. Spectral interferences are caused by elements (other than the analyte to be determined) which emit light that contains the same wavelength as the one that was chosen for detection (ICP-AES) or by species with the same m/z value as the nuclide to be detected (ICP-MS). Non-spectral interferences are caused by matrix components which are vaporized and transported to the plasma together with the analyte and as a result give rise to signal suppression or enhancement. Both kind of interferences and the possibilities to overcome them will be discussed for AES and MS.

4.6.1
Non-Spectral Interferences

Several investigations have shown that, for solid sampling ETV, non-spectral interferences occur mainly during the vaporization, the aerosol formation, the transportation and the plasma "sampling" processes. As only this last process is significantly different for AES and MS, one can expect most of the influences of matrix effects to be the comparable for ICP-AES and ICP-MS determinations. Reduction of non-spectral interferences involves applying a method which affects at least one of the former named processes or which controls the process in such a way that the effects are minimized. All four of these processes will be discussed chronologically.

4.6.1.1
Ashing/Vaporization

It has been pointed out that the use of an ETV system should enable *selective volatilization* of the matrix components. In this technique a specific temperature program is applied to the furnace to evaporate components on different moments according to their volatilization temperatures. Park et al. demonstrated the selective vaporization technique for ETV-ICP-MS with solutions. They determined Mo in a Na-matrix. Mo is known to form stable carbides, so that high ashing temperatures can be used to vaporize the matrix (sodium matrix, ashing at 1700 °C) retaining the analyte in the furnace (Mo). The Mo was released afterward with the aid of Freon [372].

This selective vaporization is, of course, only possible if the analyte of interest and the matrix elements have very distinct volatilities. If this condition is fulfilled, then the technique is mainly usable when aqueous solutions are to be analyzed. For solids, however, the complex composition of the matrix might obstruct the process of the selective volatilization. Some reactions with matrix elements might occur and change the vaporization characteristics. In some special cases, however, when the matrix is of a more easily destroyable nature, the technique can be applied very successfully. Van Berkel et al. used PDTC resins to investigate some properties of dry plasmas using this technique. The poly(dithiocarbamate) resins decompose at relatively low temperature, and at a well chosen ashing temperature (550 °C) the bulk of the matrix is volatilized before the analyte enters the plasma. In this case, a vent system of the furnace can lead the bulk of the matrix to an exhaust instead of to the plasma [392, 395].

Argentine and Barnes [474] used SS-ETV-ICP-MS for the direct determination of volatile and non-volatile impurities in semi-conductor grade organometallic materials and process chemicals. The technique showed great potential, particularly for the non-volatile impurities, since complete separation of the (volatile) matrix from the analytes could be obtained by application of a suitable temperature programme.

Mostly, the two criteria, i.e. difference in volatility and easy destroyable matrix, are not fulfilled. The matrix is mostly a very complex material containing components that are hard to vaporize. Therefore simple selective vaporization will generally not be sufficient.

However, this technique can still be used when matrix modifiers are used, i.e. changing vaporization conditions and mechanisms. This can be done in two different ways. When the vaporization temperature of the analyte and the matrix components are nearly equal a modifier may be added to the sample to promote the vaporization of the matrix (during ashing) leaving the analyte in the furnace for volatilization during the vaporization step. Alternatively, the analyte can be turned into a more volatile form (e.g. halogenide) to be vaporized before the matrix.

To overcome some of the differences in vaporization of dissimilar matrices or to facilitate the vaporization of less-volatile elements, halogenating reagents may be used as matrix modifiers. In the literature, both gaseous and solid modifiers are reported. In both cases, the halogenating reagents are used to transform the analyte elements in a more volatile halogenated form and to promote their vaporization in this way. This technique was initially introduced by Kirkbright and Snook [381] for the determination of refractory elements in solutions. These authors used a halocarbon/argon mixture to promote evaporation of Zr, B, Cr, Mo and W (0.1% CHF_3) from solutions. For the determination of B, trifluoromethane instead of trichloromethane was used because the BCl_3 tends to vaporize faster than BF_3.

The introduction of halo-forming reagents (gaseous or solid) has become a widely used technique in solid sampling ETV-ICP-AES. It has, for instance, been applied successfully to the determination of refractory elements in ceramic powders. Reisch et al. [415] investigated the use of solid halogenating reagents on graphite and ceramics. It was found that addition of $(C_2F_4)_n$ + NaF promoted the vaporization of Fe in graphite powder. It was also found that in the case of ceramics, addition of fluoridating agents did not help the evaporation of Fe. However, addition of AgCl and $BaCl_2$ as solid halogenating modifiers improved the vaporization of Fe from SiO_2, Al_2O_3 and MgO based ceramic powders [416]. Four years later the same group evaluated 7 mono-component and 6 complex modifiers (two components added, 1+1) on "difficult to vaporize carbide forming elements" in a SiC-powder matrix. They found that the complex modifier (BaO + CoF_2) decomposed almost totally, allowing the impurities of the material to be released. They later used this modifier to determine Cu, Ni, Si, Fe, Al, Ti, V, B and Ca in almost insoluble materials such as B_4C, SiC and Si_3N_4 [471]. For SiC the following mechanism was suggested, the BaO reacts with SiC to form $BaSiO_3$ whereas the CoF_2 reacts by forming $CoSi_2$. In the reaction the elements are fluoridated by the fluorine released from CoF_2 [476, 471].

A gaseous halogenating modifier has also been used very successfully by Kántor. He reported the use of CCl_4 and CHF_3 to stimulate the vaporization of a number of elements from solid matrices. Over the years, the same group continued to investigate the possibilities of CCl_4 as a matrix modifier on a great number of elements in ceramic powders, [398, 424, 402]. In more recent work they switched to Freon 12 (CF_2Cl_2) instead of CCl_4 to improve the decomposition rate of the matrix [91].

From the former section, it can be seen that halocarbon modifiers can be applied very successfully to promote the vaporization of less volatile elements from hardly soluble matrices. However, whether gaseous or solid halogenating components should be used is still a point of discussion.

Arguments against gaseous modifiers are two-fold [435]: firstly, since a heterogenous reaction is required it can be expected that the gaseous reagents may flow over the solid without reacting with the included impurities and secondly the reaction will be slow. When using complex solid modifiers, such as BaO + CoF_2, the Ba silicates formed will melt thus facilitating the contact between the sample (impurities) and the second modifier component (CoF_2) [471, 435].

On the other hand some researchers in favor of gaseous reagents argue that solid modifiers introduce large blanks. Also a premature release of the halogens is expected during the thermal decomposition of the modifier, therefore some doubt is expressed as to whether the reaction of the solid with the impurities incorporated in the core of the sample is complete. Finally, it is said that halogenation processes are far more controllable when gaseous reagents are used [90].

4.6.1.2
Aerosol Formation and Transport Effects

The second process of the sample on its way to the ICP, after the vaporization step, is the transformation of the vapor phase in a dry aerosol. This dry aerosol is necessary to have an efficient transport to the ICP. Aerosol formation and transport are interconnected and therefore will be discussed together.

The efficiency of the formation of a dry aerosol is dependent on the saturation ratio S (see Sect. 4.5.1.1). A decrease in the temperature of the sample vapor increases this ratio and promotes condensation of the vapor on germs or particles in the gas flow, resulting in a stable aerosol. From this point of view, it is clear that if a furnace is modified in such a way that a rapid cooling down of the vapor is guaranteed, a better aerosol formation will be the result. This will result in a more efficient transport of the aerosol to the ICP.

In Sects. 4.3.3 and 4.3.4, some descriptions of furnaces and their modifications are given. A very efficient modification of the furnace to promote the formation of dry aerosols was reported by Ren and Salin [465]. Although the article covers only solution analysis, the system was originally designed especially for solid sampling. As can be seen more clearly from Fig. 4.15, a flow-trough type furnace was used and 4 holes were drilled in one of the graphite contact rings. When introducing a gas flow through these holes, a fraction is led into the furnace acting as the cooling gas to promote condensation and aerosol formation and the rest of the gas flow is led through a narrow gap between the graphite contact and the Al-transport tube.

This results in a *shielding gas*, providing a thin gas layer between the wall of the transport tube and the gas mixture. In this way, the contact between particles and walls is also minimized to prevent condensation on the tubing. Thus, by using a cooling/shielding gas, the integrated signal can be increased by a factor of 22–40 for less volatile elements (Mn, Cu) and 60–120 for more volatile elements (Cd, Pb).

Once a dry aerosol is formed, it must be transported to the plasma. A shielding gas will certainly improve this transport since it was found that condensation of the particles on the wall largely occurs at the beginning of the transport tubing [423]. An other way to prevent condensation on the tubing walls consists of using a heated transfer line. The use of such a *heated transfer line* is very good practice when coupling gas chromatographic techniques. It appears that the use of a heated transfer line in the case of dry aerosols (ETV-ICP-AES/MS) is not really necessary [397]. In Fig. 4.16, the effect of the temperature of the transport tube (stainless steel) of an SS-ETV-ICP-MS system on the Pb signal from Coal Fly Ash (NIST SRM1633a) is shown. As can be seen from the graph, varying the temperature between room temperature and 420 °C does not influence the Pb

Fig. 4.15. Modified graphite furnace (**a**) and the introduction of a cooling and shielding gas (**b**) (from [465], with permission)

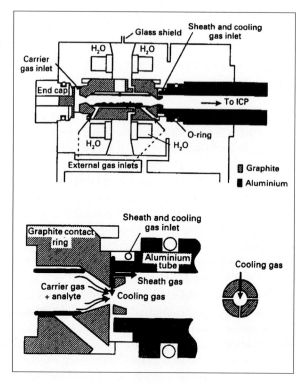

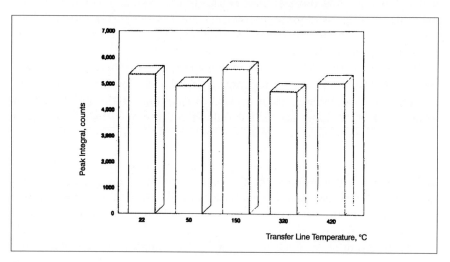

Fig. 4.16. The effect of transfer line temperatures on the Pb signal in Coal Fly Ash NIST SRM 1633a. Three replicate measurements were performed for each transfer line temperature (from [482], with permission)

signal significantly. Form this result it can be concluded that a dry aerosol was produced very successfully using the furnace described by Carey et al. [397] (see Sect. 4.3.4).

It is also claimed that the presence of a condensation site is necessary to trap the analyte before it is lost by condensation on the walls. In solution analysis addition of NaCl (0.1 µg/l) can increase the number of particles by 35% [466]. In solid sampling, no addition is necessary since large matrix amounts are already in the furnace. Even when a vent is used to lead the matrix out of the furnace, there are still enough carbon particles generated during a firing to can act as germs to produce airborne particulates [466].

Even with the use of a matrix modifier and the use of excellent furnaces, the vaporization/transportation efficiency is matrix dependent. This is of major importance for *calibration of the technique*. For this reason, several groups have investigated transport efficiencies (actually they investigated the overall efficiency starting at the vaporization step until the excitation/ionization step) for different elements in different matrices. A more detailed discussion on this overall transport efficiency and its effects on the calibration of SS-ETV-ICP methods will be given in a separate section (see Sect. 4.6.3).

4.6.1.3
Plasma Sampling

As already mentioned in Sect. 4.5.1.2, no linear calibration graphs could be obtained with SS-ETV-ICP-MS unless an internal standard was added. The signal intensity was observed becoming more and more suppressed with increasing sample mass. However, perfectly linear calibration graphs were found by several authors using SS-ETV-ICP-AES. These different observations can be explained as follows. The position of the plasma zone in which the M+ density is at a maximum is easily affected by the matrix (amount of sample introduced in the plasma). As in ICP-MS, only a very small part of the plasma is extracted by the interface, the position of the plasma is very critical and small changes of this position can result in major alterations of the signal intensity observed. Assuming one normally starts measuring under optimized conditions (maximum signal intensity), matrix effects can lead to severe signal suppression or, if non-optimized conditions are used, to signal enhancement. The plasma zone that is "sampled" in ICP-AES is much broader. This means that small changes in the plasma position, caused by the matrix, do not significantly affect the signal intensity observed, and therefore the linearity of the calibration lines.

General precautions one should keep in mind when selecting a correction method for non-spectral interferences are as follows. The addition of a matrix modifier can include the introduction of spectral interferences. Especially in ETV-ICP-MS, the use of halogenating reagents (large Cl content) will introduce large amounts of polyatomic species (e.g. $^{40}Ar^{35}Cl^+$) causing spectral interferences ($^{75}As^+$). The introduction of large amounts of modifier will also probably result in large memory effects in the ICP-MS. In this case, one should look for alternative modifiers or at least there should be a transformation, after the vaporization step, of the halogenated analytes into a more ICP-MS compatible form. If an internal standard is added its furnace chemistry (vaporization and transportation processes) should be similar to that of the analyte(s) to be determined.

Fonseca and Miller-Ihli [480] showed that different degrees of matrix suppession effects were observed when different skimmer cones were uses. the exact positon of the skimmer was shown to affect the degree of spoace charge effects.

4.6.2
Spectral Interferences

Spectral interferences are the result of an insufficient resolution of the detection system. As the detection methods for AES en MS are different, the appearance of and the correction for spectral interferences will be discussed separately for both methods.

4.6.2.1
Spectral Interferences in SS-ETV-ICP-AES

As can be understood from former sections, the solid sampling technique in AAS only became possible when an accurate and powerful background correction system (Zeeman background correction) was introduced. The same holds for ICP-AES.

Spectral interferences or overlap as specified in Sect. 4.4.1 can be interpreted using suitable software. In the same section, different approaches to collect the necessary data and to correct for a *changing background* when transient signals are used for the response aquisition, were evaluated.

As an example of a changing background due to a spectral interference, a determination of Pb in a synthetic Al matrix solution with ETV-ICP-AES using a Bell-Jar type ETV device (see Fig. 4.4) is given. This example clearly illustrates the importance of a background correction technique. Whether moving slits, fast grating drives, or lateral displacements of the light beam are used to obtain a background corrected signal, the interpretation of the data should always be the same and is of major importance. In Fig. 4.17a to d, four wavelength scans are given.

Since Pb is more volatile than Al it is detected first (a). Note that at this time the background and the Al interference are fairly small. 1.1s later Al has arrived in the plasma and interferes with the Pb peak (b). Again the attention must be drawn to the increasing background. In time slices c and d the Pb signal is disappearing whereas the Al peak is degrading much more slowly. In Fig. 4.17e, the time resolved signal of both corrected and uncorrected peak areas are shown. It is obvious that without correction of the changing background, huge errors will be made. For the actual cursor set-up shown in Fig. 4.17 a-d, an error of more

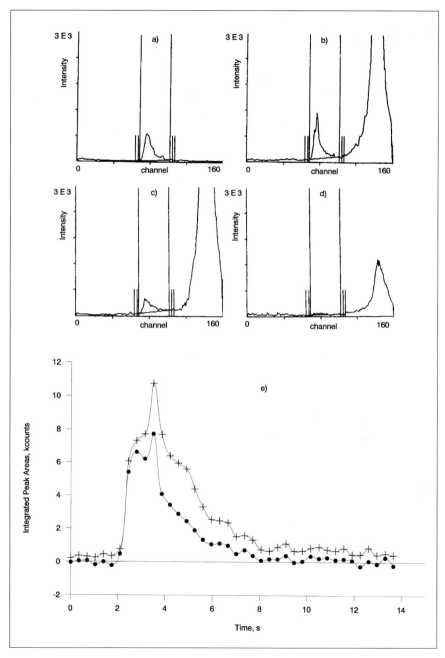

Fig. 4.17. Recording of the Pb signal in an Al-containing matrix. Spectrum of Pb (220.353 nm) with changing background caused by the wing-interference of Al (220.462 nm and 220.467 nm) obtained resp. 3.2 s (**a**), 4.3 s (**b**), 6.0 s (**c**) and 7.4 s (**d**) after the start of the volatilization step. (**e**) Both the corrected and uncorrected temporally resolved signals of Pb

then 70% was reported [171]. This result was obtained with linear background fits. A software program to calculate higher power fits (quadratic or higher), as described later by the same research group, yielded a more accurate result [171].

The example given concerned solution ETV-ICP-AES. When analyzing solid samples the background is found to change even more, probably due to the large amount of matrix elements introduced in the plasma and due to the important gas expansion. Sometimes a specially designed temperature program can be applied to selectively vaporize the matrix and have it carried out of the furnace to the exhaust.

4.6.2.2
Spectral Interferences in ETV-ICP-MS

Especially in ETV-ICP-MS with solutions, selective vaporization of the solvent (drying step) is one of the most important (and reported) advantages of an ETV used as a sample introduction device. It has been shown by many groups that the use of an ETV can reduce the levels of many *polyatomic species*. The solvent is vaporized and lead to the exhaust before the analyte enters the plasma. Figures 4.18a and b clearly shows how the use of an ETV device results in a dry plasma and allows the reduction of many isobaric interferences.

Fig. 4.18. Comparison of signals obtained from: (*A*) wet plasma and (*B*) dry plasma (from [397], with permission)

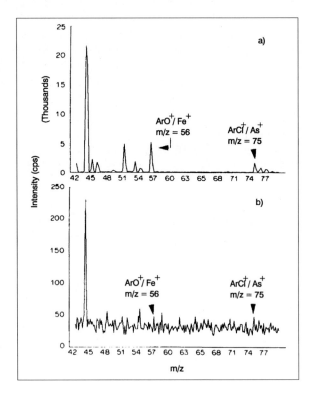

For different interferences, the use of matrix modifiers to promote the release of the matrix elements during the ashing step, was investigated. Most studies concern oxygen and chlorine containing interfering species and some are discuss materials with high sulfur [481] and calcium [373] contents.

As mentioned in the section on the reduction of non-spectroscopic interferences, either the matrix or the analyte can be selectively vaporized prior to respectively during the measurement. Darke et al. [375] were able to determine Pb in an Al matrix by vaporizing only the Pb, leaving the matrix in the furnace.

However, the above described vaporization deals with solutions. For the direct introduction of solids into the furnace, information on spectral interferences is unfortunately scarce. From the few papers found in the literature, it must be concluded that the effect of such selective vaporization in the case of SS-ETV-ICP-MS should not be overestimated. Wang et al. [482] found that interferences particularly those associated with non-volatile elements can cause problems with complex solid matrices. When introducing a solid with a high chlorine content (e.g. some biological materials) halogenation reactions will take place and introduce chlorine into the plasma.

Matrix modifiers such as gaseous or solid halogenating agents, successfully used in ETV-ICP-AES, should be replaced by other modifiers or at least should be kept from entering the plasma.

When selective vaporization is not sufficiently efficient in overcoming spectral interferences or when the matrix modifiers cannot be replaced by ICP-MS friendly reagents, a magnetic sector instrument can be the solution to the problems of spectral interferences. The resolution of these types of instruments is mostly sufficient (up to 10000) to resolve most of the polyatomic species from the analyte elements of interest. These high resolution spectrometers will probably be of particular interest for all carbon containing polyatomic species. For example it was found that in a graphite furnaces large amounts of particles are released when firing an empty furnace [466]. These particles are probably carbon particles of the furnace itself, leading to interfering polyatomic species in ICP-MS. When refractory elements such as V are to be determined, some halogenating modifier must be used to facilitate evaporation, resulting in the case of V, in a significant spectral interference ($^{35}Cl^{16}O$) on ^{51}V. A magnetic sector instrument should be able to resolve this spectral overlap at a resolution of 3000 [483].

4.6.3
Transport Efficiencies and Calibration

It was demonstrated in Sect. 4.5.1.2 that the slope of the calibration lines in SS-ETV-ICP-AES strongly depends on the composition of the analyzed material. This originates from matrix dependent differences in vaporization and transportation of the elements of interest. From the section dealing with non-spectroscopic interferences it is clear that a different matrix composition can affect every physical and chemical process an analyte undergoes on its way towards the ICP.

This has strong implications for the calibration procedure, since the influence of the matrix is reflected in a difference of the response/analyte mass ratio (=slope). Some groups have measured an "overall transport efficiency" (η). This overall transport efficiency is expressed as a analyte specific response, i.e. a signal obtained for a well known amount of analyte and has therefore a more practical significance. In solid sampling CRMs are used and the responses are related (recalculated using the certified value) to 50 ng of analyte. The overall transport efficiency, however, gives no information about the individual efficiencies of the different processes involved in aerosol formation and transport. Also there is no information as to where the losses occur.

In any case, the idea of defining the overall transport efficiency is very useful in solid sampling for several reasons. Since the overall transport efficiency is calculated as a analyte specific netto response for a certain element in a certain material, it is a measure of the sensitivity of the SS-ETV-ICP combination for the element in the material considered. In other words, it gives an idea of the slope of the calibration line which would have been obtained by varying the amount of sample. The efficiency value itself can only give information on the performance of the configuration used. However, a comparison of transport efficiencies of two experimental set-ups is more interesting. Comparing two of these sensitivities, yields very useful information for the calibration procedure since the comparison indicates how different or similar two materials are.

One of the first reports on transport efficiencies for different solid materials using ETV-ICP-AES was published by Verrept et al. [423]. They measured a *"relative transport efficiency"* as the ratio between analyte mass specific responses for ETV-ICP-AES (solutions and solids) and pneumatic nebulization ICP-AES. All signals were measured under optimized conditions. Some results are given in Table 4.1 for the elements Cd, Pb and Cu in different materials (biological and environmental).

Since the results are given as the ratio of the analyte mass specific responses of ETV to nebulizer, a relative transport efficiency exceeding 1 expresses an improvement of the transport of the analyte to the ICP and thus an increase in the sensitivity of the technique for the element in the material considered.

Dittrich and Walther [484] reported, for the relative transport efficiency of solution-ETV versus nebulizer ICP-AES, a somewhat lower improvement factor for Cd (22) and reported an improvement by factors of 26 and 32 for Mn an Ni respectively. They also reported an efficiency ratio between a solid material-ETV (BCR CRM-146) and solution-ETV; values of 0.4, 0.9 and 1.3 were reported for Cd, Mn and Ni respectively. Recalculating the efficiency ratio reported by Verrept et al. for Cd between "Sewage Sludge IO" and solution-ETV from Table 4.1 gives 0.61, which is a quite similar result.

From both papers it can be seen that, in all cases when comparing ETV-ICP to nebulizer-ICP, a factor larger than 1 is found and in some cases an improvement

Table 4.1 Transport efficiencies of ETV-ICP-AES for some elements compared to the transport efficiency of Meinhard TR-50-C3 pneumatic nebulizer ICP-AES

	Nebulizer	ETV-Reference Solution	Transport Efficiencies (relative to nebulizer technique) ETV-Solid Materials			
Cd	1	31.8 ± 1.8	Estuarine Sediment 14.1 ± 1.7	Sewage Sludge IO 19.5 ± 2.7	Incineration Ash 29.9 ± 2.0	
Pb	1	10.55 ± 0.16	Estuarine Sediment 10.64 ± 0.18	Aquatic Plant 9.18 ± 0.27	Light Sandy Soil 11.46 ±0.97	Fly Ash 7.77 ± 0.35
Cu	1	3.98 ± 0.11	Light Sandy Soil 4.06 ± 0.24	River Sediment 3.74 ± 0.41	Olive Leaves 3.58 ± 0.20	

with a factor of 30 is obtained. The magnitude of this factor depends, of course, on the experimental configuration. A comparison with values in the literature can also indicate if losses occur or e.g. if the application of a (new) modifier improves the overall transport. In the two examples given, the set-up used by Verrept et al. results in a higher overall transport efficiency for Cd.

However, even if the system does not reach the ultimate sensitivity (100% transport efficiency), this factor can be used to find how different matrices of distinct origins behave for a specific element. In other words, the relative efficiency for different materials provides information about the question whether a material is suitable as the calibrant for another material. Calibration can only be carried out if the calibrant used gives the same sensitivity as the sample, as is reflected in the value of the relative overall transport efficiency. As an example it can be seen that for Cu all 4 relative transport efficiencies (solution and 3 solids) are in good agreement. From this, it can be concluded that calibration within this group should be possible. Generally, as expected, it is found that groups of similar materials influence the processes in a similar way. Therefore, an interesting experiment was carried out by determining the Cu content of two similar materials (River Sediment and Light Sandy Soil) using a CRM also with a similar matrix composition (Calcareous Loam Soil) as the calibrant material [171]. Figure 4.19 gives the calibration lines for Cu in the three materials. As can be seen the slopes are very similar. As a result, the contents determined are in good agreement with the certified values.

From the former examples and tables, it should be clear that the composition of the matrix can actually affect the sensitivity very drastically. Therefore calibrants with similar matrices should be used.

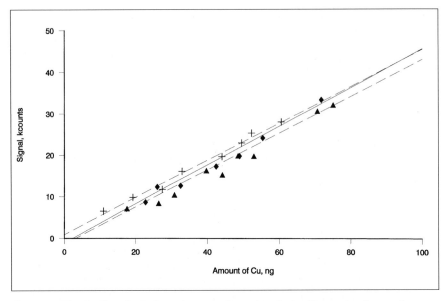

Fig. 4.19. Calibration lines for Cu in environmental samples, obtained by varying the sample amount. Only small difference is observed between the slope of Light Sandy Soil (BCR CRM 142 (+)) and Calcareous Loam Soil (BCR CRM-141 (Δ)), and a sediment; River Sediment (BCR CRM 320 (◊))

Because, in solid sampling, *calibration* is aproblem, several research groups have investigated different calibration techniques. Most of these techniques have already been discussed in Sect. 2.2. Therefore, this section will only summarize the techniques used for the calibration of the SS-ETV-ICP-AES/-MS combination.

External calibration using reference solutions is sometimes applied with little success [171, 482]. The reason for this failure is mostly the difference in relative transport efficiency. The difference is large because of the contrast in matrix composition between the reference aqueous solution and the solid powdered sample. M. Hinds et al. described in [470] that for the ETV-ICP-MS analysis of nickel alloys calibration using reference solutions, accurate results were obtained after the graphite tube had been treated with a solid Ni sample along with added diluted sea water as a physical carrier (compare discussion in Sects. 4.5.1.1 and 4.6.1.2).

In ETV-ICP-MS fairly accurate results could be obtained by adding an internal standard to correct for matrix effects [428]. Qin et al. [469] investigated the possibilities of external calibration with standard solutions when analyzing slurries with ETV-ICP-AES. These researchers found that when the slurry particles are <170 μm, intensities were identical to those obtained with aqueous solutions, and accurate results were obtained using this calibration method.

External calibration using a certified reference material (CRM) as a *solid calibrant* can be applied very successfully if the composition of the matrix of the sample is known to a certain extent and a CRM is available with a similar matrix and for which the concentration of the analyte of interest is certified. In practice, this technique has been applied most often in solid sampling [171, 103, 484, 418, 482, 428]. If a suitable CRM is not available, it is sometimes possible to produced a synthetic reference materials [392, 395, 91, 90, 421, 92]. In the future calibration using solid reference materials may, in the authors' opinion, find many applications in the industry. Once the elements of interest have been certified (possibly with more classical analytical techniques) in a product that must be monitored in the production line, this certified product can be used as the calibrant in future product control using SS-ETV-ICP-AES/MS.

Another technique that has been applied for calibration in SS-ETV-ICP is the *analyte ("standard") addition method (AAM)*. Although the transport efficiency for liquids and solids is very different (compare Table 4.1) when analyzed separately, it can be accomplished to make the responses of the reference solution and the solid additive. This can be done by drying the reference solution in the sample holder before loading it with solid material [485]. In this way most of the effects during vaporization and transport due to the difference in matrix composition are compensated. It is believed that, if no information on the unknown sample is available, these techniques give the best results [486].

Both the *single addition method*, also as the *generalized analyte addition method* have been used successfully in SS-ETV-ICP-AES [103, 486, 428, 132, 429, 430, 468, 469, 475] (see Sect. 2.2.1.3 for more theoretical information on this calibration method). In SS-ETV-ICP-MS both single and multiple analyte additions have been applied [482, 428, 132, 429].

For determinations with ETV-ICP-MS, isotope dilution can also be used as a calibration method [432, 437]. The most important requisite for successful application of this method is the availability of two isotopes which do not spectrally interfere with eachother. This must always be checked carefully since at every furnace firing relatively large amounts of C are swept into the ICP and unexpected C-related interferences can occur.

4.7
Applications

In this section, some applications are described as an illustration to the former section and as a guide for the reader interested in finding information on a particular subject. The applications are given chronologically, allowing the evolution of the research of the different research groups to be followed. An overview is also given in Table 4.2, reviewing the literature on SS-ETV-ICP-AES/MS from 1984 to 1996.

Blakemore et al. [407] determined, Ca, Cd, Cu, Fe, K, Mg, Na, P, Pb and Zn in CRM Bovine Liver NIST-1577. The sample mass used was 0.5 mg and three replicates were taken. To improve the sample homogeneity they used an embrittlement of the sample by freezing the material in liquid nitrogen prior to agitat-

Tabel 4.2 Applications on Solid Sampling ETV-ICP-AES and ETV-ICP-MS

Matrix	Elements	Instrumentation Spectrometer; ETV-system	Calibration	Special Analytical Features	Ref.
alumina	36 elements	ARL Quantoscan Spectroflame ICP-AES; boat in tube	ref. solution, solid ref. material	halogination Ar + CCl$_4$ transient signals	[398]
biological and environmental CRMs	Cu, Cd, Pb	Plasma-Therm HFP-2500D ICP-AES modified SM-30 (Grün)		comparison transport efficiency ETV - pneumatic nebulization	[423]
ceramic powders	Fe, Cd	Sequential ICP-AES (Kontron); LINN FS-7		thermochemical additives	[415]
sea water, biological materials (urine)	Cd, Cu, Pb, Zn	Jarell Ash AtomComp model 975 ICAP-AES; HGA-74 PE	ref. solution	enrichment by poly (dithiocarbamate) resin	[392]
biological: nutritional water, geological samples	Cu, Mn, Au, Hg	Spectraspan V direct current plasma ES ICP-AES, IL 455 furnace	ref. solution, solid ref. material, external calibration	addition of an ionization -suppression buffer	[419]
oil	Ca, Cu, Fe, Mg, Mn	Plasmaspec/ASS-80, ICP-AES; crucible cup	ref. solution	addition of teflon powder	[420]
waste water, Bovine liver, Blood plasma	Ca, Cd, Cu, Fe, K, Mg, Na, F, Pb, Zn	ICAP 1160 AtomComp ICP-AES; IL 555 atomizer	AAM	no ashing	[407]
silicon carbide powders	Al, B, Ba, Cd, Cr, Cu, Fe, Mg, Mn, Ti, V, Si	LABTAM 8440 Plasmalab ICP-AES;	AAM	study of signal profiles	[424]

biological materials, sea water	Cd, Cu, Pb, Zn	Jarrell-Ash AtomComp 975 ICAP-AES; HGA 74 PE		dry-aqueous plasma, thermal conductivity	[395]
botanical samples	Cd, Zn, Pb	PlasmaTherm 2500, ICP-AES; HGA 2200	ref. solution	pelletized solids	[434]
geological samples	As, Cd, Co, Cr, Cu, Fe, Mn, Pb, Se, Zn	PE Sciex Elan 5000, ICP-AES; PE Sciex 320 laser sampler	ref. solution (semiquantitative)	slurry sampling	[494]
biological and environmental CRMs	Cu, Cd, Pb	Plasma-Therm HFP-2500D, ICP-AES; modified SM-30 (Grün)	solid ref. material, external calibration	matrix effects	[171]
biological and environmental CRMs	Cu, Cd, Pb	Plasma-Therm HFP-2500D, ICP-AES; modified SM-30 (Grün)	simplified generalized AAM	automatic correction for the blank	[103]
biological: sheep's blood, organ tissue	Cd	PE Sciex Elan 5000 ICP-MS; HGA-600		separation Cd and Zn using adsorption chromatography	[495]
biological, environmental	Mn, Cu, Al, Pb, Zn	Hitachi ICP Super Scan M-306, ICP-AES; Hitachi GA-3	solid calibration	ETGF/miniature cup	[418]
ceramic powders	Fe, Cd	Plasma Spec ASS-80, ICP-AES; crucible cup	ref. solution	matrix effects	[416]
oil	Zn, Bi, Pb, Te	Plasmaspec/ASS-80, ICP-AES; crucible cup	ref. solution	modified ICP torch	[496]
biological: total diet, coal fly ash	Pb, Zn, As, Cd	VG PlasmaQuad II ICP-MS; HGA-300	ref. solution, solid ref. material, AAM	sample homogeneity influence	[482]

Tabel 4.2 Continued

Matrix	Elements	Instrumentation Spectrometer; ETV-system	Calibration	Special Analytical Features	Ref.
biological and environmental (CRM)	As, Se	PE Sciex Elan 5000 ICP-MS; modified SM-30	AAM	simultaneous determination	[429]
biological (CRM)	As	PE Sciex Elan 5000 ICP-MS; modified SM-30	ref. solution, solid ref. material, external calibration; AAM	internal standard Sb; electronic dilution	[428]
biological materials (CRM)	As	PE Sciex Elan 5000 ICP-MS; modified SM-30	AAM	Ar-dimer as internal standard and diagnostic tool	[709]
steel alloy	B	UOP-1s high resolution ICP-AES SAS-705V	ref. solution	sodium hydroxide-diammonium hydrogen	[489]
biological: drinking water, urine	As	VG PQ1 ICP-MS; PE HGA 300		phosphate as internal standard use of Pd for preconcentration halides	[426]
biological and environmental (CRM)	Cd	PE Sciex Elan 5000 ICP-MS; modified SM-30	ref. solution, solid ref. material, external calibration; AAM	matrix effects	[487]
steel	S	PE Sciex Elan 5000 ICP-MS; modified SM-30	isotope dilution	KOH as matrix modifier	[432]
drinking water freeze dried urine	As	VG PlasmaQuad PQ1, ICP-AES PE HGA 300		preconcentration of halides by trapping in the presence of Pd	[426]

Sample	Element(s)	Instrument	Calibration	Remarks	Ref.
steel	S	PE Sciex Elan 5000, ICP-AES; HGA-600	isotope dilution	KOH as matrix modifier	
steel alloy	B	UOP-1s high resolution ICP-AES; Seiko II SAS–705V	ref. solution	sodium hydroxid-diammonium hydrogen phosphate as matrix modifier	[489]
human hair	Zn, Mn, Mg	Spectroflame ICP-AES; Fisher GmbH ETV-device	AAM	vaporization temperature 2200 °C	[475]
biological: bovine liver, mussel tissue, serum	Al	ICP-AES; WF-1 heating device GF	AAM	PTFE as matrix modifier	[92]
botanical (plant) samples	B	WDG 500-LA type ICP-AES; WF-4 furnace	ref. solution, AAM	slurry sampling, PTFE as fluorination agent, slurry sampling	[469]
aqueous solution	Ag	PE Sciex Elan 5000 ICP-AES; modified PE HGA-76B		simultaneous measurement of atomic and molecular absorption signals	[441]
biological: sheep's blood organ tissue	Cd, Zn	HGA-600		separation of Cd and Zn using adsorption chromatography	[710]
HCl, HNO_3, H_2SO_4, H_3PO_4	Co, Cu, Ag, Cs, Pb, Bi, U	PE Sciex Elan 5000 ICP-AES; HGA 600		HNO_3 and H_2SO_4 cause signal enhancement	[431]
solid and liquid organometallic reagents	16 elements	HFS 5000D (RF Plasma Products) ICP-MS; IL 655 and HGA 400	ref. solution	direct vapour sampling	[474]
SiC powders	Al, Fe, Ti, V	ICP-AES; graphite crucible cup	synthetic solid ref. material	in situ processing of SiO_2, C powder, modifier and oxides of Al, Fe, Ti and V	[435]

Tabel 4.2 Continued

Matrix	Elements	Instrumentation Spectrometer; ETV-system	Calibration	Special Analytical Features	Ref.
coal, biological materials (CRM)	Ni, Cu, Cr, Pb, Mn, Co	PE Sciex Elan 5000, ICP-AES; HGA-600	ref. solution, AAM	Ar-dimer as an indicator of matrix suppression, slurry sampling	[430]
geological: basaltic lava, andesite	49 elements	PE Sciex Elan 5000, ICP-AES; modified PE HGA-76B	isotope dilution	NRCC NASS-3 Sea water as matrix modifier	[437]
inorganic matrices (oxides and carbides)	Al, Si, Zr, Ta, W	Plasma Therm 2500, ICP-AES; modified PE HGA 2200		in situ reaction with CCl_2F_2 as halogenation agent	[427]
trichloroacetic acid pentachlorophenol, o-chlorobenzoic acid, p-cloroanisole	EOCl (extractable organic chlorine)	Fisions PlasmaQuad PQ II, ICP-MS; Fisons ETV ETV Mk III	ref. solution	ethylacetate as extraction solvent	[490]
marine sediments, coal fly ash	Cu, Mn, Zn, Al, Fe	ICP-AES; modified PE HGA 2200		gradual vaporization; background correction with refractor plate	[459]
siliconcarbide ceramic powders	Al, Fe, Ti, V	ICP-AES; graphite crucible cup	synthetic solid ref. material	matrixmodification CoF_2 and BaO	[471]
nickel alloys	Bi, Pb, Te	ICP-MS	ref. solution, external calibration	sea water as a physical carrier	[470]
organometallic reagents	16 elements	ICP-AES	external calibration		[491]
biological (CRM)	Cu, Mn, Ni	PE Elan 5000A, ICP-MS; HGA 600MS	ref. solution, external calibration, AAM	supression effects from position, age and typ of skimmer, slurry sampling	[480]

ing the sample on the ball mill. Calibration was performed by standard addition.

Ohls and Hütsch [420] determined Ca, Cu, Fe, Mn, and Mg in lubricating oil. The obtained precision was 5–10% for sample masses of 5–10 mg. To promote volatility of refractory elements, teflon powder (1:1) was added. Calibration was done by CRMs, dried aqueous solutions, and organometallic reference materials dissolved in white oil.

Mitchell and Sneddon [419] determined, with an ETV-DCP system, Mn in US Geological Survey Reference Materials after diluting the reference materials with powdered cellulose (1+19). The sample mass used was 5 mg. No calibration was explicitly mentioned but external calibration with solutions was probably used.

Van Berkel and Maessen [392] investigated the use of poly(dithiocarbamate) (PDTC) in ETV-ICP-AES for the analysis of sea-water and biological materials. The resins were used to enrich the analytes and were analyzed by direct solid sampling ETV-ICP-AES. The sample mass used was about 20 mg. Calibration was done by the use of a synthetic solid reference (enrichment of reference solution on PDTC, resulting in the same matrix loaded with known amounts of the analytes). The method was used to determine Cd, Cu, Pb and Zn in a CRM urine sample.

Nickel et al. [416] investigated the determination of traces in alumina-based ceramics using SS-ETV-ICP-AES. They used halogenating modifiers (AgCl & $BaCl_2$) to determine Fe in SiO_2, Al_2O_3 and MgO – based ceramics. Calibration was done by synthetic reference materials (mixtures of Ceramics).

Ohls [421] determined Pb and Ag in iron and steel chips by analysing 5–10 mg samples with SS-ETV-ICP-AES. Calibration was done by dried reference solutions, using an empirical factor of 5.5 for matrix compensation. Detection limits for Ag, Bi, Ca, Cd, Mg, Pb, Se, Te and Zn were also given for 24 alloyed steel CRMs.

Reisch et al. [415] investigated the volatility of Fe and Cd in graphite and ceramic materials. For the graphite matrix, they found fluorides promote the evaporation of Fe, whereas chlorides are more effective in volatilizing Fe from a ceramic powder. They used a 5-mg sample weight.

Van Berkel [395] investigated some properties of dry plasmas by introducing analyte loaded PDTC resins in the plasma using SS-ETV as the introduction technique under similar conditions as mentioned above.

Karanassios et al. [434] determined Cd, Pb and Zn in some botanical samples using a pellet-ETV device. A powdered sample was mixed with graphite (1+9) and pressed into a pellet. Calibration was done by analyzing moon-element oxide-pellets (synthetic reference materials). The authors used a 0.25 g sample mass and obtained RSDs of 5–7% on 6 replicates.

Atsuya [418] used his miniature cup technique taken from SS-GF-AAS to determine Al, Co, Cr, Fe, Mg, Mn, Ni, Pb and Zn in biological and environmental materials. Calibration was done with the aid of CRMs with a similar matrix. With the miniature cup technique only 0.1–1.5 mg of sample is used.

Kántor and Záray [398] determined the detection limit of Ti in different alumina materials. They used gaseous halogenating (CCl_4) reagents to promote the vaporization. The determination of Ti itself using solutions for calibration was unsuccessful because of the difference in matrix composition.

In the same year, Kántor and Záray [424] determined background equivalent concentrations (BEC) for 10 elements in SiC powder using the chloration technique. They found an improvement with a factor of about 2–20 compared to liquid analysis. A sample mass of 10 mg was used.

The same research group determined Ca, Fe and Ti traces in SiC powders with various grain sizes using freon 12 as the halogenating agent [91]. Matrix matched calibration was used for calibration. In this technique, a boat with two sample holders is used. In a first step, SiC is heated in one of the holes at 2200 °C for 40 s leaving "purified" SiC. In a next step, an aliquot of reference solution is added in the second hole. During evaporation (2100 °C) both SiC and the reference are released. In this way calibration lines are abtained that can be used for the analysis of SiC-powders with grain-sizes changing for 0.78 μm<d_{50}<48.5 μm.

Nickel [471] reported the influence of a complex matrix modifier (CoF_2 + BaO) for the determination of impurities (Ca, B, Al, Fe, Cu, Tl, Ni, Si, Ti, V) in B_4C, SiC and Si_3N_4. The sample was mixed with the modifier (1+1+1) and the sample mass used was 5 mg of the mixture. For the determination of Al, Fe, Ti and V in ceramic powders, synthetic matrices were prepared. The results were published by Nickel and Zadgorska [92, 93].

Verrept et al. [423] reported some relative transport efficiencies for Cd, Pb, and Cu in environmental and biological materials, from which several conclusions concerning calibration possibilities were drawn and illustrated in later papers [171, 103, 486]. They found an overall improvement of the detection power of SS-ETV-ICP-AES over nebulization-ICP-AES.

Záray et al. [90] reported the use of Freon 12 as a matrix modifier to determine Ca, Fe, and Ti in SiC powders. They also found that free Si in the SiC can be determined by a selective vaporization of SiS_2 in a pure Ar-atmosphere after letting the free Si react with sulfur at 300 °C. This also shows that SS-ETV-ICP allows element speciation. The same sample masses and calibration method as in earlier papers were used. Results with RSDs of 5–20% could be obtained .

Haßler and Perzl [426] used ETV-ICP-AES with a simultaneous spectrometer for the direct analysis of SiC. For Al, Fe, Ni, Ti, V they found precision between 1–7% and limits of analysis below 1 mg/kg.

Nickel et al. [435] reported on the use of CoF_2 + BaO for the determination of carbide forming elements. The decomposition and reactions involved are the same as discussed in the Sect. 4.5.1.1, dealing with solid halogenating reagents.

Verrept et al. [171] reported the determination of Cu in two environmental CRMs using a third CRM with a similar matrix composition as a calibrant. In addition a comparison was made between CRM and reference solution calibration. Finally a discussion on whether the standard deviation on the concentration of the certified value should be included in the overall error and on the meaning of the limit of determination in solid sampling was given.

Dittrich et al. [484] also reported a comparison of transport efficiencies between SS-ETV-ICP-AES and nebulization ICP-AES for Cd, Mn and Ni. Two types of calibration were applied using BCR CRM-146 (solid calibrant) and reference solutions. The sample weight was about 10 µg–1 mg.

Boonen et al. [103] successfully demonstrated the use of the generalized standard addition method (GSAM) in SS-ETV-ICP-AES for Cu, Cd and Pb in 11 CRMs of different origins (environmental and biological). As an illustration for the second type of calibration some of the results of this paper are given in Table 4.3. The results obtained are encouraging for this method.

Table 4.3 Results for the determination of Cu, Cd and Pb in different matrices with solid sampling ETV-ICP-AES, using the simplified GSAM as a calibration method; the uncertainties are expressed as 95% confidence limits

Element	CRM	Certified value (µg/g)	95% Confidence limit (µg/g)	Found value µg/g)	95% Confidence limit (µg/g)	RSD (%)	n
Cu	BCR CRM 142 Light Sandy Soil	27.5	0.6	26.2 27.5	1.7 6.1	11.0 24.0	15 8
	BCR CRM 320 River Sediment	44.14	0.96	43.0 50.3	1.5 8.3	8.1 18.0	24 8
	BCR CRM 146 Sewage Sludge - Industrial	934.0	24.0	966.0	35.0	5.2	11
	BCR CRM 060 Lagarosiphon Major (Aquatic Plant)	51.2	1.9	54.85 54.2 57.2	0.67 1.8 5.0	4.4 6.5 9.6	49 17 8
	BCR CRM 062 Olive Europaea (Olive Leaves)	46.6	1.8	46.26 45.7	0.66 1.0	3.2 2.4	22 8
Cd	BCR CRM 176 City-waste Incineration Ash	470.4	8.5	486.0	17.0	7.2	19
	BCR CRM 146 Sewage Sludge – Industrial	77.7	2.6	73.3	2.7	7.1	17
	BCR CRM 277 Estuarine Sediment	11.88	0.42	12.35	0.78	11.0	16
Pb	BCR CRM 146 Sewage Sludge - Industrial	1270.0	28.0	1227.0	34.0	4.6	14
	BCR CRM 176 City-waste Incineration Ash	10870.0	170.0	10320.0	730.0	9.4	10

Wang et al. [482] reported the determination of Pb, Cd and As in Coal Fly Ash and Total Diet using SS-ETV-ICP-MS. For this they used reference solutions, resulting in poor accuracy of the results.

Argentine and Barnes [474] used ETV-ICP-MS for the direct determination of volatile and non-volatile impurities in semiconductor-grade organometallic materials and process chemicals. The technique showed great potential, particularly for the non-volatile impurities, since complete separation of the (volatile) matrix from the analytes could be obtained by application of a suitable temperature programme.

Vanhaecke et al. [428] and Boonen et al. [429] determined trace amounts of As and Se in CRMs of biological and environmental origin with SS-ETV-ICP-MS. Although important non-spectral interferences could be established, several calibration methods were shown to offer possibilities for accurate quantification.

Vanhaecke et al. [706] showed that in SS-ETV-ICP-MS the Ar^{2+} signal can be used as an indication for non-spectral interferences. By using the ($^{75}As^+$/ Ar^{2+}) signal ratio, As could be accurately determined in BCR CRM 279 Sea lettuce.

Galbács et al. [487] determined Cd in a number of CRMs of biological and environmental origin with SS-ETV-ICP-MS. Different calibration techniques were successfully used for quantification.

Byrne et al. [488] investigated the mechanism of volatilization of tungsten with ETV-ICP-MS. The signal profiles showed two separate peaks: one below 1100 °C from WO_3 and one above 2500 °C from WC. Associated with this WC-peak are both memory effects and high blank values. Therefore optimum signal to background ratios and limits of detection and minimum memory effects are obtained at low vaporization temperatures.

Naka and Grégoire [432] determined trace amounts of sulfur in steel by ETV-ICP-MS. Isotope dilution was used for calibration. Fe was removed by solvent extraction with 4-methylpentane-2-one. KOH, used as a matrix modifier, could only partially eliminate interference effects caused by concomitant elements such as Cu, Al, Si, V, Co, Ni, Cr, Mo, Ti, Mn and W.

Grégoire and Lee [707] successfully used ETV-ICP-MS for measuring isotope ratios of Cd and Zn in blood and tissue samples at naturally occurring concentration levels. The precision obtainable was 30-150 times greater than the variation in isotope ratios observed. Cd and Zn were separated from the matrix components using adsorption chromatography.

Okamoto et al. [489] managed to delay the volatilization of B using sodium hydroxide-diammonium hydrogenphosphate as a chemical modifier. In this way, interferences of major steel alloy components (Fe, Ni, Cr) could be drastically reduced.

In the first part of their study, Lamoureux et al. [441] developed an ETV device that can operate as an efficient sample introduction system for ICP-MS and at the same time for the measurement of atomic and molecular absorption signals. For this purpose, the sample vapor was extracted from the dosing hole as opposed from the tube end so that the modified ETV could be used in normal

ETAAS conditions. In this way, the interference of a $MgCl_2$ matrix on the determination of Mn by ETAAS could be shown. No ETAAS signal could be obtained during the pyrolysis step. The ICP-MS signal, however, showed a large Mn signal, which indicates that Mn has been lost in a molecular form during the pyrolysis step. In a second part, Lamoureux et al. [437] established detection limits for Co, Cu, Mn, Nd, V, Y, Yb and Zn which varied from 0.075 pg for Co to 0.617 pg for Zn. Further, the sensitivity of 49 elements was determined. NASS-3 Sea water was used as chemical modifier.

Grégoire et al. [430] used slurry sampling ETV-ICP-MS for the determination of Ni, Cu, Cr, Pb, Mn and Co in three reference materials of biological origin. Due to matrix effects, standard addition was shown to be superior to calibration using aqueous standard solutions. To obtain linear calibration graphs, a matrix modifier (NRCC NASS-3 Sea water) was added. The Ar-dimer was successfully used as an indication of plasma loading effects.

Hu et al. [468] determined Al in biological reference materials using PTFE as a chemical modifier. Due to the formation of AlF_3, Al was vaporized at lower temperatures (1300 °C). Ca, Cu, Fe, K, Mg, Na, Si and Zn (2-5 mg/ml) were shown not to affect the determination of Al.

In a first part of their study, Ren and Salin [427] improved the vaporization of oxides (Al_2O_3, SiO_2, ZrO_2) and carbides (TaC, WC) from solid samples to almost 100% using Freon-12 (CCl_2F_2) as a halogenation agent and 2400 °C as a vaporization temperature. For WC, the addition of $BaCl_2$ was required. In a second part, the linearity of the mass/response curves were investigated for Cr, Cu, Mn, Zn, Al and Fe in 4 marine sediments and a coal fly ash [459]. Gradual vaporization was used to minimize spectral interferences. Background correction was performed by means of a refractor plate placed right behind the entrance slit.

Manninen [490] used ETV-ICP-MS for the determination of the extractable organic chlorine (EOCl) content in solid and liquid samples. The results obtained are comparable with NAA results. The precision, however, was rather poor (RSD >20%).

Qin et al. [469] investigated the effect of the particle size in the analysis of botanical samples by slurry sampling and fluorination-ETV-ICP-AES. Accurate results were obtained for B. The recovery (intensity obtained from the slurry / intensity obtained after digestion) was shown to be 100% for a particle size <170 μm.

Marawi et al. [369] used Pd-coated (electroplated / sputtered) metallic platforms (W, Ta, Mo, Re) in order to increase the linear dynamic range. A sputtered W platform was shown to retain the Pd up to 80 firings and increase the Pd surface available for analyte trapping As hydrides, improving the linear dynamic range with an order of magnitude.

Grégoire et al. [431] investigated the vaporization of some acids and their effect on analyte signals in ETV-ICP-MS. HNO_3 and H_2SO_4 were shown to enhance the signal. H_3PO_4 caused a signal decrease for the elements Ag and Bi and HCl did not alter the signal intensity significantly.

Plantikow-Voßgätter [475] and Denkhaus determined Zn, Mn and Mg in human hair with ETV-ICP-AES. The temperature programme was optimized by means of differential thermal analysis (DTA) and thermal gravimetry (TG).

Argentine et al. [491] investigated the vaporization of 16 elements from liquid and solid organometallic reagents with ETV-ICP-AES. They used external calibration and direct vapor sampling for calibration.

Hinds et al. [470] determined the volatile elements in nickel alloys by ETV-ICP-MS using sea water as a physical carrier. Bi and Pb could be determined using a external calibration using reference solutions, better results in terms of accuracy were obtained when solid RMs were used for calibration. The determination of Te was not successful using either solution or solids calibration. The authors suspect an interaction between the Te and matrix components that alters the release mechanism.

Ming-Jyh et al. [714] applied slurry sampling ETV-ICP-MS to the determination of Cu, Cd and Pb in several sediment samples. The influence of instrument operating conditions, slurry preparation, and non-spectroscopic and spectroscopic interferences on the ion signals of isotop ratio determination was investigated.

In 1994 Darke and Tyson [492] published a review on solid sampling plasma spectrometry. In this article the possibilities and limitations of laser ablation, electrothermal vaporization and slurry nebulization are compared.

In1995 Moens et al. [493] published a review on direct analysis of solid samples with ETV-ICP-AES and -MS.

CHAPTER 5

Introduction of Slurry Samples into the Graphite Furnace

MARKUS STOEPPLER AND ULRICH KURFÜRST

5.1
Introduction

With the introduction of modern furnace technology and improved background correction such as continuum source, Zeeman and Smith-Hieftje systems, slurry sampling graphite furnace AAS (GF-AAS) became, in a relatively short time, a convenient and increasingly used technique for the fast and relatively easy introduction of a number of quite different solid materials into atomizers. The technique thus combines, to some extent, the properties of both liquid and solid sampling for numerous materials and elements. Slurry techniques have similarities and to some extent overlapping working areas on the one side with the analysis of liquid (decomposed) samples, on the other with direct solid sample introduction into furnaces.

Slurry sampling is similar to direct solid sampling in that the sampling error might be significant due to the relatively low absolute masses handled and that the method of calibration applied must be chosen more thoughtfully. Slurry sampling differs, however, from the direct solid sampling approach because (a) of volumetric dosing and quantification by pipetting, (b) the possibility of using matrix modification for a number of elements and matrices and (c) the applicability of all types of commercially available and laboratory made atomizers. From this point of view, the slurry technique is to some extent similar to the classical approach using liquid samples.

Slurry sampling GF-AAS has, like all analytical methods, advantages as well as limitations, but has already proven, particularly in its most sophisticated automated versions, to be a very useful and versatile method which is complementary to, rather than competitive with direct solid sampling and liquid sampling.

Slurry sampling has broadened the potential of GF-AAS techniques to samples that were previously difficult to analyze or in some cases almost unanalyzable by the classical "wet" approach after decomposition as well as often being problematic for direct solid sampling. Its relatively ease of use, the subsequent commercially available automation, and the feasibility of adapting laboratory systems to nearly all commercial autosamplers for GF-AAS systems have been the reasons for its increasing application in numerous analytical tasks.

The use of slurried samples in atomic absorption spectrometry with flames and furnaces was reviewed in some detail earlier [34, 17, 175]. There are a number of recent and comprehensive papers either including this topic or exclusively dealing with slurry sampling [e.g. 499, 157, 500, 16, 501, 502, 503].

In particular, it has been Miller-Ihli who, since 1988, has significantly contributed to the design and construction of basic automated instrumentation, further evaluation, and improvement of the technique, its theory and application in the broad field of trace analysis. There are, in addition, many very valuable contributions by other workers dealing with comparative studies with direct solid sampling GF-AAS, conventional flame-AAS, GF-AAS and ICP-AES, in some cases also including radiotracer experiments. Almost all of these papers confirmed principally the practicability of slurry sampling.

This chapter deals with the use of GF-AAS for slurries, but not with slurry methods for flame AAS, FIA-flame AAS and ICP-AES. This is because the introduction of slurries by nebulization into flames [e.g. 504, 505, 506, 507] and plasmas [e.g. 508, 509] as well as into hydride and cold vapor systems [e.g. 510, 511, 512] requires extremely small particle sizes, typically <3 μm, often only achievable by vigorous milling procedures [509]. The slurry technique for GF-AAS, can be performed with significantly larger particle sizes that are close or even similar to those which are obtained by conventional grinding techniques commonly applied prior to decomposition. The sample sizes used for the slurry method are also similar to those for direct solid sampling GF-AAS.

5.2
Preparation and Characterisation of Slurries

This section deals with the prerequisites for the preparation of the slurry. It commences with a general working scheme based on recent experience, which is displayed in Fig. 5.1. This is followed by a discussion of sources of errors, which occur typically in slurry analysis (see also the general discussion of errors and the resulting uncertainty in the result in solid sample analysis in Sect. 2.4.1).

5.2.1
Slurry Preparation

5.2.1.1
Particle Size, Grinding

The mean particle size of a slurry is of particular importance for the *representativity* of the analytical results (see Sect. 5.2.2.1) as well as for the *uninterrupted passage* of the slurry through tubing and pipette orifices.

The data reported for particle sizes cover a wide range in the earlier but also in the more recent literature, i.e. from particle sizes around and even below 20 μm [e.g. 143; 142; 109; 513; 514; 515, 516; 517, 518, 238] around and below 50 μm [e.g.

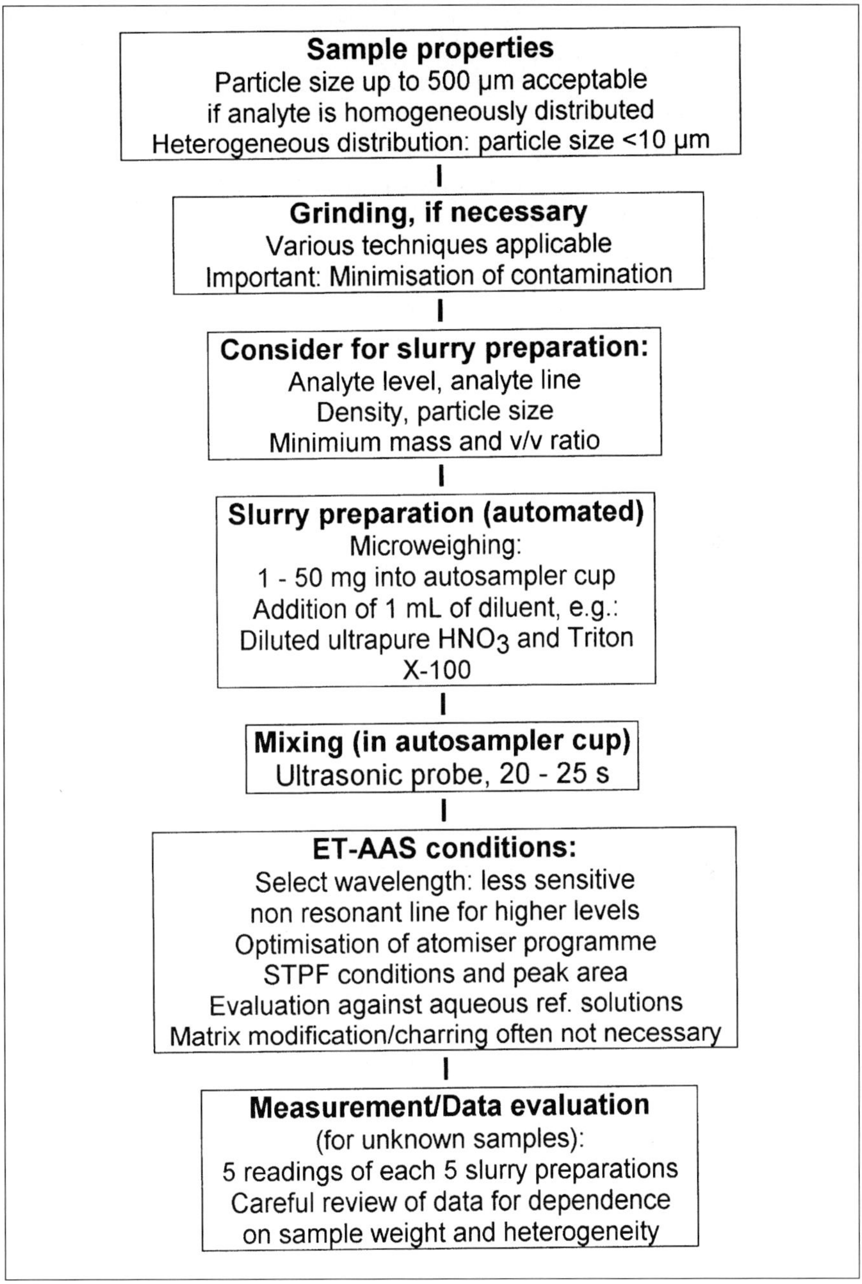

Sample properties
Particle size up to 500 μm acceptable
if analyte is homogeneously distributed
Heterogeneous distribution: particle size <10 μm

Grinding, if necessary
Various techniques applicable
Important: Minimisation of contamination

Consider for slurry preparation:
Analyte level, analyte line
Density, particle size
Minimium mass and v/v ratio

Slurry preparation (automated)
Microweighing:
1 - 50 mg into autosampler cup
Addition of 1 mL of diluent, e.g.:
Diluted ultrapure HNO_3 and Triton
X-100

Mixing (in autosampler cup)
Ultrasonic probe, 20 - 25 s

ET-AAS conditions:
Select wavelength: less sensitive
non resonant line for higher levels
Optimisation of atomiser programme
STPF conditions and peak area
Evaluation against aqueous ref. solutions
Matrix modification/charring often not necessary

Measurement/Data evaluation
(for unknown samples):
5 readings of each 5 slurry preparations
Careful review of data for dependence
on sample weight and heterogeneity

Fig. 5.1. Flow-chart of sample preparation for slurry-GF-AAS, based on automated operation and recent experience (from [501], with permission)

519, 520, 521, 522, 523, 524, 525, 526] to particle sizes of around 100 μm and even larger [527, 157, 16, 501, 502, 528].

Extremely small mean particle sizes around 10–20 μm might improve the precision of the analytical results, but are only mandatory if the elements to be determined are not homogeneously distributed in the solid material [500, 16]. Thus, the degree of heterogeneity must be known or determined by the analyst, e.g. by replicate analysis of an appropriate number of samples (i.e. preparations) for the elements sought, if precision analysis, which means small uncertainty in the analytical result, is required (compare Sect. 2.3 and 2.4).

Another frequently observed property of slurries with very small particles is an increased analyte dissolution (extraction) during slurry preparation, particularly if the medium is acidic (see Sect. 5.2.2.1), which might, on the one hand, additionally and positively influence the precision of the measurements of slurry test samples. On the other hand, with very small particle sizes below 20 μm, problems might be encountered with static electric charge during weighing [502].

If the analytes are quite homogeneously distributed in the solid material, significantly larger particle sizes up to about 500 μm can be used. This has been reported for zoological, botanical, and food samples, whereas fine grinding obviously improved the precision of metal determinations in soil, sediment and coal fly ash [137]. Even larger particle sizes well above 100 μm do not negatively influence the analytical results for relatively homogeneous materials, if tubing diameters and pipette orifices are adapted to those particle sizes. This can be achieved by increasing the diameter of pipette orifices from the usual 300–400 μm to around 800 μm and replacing the tubes by those of larger diameters [16, 501].

Many of the materials to be analyzed, e.g. reference materials as well as a number of technical materials already have mean particle sizes in the optimal range of <50–100 μm for slurry sampling or even below. Thus it was often possible to use these materials directly for slurry preparation without prior grinding [e.g. 322, 523; 527, 502; 524, 529; 530; 531, 532, 533; 534, 175, 535, 536, 537, 538; 533; 539, 239, 540 541, 542; 517, 518, 543, 238, 526].

If the material has to be ground prior to slurry preparation, the grinding techniques vary, depending on the properties of the sample matrix, especially on its hardness. It is also very important for all materials that care must be taken to minimize contamination from the grinding devices, and that, whenever possible, sieving and discarding larger particles should be avoided in order to achieve results that are representative for the whole material.

Biological materials may be ground by using PTFE beads in polyethylene bottles [501]. As alternatives, mortar grinding [16], grinding in porcelain [523], agate ball mills [544, 525] and dental mills [545] might be used.

More general grinding techniques suitable for a wide variety of materials, mainly after careful drying procedures of fresh samples, including grinding and pulverizing mills are e.g. the Retsch Spectro Mill or the Fritsch planetary, centrifugal or jar mills. For these devices, stainless steel parts should be avoided if

Fe, Ni, Cr etc. have to be determined. Silicon nitride and boron carbide offer good abrasion resistance and minimal contamination if these metals are involved [16]. The use of zirconia beads, proposed by some workers [e.g. 546], however, was shown to introduce significant contamination levels of Al, Cr, and Fe [157, 16], thus should not be used if these elements have to be determined.

For most elements of interest, *cryogenic grinding* at the temperature of liquid nitrogen (–196 °C) is a straightforward and almost contamination free grinding technique if devices made of PTFE and titanium are used. These techniques have been developed during the last decade and successfully used for grinding and homogenizing of various fresh and dry biological (PTFE) and environmental materials (predominantly Ti, less frequently highly resistant stainless steel) in environmental specimen banking and for the preparation of heat sensitive biological reference materials. Typical particle sizes achievable by these techniques are, depending on the processed material, around and often below 100 μm [547, 548, 549, 550].

In some cases, either with or without previous grinding, the samples are *ashed* (calcinated) at relatively low temperatures of 300–500 °C prior to the preparation of the slurry. This is performed mainly to destroy difficult to grind materials such as plant fibers or hair [551, 498, 552, 553, 554]. However, pre-ashing might also be a means for pre-concentration of elements present at relatively low levels as was recently reported prior to solid sampling GF-AAS [186].

5.2.1.2
Subsample Intake

The amount of subsample, typically 1–50 mg, taken for preparation of the slurry depends on the final concentration of the element(s) to be determined, on its (their) homogeneous or heterogeneous distribution in the material, but also on its (their) particle size and density. Density and particle size should be considered for a calculation of the number of particles needed for the preparation of a 1 ml slurry to ensure that there is a minimum number of 50 particles (if no significant heterogeneity occurs) in each 20 μL injection (which is the predominantly used amount) in order to reduce sampling errors (see also Sect. 5.2.1). For lower or higher injection volumes, corresponding conditions have to be ensured.

Table 5.1 is very useful for planning and performing slurry preparation [16].

Table 5.2 informs about typical densities of some common materials based on data given by Majidi and Holcombe [555] as well as on those experimentally determined by Miller-Ihli [500].

For final optimization, each of the criteria mentioned has to be balanced by the analyst to ensure that a representative sample is weighed out and a slurry as homogeneous as possible is achieved. It is also important that the concentration of the analyte is within the analytical range for the selected wavelength, which is similar to the proper selection of instrumental conditions and wavelengths discussed above for direct solid sampling in AAS (see chapter 3) and that the slurry is not too viscous as already mentioned above.

Table 5.1 Minimum mass needed for the preparation of a 1ml slurry with 50 particles in a 20 µl injection depending of particle density and particle size (from:[16] with permission)

density (g/cm³)	particle diameter (µm)	minimum mass/1ml (mg)
0.5	25	<0.05
	50	0.08
	100	0.65
	250	10.2
	500	81.8
1.0	25	<0.05
	50	0.16
	100	1.30
	250	20.4
	500	245
1.5	25	<0.05
	50	0.24
	100	1.95
	250	30.6
	500	245
2.0	25	<0.05
	50	0.32
	100	2.60
	250	40.8
	500	327
2.5	25	<0.05
	50	0.40
	100	3.25
	250	51.0
	500	408

Table 5.2 Results of the determination of sample densities

material	density (g/cm³)
NIST SRM 1566a "Oyster Tissue"	0.28
Paper	0.7–1.2
NIST RM 8431 "Diet"	1.09
Coal (bituminous)	1.2–1.5
Coal (anthracite)	1.4–1.8
Bone	1.7–2.0
NIST SRM 1400 "Bone Ash"	1.95
Clay	1.8–2.6
NIST SRM 1646 "Estuarine Sediment"	2.47
Glass	2.4–2.8

If slurry preparations have to be diluted to allow easy pipetting or to reduce the amount of sample injected into the furnace, care must be taken to ensure that a representative number of particles according to the data given in Table 5.1 is injected. If this is not the case, the precision of the measurements can significantly decrease as has been reported for the determination of Ti in very dilute slurries of plant materials [556].

Weighing should be performed on a commercially available electronic microbalance that can weigh up to 150 mg with a readability of 1 µg. Static electricity may be controlled by using a commercially available ion source [16] (e.g. Staticmaster, NRD Inc., Grand Island, NY, USA).

5.2.1.3
Dilution, Reagent Addition

Depending on the subsequent analytical procedure the reagents described for the preparation and stabilization of slurries are different. However, the former frequent use of thickening agents as slurry stabilizers, described below, was and is only useful for manual introduction or special cases if autosamplers are applied. More recently and for automated injection, much simpler recipes have been reported based on the addition of dilute acids or just water.

The procedures for preparation of slurries using different *thickening agents* (Viscalex, glycerine, glycerol) are quite similar (see Sect. 5.2.3). Typically, a thickening agent, e.g. the originally acidic and low viscous Viscalex HV30 [519] is mixed with the sample material, and the mixture mechanically, magnetically or ultrasonically agitated.

It is often necessary to add a defoaming agent in order to avoid bubbling upon agitation. Hereafter, the sample is diluted with water and the pH adjusted with ammonia to a range of 6–10 in order to achieve a stable viscous gel. An advantage of procedures of this kind is the long-term stability. A disadvantage is that the high viscosity might pose problems if an autosampler is used for introduction of the viscous droplet into the graphite tube [521]. Additional drawbacks are the possibility of contamination from the various reagents used at relatively high concentrations as well as the rather complex technique compared with the simplicity of agitation alone [533].

The preparation of slurry is quite simple by the addition of water alone or aqueous solutions with only very small amounts of added chemicals so that contamination is minimal. This, however, requires powerful agitation (see Sect. 5.2.1.4), followed by quick injection into the autosampler tray. Also, in most cases, the desired extraction of analytes into the liquid phase is less than with acidification.

A number of workers added water alone to the powdered material [e.g. 545, 142; 109, 522, 551, 557, 533, 508, 546, 534, 541, 542, 501].

Another version, which is relatively often reported, is the addition of various modifiers and/or Triton X-100 to the aqueous slurry suspension. These are, e.g., Mg and phosphate modifiers [558, 514, 513], Pd and phosphate modifier [608], Ni nitrate [551], Pd-Mg modifier [610, 611] with aqueous ammoniacal solutions [524], water and Mg modifier [559] or with water and various (including Pd) modifiers [538; 579; 524, 538], 0.04% or 0.01% aqueous solutions of Triton X-100 [137, 322], a 1% aqueous solution of Triton X-100, with Ni nitrate [528], water and 0.03% hexametaphosphate [556], water and Mg nitrate [518], water and Ca nitrate [543].

The hitherto very frequently used dilution/dispersion reagent, which is also the extraction reagent for many elements, is nitric acid at various concentrations and with various additional reagents.

Just nitric acid, at concentrations ranging from 0.1% up to 5% has been used by a number of workers, e.g. [560, 561, 498, 552, 531, 532, 497, 535, 562, 537, 539, 239, 517].

Other workers have applied mixtures of dilute nitric acid with Triton X-100, in some cases also with Triton X-100 and modifiers (for details see Sect. 5.4.1) e.g. [515, 563, 564, 529, 142, 497, 535, 240, 516, 540, 565, 567, 501, 502, 517].

Typically, an appropriate amount of the ground sample material (grinding is unnecessary if the sample is already fine enough) is weighed into larger vessels or autosampler cups and ultrapure water or ultrapure dilute nitric acid and frequently also various matrix modifiers are added. In order to moisten the material and to assist in particle dispersion, low amounts of, e.g. 0.005%, Triton X-100 may be added [502]. Slurry agitation is then performed by different systems which will be discussed below.

5.2.2
Accuracy of Slurry Analysis

5.2.2.1
Representativity in Slurry Analysis

The slurry method requires two sampling steps in the laboratory (see Fig. 5.1). Correspondingly a sampling error may occur with the preparation of the slurry (external, respectively in cups) and with the pipetting of the suspension.

If n_s slurry preparations of an unknown sample (in different cups) are independently analyzed, the sample content is calculated from these results. Accordingly, the precision is calculated from n_s individual results.

For the analysis of n_s slurries based each on the test sample mass m_s the representative mass M_r can be simply calculated by

$$M_r = n_s m_s \qquad\qquad\qquad\qquad \text{Eq. 5.1}$$

For the validity of Eq. 5.1, it must not be assumed, that all of the m_s is introduced into the furnace, if for each slurry preparation, the analysis is performed with a sufficiently large number of pipetted samples (i.e. a large number of test samples n_{ss}). However, if the repetitive analysis is performed only out of one slurry preparation, the estimated mean content is given by

$$M_r = n_{ss} m_{eff} \qquad\qquad\qquad\qquad \text{Eq. 5.2}$$

where
n_{ss} = number of replicate measurements (test samples) from the slurry

The effective (sample) mass m_{eff} is the mass of the solid sample which is introduced into the furnace by one pipetted slurry aliquot, calculated by

$$m_{eff} = \frac{V_{ss}}{V_s} m_s$$ Eq. 5.3

where
V_{ss} =injected volume of the slurry
m_s =sample mass used for preparation of one slurry
V_s =volume of the slurry preparation

(note: m_s/V_s="slurry concentration", further shortened as m/V, e.g. in the unit of mg/ml; this value is often given as a ratio in %, incorrect because this term is obviously not dimensionless, see discussion below.)

Example

If an aliquot of 20 µl is pipetted from a slurry, which is prepared from 10 mg of the sample in 1 ml water ("1% m/V"), the effective mass of the solid sample introduced into the furnace is 200 µg.

The effective sample mass is usually small. However, some workers have reported that even in aqueous solution, a more or less significant fraction of the analyte may be extracted from the solid particles [e.g 125, 568, 569]. Only in the case of total extraction into the liquid phase, is the analysis of n_{ss} slurry aliquots representative of the total mass used for the slurry preparation (M_r =m_s).

If the analyte is partly extracted, the representative mass M_r for the analysis out of one slurry preparation may be calculated by the relation [502]

$$M_r = m_s \left[\frac{V_{ss}}{V_s}\left(1-f_x\right)+f_x \right]$$ Eq. 5.4

where
f_x=fraction of analyte in solution

Example

Figure 5.2 shows the consequence of this situation for the value of M_r calculated from Eq. 5.4 depending from the volumes V_s at a slurry preparation of 1 mg/ml. Table 5.3 shows the calculated representative sample mass M_r for different slurry preparations, depending on the fraction of analyte extraction.

The slurries used for in Table 5.3 are apparently of equal mass/volume (m/V) concentration. It illustrates the benefit of using larger masses of solid with larger final slurry volumes, in order to increase the mass of sample represented by the analysis [712]."This clearly illustrates the problem when discussing slurry

Fig. 5.2. Representative mass depending from the volume of the test samples (slurry preparations) for different fractions of analyte extraction, considering a fixed slurry concentration of 1 mg/ml and an injected volume of 20 µl (from [502], with permission)

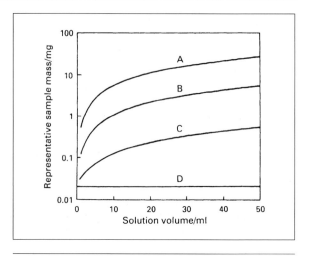

Table 5.3 Representative sample mass for different fractions of analyte extraction considering a fixed sample injection of 20 µl (from [502], with permission)

extraction (%)	representative mass (mg)		
	slurry preparation		
	2 mg/1ml	20 mg/10ml	200 mg/100ml
0	0.04	0.04	0.04
15	0.34	3.43	34.3
100	2.0	20	200

preparations using terminology such as 'a 0.2% slurry' which is often used in the literature" [502].

Concluding, Miller-Ihli pointed out "although M_r is a reasonable indicator..., it is only that. Because calculations of M_r are based on the assumption that the percentage of analyte in the liquid phase of the slurry is proportional to the amount of solid dissolved, it is not absolute. There is no easy way to routinely determine the actual amount of material dissolved for small slurry preparations. There is a strong possibility that when working with materials such as sediments where analyte is adsorbed onto the surface of the solid that only 2% dissolution could lead to 50% of the analyte being extracted into the liquid phase. In terms of how representative the measurement is, a safe assumption is that the analysis is representative of a mass *not larger than* the calculated M_r value" [502]

The analyte fraction which is extracted can be determined as described in [157]: "The mixed slurry was analyzed and then allowed to stand for several hours so that the solid material could settle out and then the supernatant (liquid fraction) was analyzed. The amount in the liquid fraction was then rationed to the total amount of analyte in the slurry and expressed as a percent."

For a spinach material extraction of 98% for Cu and 74% for Cr were found [501], these are typical values for biological samples. For inorganic samples, the extracted fraction is generally significantly smaller [125, 565]. The addition of acids to the slurry suspension usually increases the analyte extraction. For a river sediment, the extracted fraction of Cr was <1% after 48 h, whereas with 2.5% nitric acid 27% of the Cr was extracted [502].

Leaching experiments for various trace elements in graphite powder are described in [571]. It was found that with pure water, the extraction ranges from almost zero for Cu to 80% for Na. By using dilute nitric acid, the leaching efficiency increases for Cu to 15 % (0.2 mol/l) and to 50% for Mg and Zn (0.08 mol/l) with no considerably additional benefit arising from higher acid concentrations.

It was suggested that, in the case of a high extraction efficiency, the sampling error diminishes due to a more homogeneous distribution of the analyte over the slurry suspension [125, 565, 571]. Actually, this effect occurs if the assumptions for the extraction process – homogeneously distributed analyte in the particles and the extracted analyte fraction proportional independent of particle size – are met with the particular sample (see discussion in Sect. 6.6.3)

5.2.2.2
Random Effects

In Sect. 2.4.1.2, effects which contribute to the analytical precision ("random errors") in solid sampling are discussed. Additionally in slurry analysis the *volumetric error* s_V has to be considered. This term comprises several different sources for random effects. These are the effects associated with the variation in the pipetted sample volume (V_s), in the number of particles extracted in the sample volume (i.e. m_{eff}), and in the mass of the individual particles. Because the contribution of each source of error cannot be isolated from each other, the sampling process of a slurry is generally not under statistical control.

Holcombe and Majidi [572] have theoretically analyzed the contribution of this effect to the total precision (RSD). They concluded that the volumetric error might be minimized by working with smaller particle sizes, larger total masses of material and narrower particle size distributions.

For controlling the volumetric error, Miller-Ihli propose to use a minimum sample mass for the preparation of the slurry (m_s) to ensure that an average number of 50 particles are pipetted [16, 501]. In this case the respective uncertainty is limited to 2%, assuming that the uncertainty is ±1 particle. Exemplary calculations for the minimum sample mass for slurry preparation are given in Table 5.1 with respect of material density and particle diameter. Typical data for the density of some commonly analyzed materials are listed in Table 5.2.

If the determination is based on the analysis of several slurry preparations, the volumetric error is of minor importance. The contributions to the calculated between-slurry precision which are associated with the preparation – mainly the weighing and the sampling error – are commonly known or determinable. Thus, this analytical process might be regarded as being under statistical control. In [239] illustrative examples of slurry data for in-slurry and between-slurry precision are given. For a statistical evaluation ANOVA (analysis of variances) can be a suitable method to study the contribution of the various effects [125, 501].

5.2.2.3
Systematic Effects

Additional to the general discussion in Sect. 2.4.1.1 other sources for a bias ("systematic errors") must be reckoned with in slurry sampling.

Inherent to slurry sampling is the occurrence of a *sedimentation error*, which may lead to a biased result. If the the agitation of the suspension is stopped before sampling an aliquot of the slurry, the particles will begin to settle. Depending on the time elapsed between the end of agitation and pipetting, a reduced recovery, leading to a bias, may appear.

In a basic theoretical study, Majidi and Holcombe [555] illustrated the influence of the main parameters to the sedimentation rate and by this to the recovery of the analyte content. With their conclusion, they outline the interdependence of these parameters as follows: "In order to maximize the sampling time window for a given matrix, the medium should be selected to have high viscosity and a density similar to that of the particles, and the sampling depth should be maintained as deep as possible without reaching the bottom of the container. Sample recovery for a range of particle size distributions has confirmed that samples with a narrower range of distribution have a longer time window…". Figure 5.3a and b visualize sampling time windows of 100% recovery depending e.g. from particle size and viscosity.

An extreme case was reported in [336], where a very large particle density up to 10 g/cm^3 (powdered molybdenium) prevented the preparation of a stable homogeneous slurry. In the discussion below (Sect. 5.2.3), appropriate measures (actions) are discussed for stabilizing the slurry.

In [573] the use of the slurry technique for overcoming the problem of sedimentation of particles of natural origin or due to processing (e.g. bleaching earth, catalysts or rust) is described. Thus, a separate analysis of the solid fraction of edible oils and fats could be avoided and the precision of the analysis was somewhat better.

A sampling efficiency (recovery) of only 69% was achieved if a slurry of a titanium dioxide powder with larger particle size was (discontinuously) stirred by an ultrasonic probe, but near 100% was obtained with (continuously) magnetic stirring. However, for a finer powder with both agitation techniques more than 97% was obtained [574].

How a combination of unfavourable conditions can yield a bias in the analytical result was shown in a collaborative study performed and evaluated by Miller-Ihli [30]. When a very low slurry concentration was prepared from a sediment sample with a very high content of Pb in order to use the sensitive line at 383.3 nm, the determined content was found to be significantly too low. However, as Fig. 5.4 shows, a slurry concentration that provides an effective sample mass of >0.04 mg yielded good results. Such a preparation was recommended as a consequence of a "systematic approach" ([16], see Sects. 5.2.1.2 and 5.2.2.2). In this case the unsenitive line at 361.4 nm was required for the analysis (see Sects. 3.2.1

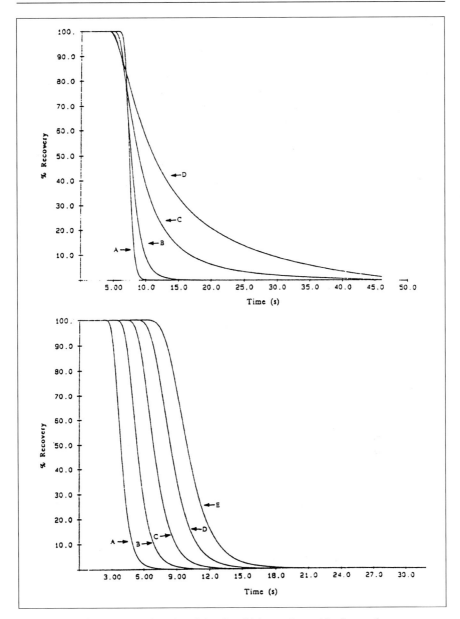

Fig. 5.3. Plots of recovery as a function of time for which sample particles from a slurry are "removed" by setting; parameters are chosen as e.g. density of particles 2 g/ml, density of liquid medium 1 g/ml, radius of particles 100-330 μm, depth of sampling 10 mm (from [555], with permission).

a Curves *A,B,C,* and *D* correspond to 10%, 20%, 40% and 80% relative standard deviation respectively of the distribution of the term (radius of particles) ·(difference in density of particle and medium) in a medium with a viscosity of 1 mPa s.

b Curves *A,B,C, D* and *E* represent the time profile for the mediums with viscosities of 0.5, 0.7, 0.9, 1.1, 1.3 mPa s respectively

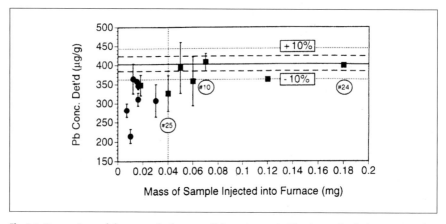

Fig. 5.4. Comparison of slurry results from a collaborative study. The determined Pb-contents of a Sediment (PACS-1) are plotted as a function of the effective sample mass injected into the furnace. Circles represent analyses at 283.3 nm and squares represent analyses done at 261.4 nm (from [30], with permission)

and 3.2.2). The author explained this finding by the "nugget-effect", i.e. rare particles containing a large analyte amount are not "captured" with the low number of particles in the test samples (or if included in a test sample the respective test result was discarded as an outlier, see Sects. 2.3.2 and 2.4.3). Thus the volumetric effect that is basically of random character has yield a systematic bias.

As it is with conventional digestion techniques the contamination by blank values of the diluent and the reagents must be considered to some extent with slurry analysis [157].

Finally, if only one slurry is prepared and analyzed, the resulting mean content from the pipetted test samples might be biased due to an error occured with the slurry preparation process.

5.2.2.4
Uncertainty in Slurry Analysis Results

In this section only the particularities in the estimation of a combined uncertainty in slurry results are outlined, whose general characteristics are discussed in detail in Sect. 2.4.2. The concrete uncertainty components for slurry analysis are included in Table 2.12.

Commonly a calibration curve "response versus concentration" based on reference solutions is established with the slurry method, accordingly the respective terms must be estimated, considering the uncertainty components which represent the AQC-process (uncertainty of the measurement and the certified content of the CRM, compare Sect. 2.4.2.4).

The final content of the unknown solid sample c_s is calculated from the determined concentration of the slurry c_{ss} by considering the sample weight and the volume of the slurry preparation, thus the respective factor appears in the uncertainty terms as well

$$c_s = \frac{V_s}{m_s} c_{ss}$$ Eq. 5.5

If the solid test samples, which are used for the preparation of the slurries, are not constant, but differ because in the preparation it is often tedious to adjust the weights to an exact value, these must be transformed in order to achieve a representative uncertainty in the test sample measurements. The response R^*, which is representative for the mean of the used test sample mass can be calculated according to $(i=1,\ldots,n_s)$

$$R_{s,i}^* = R_{s,i} \frac{m_{s,i}}{\overline{m}_s}$$ Eq. 5.6

Usually, the uncertainty in the calibration, if it is based on reference solutions, is small compared to the other contributions, so that it can often be ignored. Hence, the combination of the remaining uncertainty contributions can be expressed in terms of "content units" instead of "response units", hence the calculation becomes simpler (see Table 2.12).

It should be stressed again, that for the uncertainty in slurry analysis, the *standard deviation between different slurry preparations* must be determined. If the standard deviation of repeated measurements from only one slurry preparation is used, the uncertainty due to sampling and preparation are not considered. As a result, the calculated uncertainty may be underestimated.

Example

In [502] for the analysis of an unknown spinach sample a mean content for lead c_s=0.21 mg/kg and $s(c_{s,i})$=0.01 mg/kg was found, based on the analysis of 3 slurry preparations. The AQC-analysis of the NIST-SRM1570 Spinach has a yield of c_q=1.23 mg/kg and $s(c_{q,i})$=0.07 mg/kg, based on 3 slurry preparations as well. The certificate for the SRM specifies a lead content of c_r=1.2±0.2 mg/kg from which an uncertainty in the certified content of $u(c_r)$=0.077 mg/kg may be supposed (see example in Sect. 2.4.2.3). Thus, the uncertainty components are estimated from the respective equations from Table 2.12: $u_s(c_s)$=0.0058 mg/kg, $u_q(c_s)$=0.0069 mg/kg, $u_r(c_s)$=0.0135 mg/kg (the calibration uncertainty component is supposed to be negligible). With these values the combined uncertainty is estimated as $u(c_s)$=0.016 mg/kg. With df_{eff}=5, the coverage factor is chosen as k_p=2.5, thus the extended uncertainty for the determined sample content amounts to ±0.04 mg/kg (±19%) instead of ±0.025 mg/kg (±12%) if only the sample measurements are considered.

5.2.3
Slurry Stability

As discussed above, for reliable results it is mandatory to achieve a sufficient stability of the slurry until the time of injection into the graphite furnace in order
to avoid a significant sedimentation error. For many materials, this is a real problem because of their rapid sedimentation rate from an unagitated and non-vis-
cous or less effectively agitated sample due to particles of different sizes and density [555], see also Sect. 5.2.1.1. Therefore, a representative sample can only be
obtained if either the viscosity of the slurry medium is changed to some extent
to minimize the sedimentation rate after terminating the agitation, or if the slurry is continuously and effectively agitated during the sampling procedure.

The "classical" approach to achieving an appropriate slurry stability is raising the viscosity by adding agents such as Viscalex or Glycerol. Fuller and
Thompson [519] and Fuller et al. [143] were the first to apply this technique
for the determination of several elements. They stirred various rock samples
mechanically, ground them to pass through a 325-mesh sieve (particle size
«44 μm) for 1–2 min with Viscalex HV30, and a de-foaming agent, adjusted to
pH 6–10. The resulting suspension of around 2% (V/V) Viscalex gel (1% m/V)
was reported as being stable for several days. Ottaway and coworkers [544, 521]
applied the same technique for the determination of Pb in spinach using an
autosampler. Because the suspensions based on manually ground material
were initially only stable for short periods, especially at higher spinach concentrations, the samples were ground in an agate ball mill for 1.5 h. After this
procedure, the size of the particles was well below 50 μm. With this procedure,
material slurries in the 1–10% (m/V) range were prepared that were reported
as being stable for several hours. More recently, Yu et al. used a similar stabilization technique for lead determination in various NIST plant SRMs by manual injection [536]. The material was ground to particle sizes below 50 μm. For
a Citrus Leaves SRM slurry (0.3% m/V) containing Viscalex HV30 (1.5% V/V)
a stability of 3 h was reported.

Another stability agent is glycerine. Hoenig and coworkers [575, 576] used
mixtures of glycerine, methanol, nitric acid and matrix modifiers for the preparation of slurries from different geological and biological materials for the subsequent determination of various elements using an autosampler. Best results for
stable slurries were obtained with 1+1 glycerine-demineralized water for environmental samples and glycerine-methanol for lyophilized animal tissue samples. It was found that these suspensions were stable for up to 1 h thus permitting introduction via an autosampler.

Bendicho and De Loos-Vollebregt tried to stabilize slurries of glass materials
with solutions containing 5–80% glycerol in the autosampler cup: the slurry was
stirred and left in the autosampler cup without further mixing for automated
injection [497, 535]. The authors reported that these slurries were very unstable
at low concentrations of glycerol, whereas the high concentrations of glycerol

caused injection problems with the autosampler. Thus, they concluded that for high density materials such as glass powders, glycerol was unsuccessful as a stabilizing agent because the distribution of the particles was not homogeneous unless the slurry was continuously stirred.

For any mode of ultrasonification, the stability of the slurry appears, from the data available from the recent literature, to be of minor or even no importance, if the agitation is performed directly on the sample tray and consequently continued during sampling. However, in some cases, there is some elapsed time between the termination of ultrasonification and injection because of the different techniques applied. Therefore, some studies have been performed on the stability of slurries after ultrasonification. These studies also provided additional information about the extent of element extraction from the solid into the liquid phase of the slurry.

Dobrowolski and Mierzwa reported, from their studies with plant and silica gel samples, that ultrasonification in comparison to mechanical mixing improved the stabilities of the slurries, especially a few minutes after termination of the 20 s mixing [240, 516, 517]. Use of nitric acid with increasing concentration as a diluent, leads to a further significant extraction of elements into the liquid phase. This was particularly observed for Cd in contrast to Pb from NIST SRM1571 Orchard Leaves and for Cd, Cu and Pb in contrast to Al, Fe and Ti from two silica samples.

5.3
Instrumentation and Procedures

Numerous approaches have been reported up till now for manual and automated slurry agitation and introduction.

5.3.1
Agitation

Important for final homogeneity of the slurry and reliable results is effective agitation after addition of the various reagents mentioned above to the material so that sedimentation, at least during the time of sampling, can be avoided or as far as possible minimized.

A number of agitation techniques were used mainly during earlier work performed with the slurry approach. These techniques were manual or rod (electrical) stirring or mixing. These were often sufficient if thickening agents were used but they were also applied by several workers in other procedures [519, 143, 544, 521, 576, 575, 531, 532, 577, 536]. Other techniques are Ultra-Turrax agitation [e.g. 560, 561] magnetic agitation with reported mixing times of up to 20 min [e.g. 520, 142, 109, 558, 514, 522, 513, 578, 535, 523, 579. 524, 559, 557, 528, 538, 556, 580, 581, 541; 568, 525], however, with the disadvantage of problems if magnetic samples, e.g. ores and some soils have to be analyzed [559] and vortex mixing [545,

557, 239]. Less frequently used techniques were agitation in water with zirconia spheres as a combination of grinding and agitation prior to sampling [551, 546] and passing an Ar stream through a narrow capillary tube introduced into the slurry medium, allowing sample preparation directly in autosampler cups [497, 535].

The recently most frequently and (with automated slurry introduction) almost exclusively applied agitation technique is ultrasonification with appropriate power settings applied for a relatively short time of, typically, 15–25 s [e.g. 322, 527, 157, 500, 16, 501, 502, 498, 552, 563, 564, 529, 530, 534, 562, 537, 137, 539, 240, 516, 239, 540, 568, 525, 567, 518, 517, 238, 526]. Occasionally also somewhat longer sonification times, for up to a few minutes, were reported as being necessary for complete homogenization. The advantages of this elegant and versatile approach for manual as well as automated methods were confirmed by a recent carefully performed study by Sandoval et al. [239]. The authors compared different techniques for the preparation of sediment slurries for Cd and Pb determination. The studied techniques and apparatus were: Glass impact beads using a Braun cell homogenizer (B. Braun Melsungen), mechanical stirring with both propeller type and foamless generators (Omni-Mixer, Servall) vortexing (Vortex-Genie mixer, Scientific Products) and ultrasonic treatment by a hand held high intensity ultrasonic processor (50 W-mode, Sonics & Materials Inc.) with a 3.2 mm diameter probe. The authors concluded that "ultrasonic homogenization on the basis of precision and accuracy, speed and ease of operation was clearly the method of choice".

5.3.2
Manual Slurry Introduction

Former as well as more recent approaches were based on manual introduction of slurried samples by commercially available micropipettes with acceptable results for precision and accuracy for a number of solid matrices. The authors mentioned, in most cases, the simplicity and speed compared with the conventional approach after decomposition. Most authors reported that calibration, either against aqueous reference solutions or by the analyte ("standard") addition method (AAM), frequently achieved acceptable results. Some examples of the hitherto applied very different approaches for the preparation of slurries from numerous materials for manual slurry introduction are given below.

Brady et al. analyzed plant leaves [545]. The material used was NIST SRM1571 Orchard Leaves. It was dried in an oven at 105 °C overnight and a subsample ground in a dental mill just prior to analysis. Samples of 2–20 mg were weighed out and slurried in 2–8 ml of water using a Vortex mixer. After 20 s of slurrying, a 10 µl sample was withdrawn with an Eppendorf micropipette while mixing, and injected into the HGA.

Fuller and Thompson analyzed rock samples [519]. The slurry was prepared by addition of Viscalex HV30 and a de-foaming agent to the ground material with a particle size well below 44 µm. Injection into the atomizer was performed with a 50 µl micropipette.

Stoveland et al. analyzed sewage sludge [561]. The sludge, diluted fifty-fold with 1% HNO_3, was homogenized with an Ultra-Turrax that had a shaft and rotor made from 99,7% pure Ti. In most cases, a satisfactory dispersion was achieved after 5 min at 8000 rpm. Sludges that are not readily dispersed by this procedure required initial treatment with a coarser homogenizer so that a combined treatment was necessary for 10 min at $8000min^{-1}$. Injection into the atomizer was performed with 10 µl micropipettes.

Fuller et al. compared flame AAS, GF-AAS and ICP atomization techniques for the direct analysis of several elements in slurries of ores and silicate rocks [143]. The procedure for preparation of the slurry for GF-AAS was that already described above [519]. The injected volume into a HGA was 20 µl. Acceptable results could be achieved if the samples were ground to a particle size <25 µm.

Ebdon and coworkers contributed significantly to developments covering all aspects of slurry preparation and application with flame AAS, GF-AAS [520, 515, 546, 578, 553] and plasma emission spectrometry using manual sample injection [570].

Ebdon and Pearce analyzed coal samples [520]. The slurry was prepared as follows: 0.5–1 g coal was ground to less than 45 µm (Retsch Spectromill) and weighed into a 50 ml polypropylene bottle. Then a reagent solution (containing 10 g/l each of Ni and Mg nitrate, 50 mg/l of HNO_3 conc. and 100 ml/l of ethanol) was added. A magnetic stirrer bar was then inserted, the bottle sealed and the contents stirred for 5 min. Then, 5–20 µl aliquots were taken and manually injected into the atomizer while being stirred continuously.

Another approach was applied for the determination of As in coal in order to eliminate background interferences from Al [578]. Slurry preparation was similar to that previously described [520]. Studies with a thixotropic medium were reported not to improve the dispersion of the coal under the given conditions. Background problems were able to be overcome by Smith-Hieftje (S/H) background correction.

Using the same instrumentation and slurry preparation technique, the determination of Se in coal was investigated [515]. In this case, in addition to Smith-Hieftje background correction, the complete elimination of interference was only possible by introduction of an air ashing step. Air ashing was subsequently studied in more detail for the determination of a number of elements in slurries of biological and environmental reference materials, including hair [553]. The preparation of the slurries was somewhat different. 1% (m/V) slurries of the certified reference materials (except hair) were made by weighing about 200 mg of the material into a 30 ml polypropylene bottle. Zirconia beads (2 mm diameter) and 3 ml of water were placed in the bottles and the bottles mechanically shaken for 1–4 h. The slurries were then diluted to 20 ml by addition of water. A 1% (m/V) slurry of hair was made by weighing ca. 250 mg into a clean platinum dish and the dish heated at 300 °C for 3 h. The majority of the ash was then weighed into a polypropylene bottle and slurried as before. The final volume of the hair slurry was 20 ml.

Jackson and Newman analyzed soil samples based on previous work [142]. The dried soil was ground by hand first and then amounts of 2 g ground in a miniature ball mill (Rotomill) for 10 min. 5–20 mg of this soil powder were weighed into a 50 ml beaker and 25 ml distilled water added. Slurries were obtained by magnetic stirring for 3 min at maximum speed. While the suspension was stirred, 20 µl aliquots were taken by a micropipette and injected into the graphite furnace.

Despite the fact that, from approximately 1985, there was a significant increase in automated methods, manual introduction of slurried samples was and still is frequently applied.

De Benzo et al. analyzed slurries of four different ashed plant materials with GF-AAS [498]. Additional experiments were performed with flame AAS. The slurries were prepared after milling and heating up to 500 °C by suspending 100 mg of the calcinated material in 100 ml of 0.1% nitric acid. It was reported that these slurries, if a ratio of 2 mg/ml was not exceeded, showed constant signals for 60 min without stirring.

Barnett et al. determined B in certified plant reference materials [551]. Portions of the reference materials were dried and samples of 0.5–0.7 g charred at 250 °C for 6 h, then shaken by addition of zirconia spheres and Ni nitrate solution as a matrix modifier for 1 h to reduce the particle size and the bottles shaken vigorously before each sampling.

Carrión et al. determinated several elements in pine needles. The preparation of slurries for manual preparation was as described above either by calcination (heating up to 400 °C in a furnace) followed by stirring with 0.1% nitric acid or by direct stirring of the sieved material (particles of <160 µm) [504]. Injected volumes were either 5 µl (CRA-90) or 20 µl (HGA-2100) after 1 min stirring.

A series of studies on slurry methods in various materials was performed by López-García; Hernández-Córdoba, Viñas et al. (see Sect. 6.3.2). E.g. this group developed a slurry method for As determination in Fe(III)oxide pigments[528]. The pigments had relatively low particle sizes, thus the samples were only sieved using a 100 µm sieve and the small fraction containing larger particles discarded. To 10–250 mg of the sample 25 ml of a solution containing 0.1% (m/V) Triton X-100 and 0.1% (m/V) Ni nitrate as matrix modifier was added. The suspension was magnetically stirred for at least 10 min and 25 µl aliquots taken while stirring for manual introduction into the atomizer.

The same authors published a slurry procedure for the determination of Pb in paprika [554]. The preparation of slurries started with ashing (calcination) of 1 g oven dried material at 350 °C in a porcelain crucible. The sample ash was then ground in an agate ball mill for 10 min and sieved with a 325-mesh sieve. The small fraction containing larger size particles was discarded. To 10–100 mg samples 25 ml of a solution containing 0.1% Triton X-100 and 0.1% ammonium phosphate were added. Then the solution was magnetically stirred and 20 µl aliquots taken while the suspension was being stirred and manually injected into the atomizer.

Another slurry procedure was described for the determination of Ti in lettuce, spinach, peas and paprika using the same instrumentation as above [556, 506] as well as the calcination prior to slurry preparation.

The same authors studied a fast-temperature approach in that the drying and ashing steps were replaced by a modified drying stage with optimized temperature and hold time for several elements in samples with high silica content (diatomaceous earth) [568]. The materials were first ground for 10 min with an agate ball mill and the size distribution checked by passing the ground sample entirely through a 30 μm sieve. 10–1000 mg of the ground material was suspended into 25 ml of a suspending solution in 50 ml plastic beakers. The suspending solution contained mixtures of ammoniumdihydrogen-phosphate and variable percentages (0.1–3% for slurries in the 0.02–2% range) of HF for the determination of several metals, and only HF for the determination of Mn. When slurries more dilute than 0.04% (m/V) were required, the volume of the suspending solution was increased to 50–100 ml. From all suspensions 10–25 μl aliquots were manually injected while being magnetically stirred. Accuracy was checked by analyzing appropriate certified reference materials.

Yu et al. analyzed plant reference materials [536]. First, the coarse powders were ground in a microdismembrator or in an agate ball-mill to obtain particle sizes <50 μm. The slurries were prepared by adding Viscalex HV30, antifoam agent and ammonia to adjust the pH value of the suspensions to about 9.0. The injected volume was 15 μl.

Lynch and Littlejohn developed a slurry method for the analysis of food samples [577]. The slurries, prepared from various finely ground materials by addition of modifiers, antifoaming agent, and concentrated ammonia solution were vigorously shaken for 15 min to achieve thorough dispersion. Hereafter 20 μl samples were injected either manually or by using an autosampler with an integrated magnetic stirring device. This is mentioned below for additional work of this group dealing with automated sample injection.

Ohta et al. applied a Mo tube atomizer for the determination of Cd in slurries of biological material [534]. The slurry was prepared by dispersion of 1–10 mg of powdered material in water and the slurry ultrasonically agitated for 5 min. Hereafter, 1 μl aliquots were injected into the metal tube.

Fernández et al. determined a number of elements in atmospheric particulates by GF-AAS and flame AAS [537]. The particulates, usually having low particle sizes, were suspended in 1% nitric acid (50 mg/100 ml m/V) and then homogenized by ultrasonic agitation. It was reported that slurries could be prepared up to 0 4 mg/ml in 1% nitric acid without agglomeration of solid particles.

Hinds and Jackson determined Pb in soil [533]. Since the magnetic stirrer applied for the preparation of slurries showed problems with magnetic soil particles, vortex mixing was used for this material. To weighed amounts of soil, typically 45 mg, in a 50 ml conical centrifuge tube 20 ml of distilled water was added. The sample was agitated by a vortex mixer. The slurry was then continuously agitated at a mixer speed that suspended the soil without splashing and permitted the manual introduction of 20 μl slurry aliquots into the atomizer.

Hinds et al. studied Pb determination in soil applying a fast temperature program [538]. Slurry preparation was as described above but only with magnetic stirring. The authors concluded that temperature programs for the direct determination of Pb in soil slurries might be significantly shortened by drying at higher temperatures, thus omitting the charring step.

Dolinsek et al. studied the direct analysis of Cd and Pb in IAEA and NIST environmental and botanical CRMs either by direct introduction or by a slurry technique [137]. The investigated materials were dried for 4 h at 80 °C prior to analysis and appropriate samples ground in a vibrational mill equipped with either a stainless-steel or zirconia grinding jar and balls. Slurries were prepared by diluting 5–200 mg of powdered material in 1–25 ml of doubly distilled water containing 0.04% Triton X-100. If necessary, a corresponding amount of $(NH_4)_2HPO_4$ was added directly to the slurry. Prior to sampling, the slurries were homogenized using an ultrasonic device with a 3 mm diameter Ti probe. From this suspension aliquots of 10–20 μl were taken and placed directly in the inner graphite cups.

5.3.3
Automated Slurry Introduction

5.3.3.1
Separate Slurry Agitation

The generally very promising, relatively inexpensive and sensitive GF-AAS technology was boosted and sample throughput increased by the introduction of AS-1, the first commercially available autosampler [582]. This was able to compensate to some extent the rather lengthy furnace programs and was a remarkable contribution to the the rapid introduction and further extension of GF-AAS into nearly all analytical branches. Therefore, it is obvious that many workers tried to combine automated sample introduction, meanwhile offered by almost all manufacturers of AAS instruments, with the slurry technique and to overcome the disadvantages of the potential instability of suspensions by different, often ingenious, approaches.

Ottaway and coworkers, and after his early decease, Littlejohn et al. studied in much detail the application of slurry techniques in combination with automated sample introduction.

Stephen et al. [544] and Littlejohn et al.[521] evaluated a slurry technique for the determination of Pb in spinach. The preparation of the sample was similar to that described previously [143]. The freeze-dried powdered spinach was ground in an agate ball-mill for 1.5 h to reduce any small stalks and larger particles to obtain a uniform powder. After this procedure all the particles were 50 μm in diameter or less, with 75% of these below 17 μm. Weighed 0.5–10 g aliquots of this material were transferred into a Jencons Uni-Form homogenizer/grinder and homogenized with 10–20 ml of distilled water. Then the slurry was transferred into a 100 ml calibrated flask where 2–3 ml of Viscalex HV30 were added, and the mixture neutralised with ammonia solution. It was found that the 1% (V/V) concentration of Viscalex reported earlier [546] was insufficient to produce a stable suspension. Thus, higher concentrations of Viscalex were employed depending on the concentration of sample material. For example, a stable suspension of 10% (m/V) spinach required a 3% (V/V) Viscalex solution. The flask was continuously agitated during the formation of the slurry and the thickened sample was diluted to 100 ml with distilled water. If foaming occured, a few drops of an antifoaming agent were added before dilution of the slurry. Calibration was performed by analyte addition. 10 ml aliquots of the suspension were transferred into four acid-cleaned glass test tubes and appropriate amounts of Pb stock solutions added to each test tube. Small volumes of these mixtures were then transferred into autosampler cups. A difficulty was initially encountered in dispensing the slurries with the autosampler. Droplets of the Viscalex solution frequently remained attached to the autosampler capillary and dropped onto the inside wall of the tube near the injection hole, with consequential impairment of analytical precision. This problem could be overcome by careful adjustment of the capillary delivery position and frequent cleaning of the tubing with acetone. The atomizer program was improved by an oxygen ashing step so that, with higher powder concentrations, lower detection limits could be achieved.

Further studies applying the same slurry procedure were performed by this group for the determination of several elements in foodstuffs and the BCR certified reference material single cell protein [144]. The authors stated that, for most of the investigated materials, milling for 45–60 min in the agate ball mill was sufficient to produce a fine powder suitable for the preparation of stable slurries up to 10% (m/V). However, particle size depended strongly on the material used. Liver required, for instance, longer grinding times. The authors mentioned that the use of a more efficient background correction system than the D_2 system used would improve precision and accuracy further.

Hoenig and coworkers developed and applied an automated slurry procedure. The first contribution for the determination of Cd and Pb in different plant tissue, sediments, lyophilized animal tissue and CRMs was based on the use of a programmable sample dispenser [576]. Aliquots of the materials, typically 3–30 mg, were weighed directly in 2 ml polyethylene microvials of the autosampler. Then 1 ml of a liquid medium consisting of glycerine, methanol, $(NH_4)H_2PO_4$, $Mg(NO_3)_2$ and nitric acid was added and aliquots of the samples were suspended and homogenized with a Mini-Mix stirrer prior to analysis. Aliquots of the suspended samples, typically 5–10 µl, were dispensed on a preheated platform via the autosampler. These volumes represented an initial solid sample amount between 15 µg and 300 µg. The authors also performed repeatability tests with different sample types.

In another study, a Zeeman AAS instrument with the programmable sample dispenser was used for the determination of a number of elements in sediment samples [575]. The preparation of the suspensions was similar to that previously given, however, a mixed modifier with Pd was applied in this study. The authors proposed for better results, a partial pre-digestion with nitric acid for an almost complete mobilization of Cd, Cu and Pb. More recently they applied the same instrumentation and procedure for the determination of high Al concentrations in suspended matter samples collected in natural waters [562]. Since the concentration of Al in the analyzed material was very high, the slurry used for the determination of a number of other elements was diluted further by a factor of 10–30. Intensive ultrasonic mixing before taking a portion of this dilute sample was necessary. Test samples of 5 µl of these highly dilute samples were injected into the atomizer.

Olayinka et al. determined Cd in dried pig kidney, seafood and vegetable soup samples and the NIST certified reference material (SRM) Bovine Liver [522]. The samples were ground in an acid-washed mortar and pestle and sieved to the required particle size <44 µm. Aliquots of 10–50 mg were weighed into a beaker, 25 ml of water added and magnetically stirred for 10 min. Analyte additions were prepared by adding appropriate amounts of Cd reference solutions to four samples. Small volumes of the slurries were then injected with a Fastac 254 autosampler into the atomizer. The furnace program included an oxygen-ashing step.

5.3.3.2
Agitation of Slurries on an Autosampler Tray

Lynch and Littlejohn used a modified AS-40 autosampler for food slurries with magnetic stirring [524]. The stirrer was placed under the sample cup being sampled and stirring performed by a small PTFE coated magnet placed in the sample cup. In this study, only an antifoam solution and Pd nitrate solution as a matrix modifier were added to the slurried material and a furnace program exclusively with nitrogen applied. The authors concluded that Pb in slurries could be directly calibrated by this procedure with aqueous reference solutions up to 4% (m/V). Above this, systematic effects in sample introduction and interference necessitated the use of the analyte addition technique. Some measurements with Cd and Sn indicated that Pd might be a general modifier for the analysis of slurries in graphite furnaces. The determination of Cd in food either by manual or automated sample introduction using the same technique has been already mentioned above [577].

Haraldsen and Pougnet used a modified AS-40 autosampler for the determination of Be in coal slurries [559]. The autosampler was modified for magnetic stirring as shown in Fig. 5.5a. The stop switch for the vertical sampler travel was modified in such a manner that the pipette tip could sample the slurries in glass vials, 50 mm high. Figure 5.5b shows the arrangement of sample vessel, the motor, rinse container and matrix modifier container. The samples were placed manually in the sample container holder. The operation then proceeds as for liquid samples. Sideways motion allowed rinsing of the pipette tip between injections and the addition of the matrix modifier.

Fig. 5.5. Adaption of magnetic stirring in the PE AS-40; on the left side magnetic stirring mechanism, **b** position of sample container during pipetting (from [559], with permission)

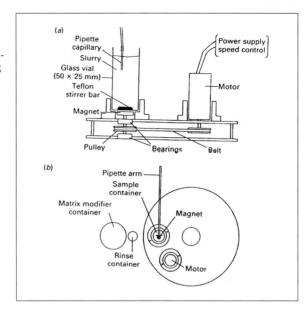

Sample preparation was as follows: coal samples were oven dried and then ground in an agate ball mill for 2 h. Slurries were prepared by weighing an appropriate mass of sample into acid-washed glass vials. A 15 ml volume of the matrix modifier ($Mg(NO_3)_2$ in acidic (HNO_3) Triton X-100 solution) was pipetted into the vial. The sample mass was so that the final concentration of Be in the slurry produced an absorbance value that lay on the linear portion of the calibration graph. Hereafter, the slurries were ultrasonicated for 5 min in order to disperse the particles and the sample vials inserted in the sample tray for determination.

Bendicho and De Loos-Vollebregt used an AS-40 autosampler for the determination of various metals in siliceous certified reference materials and a quartz sample [497, 535]. The slurries were prepared in different ways after weighing typically 1–50 mg of the material in the autosampler cup:

a) 1.75 ml aliquots of a solution containing increasing concentrations of glycerol was added to the autosampler cup. The slurry was prepared and stirred and was left in the autosampler cup without further mixing.

b) 1.75 ml of water containing 0.2% nitric acid was added into the autosampler cups. The sample was then homogenized by a stream of Ar gas (100 ml/min) using a narrow polythene capillary for 30 s through the cup. The gas flow was stopped just before the sample was to be withdrawn.

c) 1.75 ml of water containing 3% HF was added into the autosampler cup and the Ar introduced into the sample cup for 3 min in order to remove the volatile silicon fluoride from the slurry. Similar to procedure b), the capillary was removed and the gas flow stopped when the sample was to be withdrawn.

An AS-1 autosampler was modified similar to the approach of Haraldsen and Pougnet [559] by placing magnetic stirrers besides and under the sampling tray for stirring of up to 10 sample vials using PTFE coated small magnetic stirrers for the determination of boron in cell suspensions [580].

A simple remote-controlled magnetic stirring device for continuous homogenization of slurries during sampling with a PE AS-70 autosampler has been designed by Docekal as a small turbine with small rotating PTFE coated magnetic bars [581]. This device can be easily incorporated into an autosampler tray and can be driven by compressed air or cooling water (Fig. 5.6). It was used by Docekal and Krivan for the determination of a number of trace elements in high-purity Mo trioxide [541]. The slurries were prepared as follows: 0.1–0.25 g of the Mo trioxide powder was added into a 20 ml polystyrene vessel with 10 ml of doubly distilled water. Suspensions were pretreated in an ultrasonic bath for 15 min in order to disintegrate particle agglomerates. Under continuous stirring with the magnetic stirring device inserted into the AS-70, 20 µl aliquots were automatically dispensed into the tube from the alternative position of the autosampler usually assigned for the vessel containing the modifier solution.

Slurries of a number of plant materials, including CRMs were prepared using ultrasonification by Dobrowolski and Mierzwa [240]. The preparation of the slurries started with grinding in a vibrating mill with chambers and balls made from tungsten carbide for all elements studied, except for Co. For Co, the material of the mill was changed to Zr oxide. After 15–20 min of grinding, approximately 85–95% of the particles had sizes <20 µm. The ground material was then weighed and weighed portions of 0.05% nitric acid in polyethylene vessels

Fig. 5.6. Schematic drawing of the stirring device (from [581] with permission); (a) cover body with input and output fittings for the connection of tubings with the driving medium, (b) two magnetic pieces mounted in an opposite magnetic direction, (c) rotor with holes for fixing of magnetic pieces, (d) stator main body with tangentially drilled channels

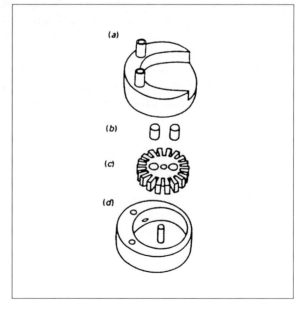

Fig. 5.7. Ultrasonic probe fitted on a fully programmable Gilson 221 sample changer, slightly adapted of the Varian SpectrAA 400 Zeeman AAS system (from [583], with permission)

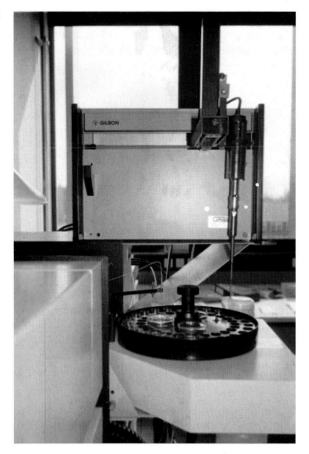

added. Ultrasonic homogenization was performed in vessels of the autosampler directly prior to slurry sampling of 20 µl aliquots into the graphite furnace.

Hoenig and Cilissen developed an automated device for use with all Varian SpectrAA systems fitted with a graphite furnace and programmable sample dispenser [583]. The slurries were mixed with an ultrasonic probe, and monitored by a programmable sample changer (Fig. 5.7). Amounts of 1–20 mg of the finely powdered samples were weighed directly into the autosampler microvials and treated with 50 µl of concentrated nitric acid overnight before the addition of 950 µl of demineralized water as already previously described by the authors [575, 562].

For each working cycle, the Gilson sampling system is started by a remote signal, supplied at the end of the dispensing cycle by the rinse valve of the autosampler. Programming of the Gilson exchanger is performed in order to skip a certain number of these remote signals, which are generated even when the mixing is not necessary, i.e. for blank and reference solutions. Another remote signal, supplied by the sampling device to the ultrasonic probe, performs the mixing of the samples for a distinct time just before their uptake by the autosampler capillary. After each mixing of the sample, the probe is automatically rinsed in an external vessel. During an automatic run cycle, the ultrasonic probe must perform the mixing in four different autosampler carousel positions. The program is valid for the carousel peripherical row only, i.e. sample positions 1–25. A detailed description of the whole program cycle is given in the mentioned paper [583]. The system was used for a detailed comparative study on the use of conventional and fast temperature programs for a number of elements in CRMs [584].

5.3.3.3
Commercial System

A breakthrough for automated analysis of slurried samples was the presentation and commercial introduction of ultrasonic agitation in combination with autosamplers. Miller Ihli performed first studies with ultrasonification and an autosampler[527]. The analysis of certified reference materials with a prototype system (SSI) by Epstein et al. in combination with GF-AAS and Zeeman background correction was the second step in this direction [563]. This was followed by the introduction of the Perkin-Elmer USS-100 (Ultrasonic Slurry Sampler) the first commercially available system [585]. It is based on a patented design [586], basic laboratory experiments [587], and offered the potential for use in combination with several AAS instruments (in the first instance from the same manufacturer).

The USS-100 (Fig. 5.8a) consists of the following hardware:

A tower assembly containing a 2 mm stepped titanium probe and a pneumatic cylinder, an ultrasonic power supply to provide power to the probe, a control assembly that monitors autosampler arm and travel movement, provides logic for the operation of the pneumatic cylinder, activates ultrasonic mixing, and sets the mixing time, and an adapter box to interface the control assembly with the AS 60/70 electronics and to permit the bypassing of the USS-100 for normal solutions sampling.

Because of the importance of this system, its versatility, and increased use in research and routine analysis, some more details will be given. In Fig. 5.8b, the timing diagram is shown how the USS-100 interfaces with the normal operation of the autosamplers AS-60 or AS-70 to which it fits (see also [587]). The USS-100 must monitor and interrupt, where appropriate, routine autosampler processor control. It achieves this by monitoring several signals sent from the autosampler processor to the autosampler. The figure shows the signals monitored by the USS-100 and the resulting responses made by the USS-100.

Prior to the start of a cycle, both the ultrasonic probe and the sampler capillary reside in the wash port. As the cycle starts, the sample tray rotates if required. A signal is then sent from the processor to raise the sampler arm and the attached capillary from the wash port. When this signal is detected by the USS-100, a sig-

Fig. 5.8. a The USS-1 attached
to a commercial Zeeman-AAS
instrument (PE Zeeman
4100ZL), **b** timing diagramme
of the USS-100, see text (from
[585], with permission)

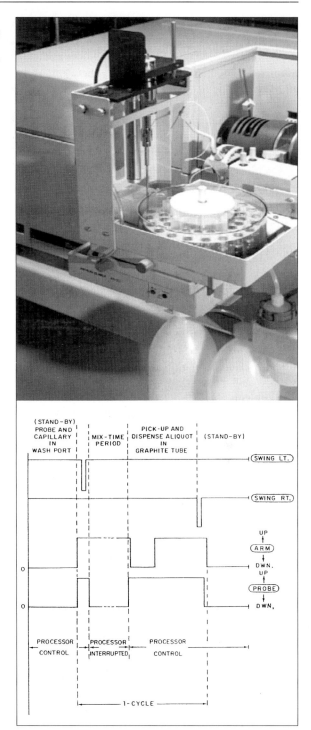

nal is sent to evacuate air from the pneumatic cylinder that raises the probe. A signal from the autosampler processor then directs the autosampler table to swing to the left, so that the desired sample cup is in the correct sampling location. This "swing left" signal is to the operation of the USS-100 (Fig. 5.7b). When it is directed by the USS-100 four responses are initiated:

a) normal processor control of the autosampler is interrupted, leaving the autosampler arm in the "stand-by" position,
b) air is returned to the cylinder that lowers the probe into the sample cup,
c) the ultrasonic power supply is turned on, and
d) a timer is started.

When mixing is completed, the sample is withdrawn from the sample cup and the processor control is returned, the autosampler continues its operation and dispenses an aliquot of the sample into the graphite tube. The cycle is completed by a processor sample table "swing right" signal and subsequent "arm down" signal. The USS-100 monitors these signals and responds by lowering the probe into the wash port.

The preparation of slurries for this system was as follows. 1–2 mg of the material to be analyzed were either weighed directly into the polyethylene autosampler cups or sampled from a vortexed suspension of 10–50 mg [527]. If no significant difference could be observed between these approaches, the simpler method using just weighing into the autosampler cups was used. After weighing a diluent solution (5% HNO_3 and 0.04% Triton X-100) was added to the samples in a weight ratio of approximately 1 part to 500 parts diluent. Matrix modifiers were also added directly to the slurries. It was stated that ultrasonic agitation can prevent larger particles from deposition and that these might be disintegrated in some cases as well [529]. Also the extraction of a number of analytes from the particles was facilitated by ultrasonification.

The use of the USS-100 in combination with Zeeman AAS instruments applying various sample preparation and graphite furnace programs for the determination of a number of elements in environmental and biological materials was shown by several workers [e.g. 530, 564, 588, 539]. Further applications of this system were reviewed and discussed by Miller-Ihli [157, 16, 501, 500].

Richter and Dannecker evaluated the slurry sampling technique for elemental analysis in inorganic solid samples (CRMs from NIST and BCR) in comparison to digestion-GFAAS [565]. The preparation of slurries was as follows. Direct weighing of appropriate amounts of samples into the autosampler cups and addition of 2.5% HNO_3. If the particles were difficult to moisten or not heavy enough additionally a 0.03% Triton X-100 solution was added.

Bendicho and Sancho used a Perkin-Elmer 4100 ZL instrument, fitted with longitudinal Zeeman background correction and a transversely heated graphite atomizer (THGA) and an AS-70 equipped with the USS-100 for the determination of Se in wheat flour [540]. The authors reported that with this instrument under the chosen working conditions the problems associated with vapor phase and spectral interferences could be eliminated.

Krivan and coworkers showed in a series of investigations with high purity industrial (e.g.semiconductor) materials using the 4100 ZL with THGA and the USS-100 that the slurry technique achieves, because of very low blanks, much lower detection limits for a number of important trace metals compared with solution GF-AAS and other trace analytical methods [518,543,607,526,589,571].

5.4
Interference and Overcoming it

Despite the fact that all furnace programs are similar to that used for solutions, it is not possible in many cases to adapt them in exactly the same manner. The reasons are:

- In most cases the slurry contains the elements to be determined in the dissolved as well as in the particulate phase with, at the beginning, unknown data for the extracted part, so that the atomization curves are different to that for the same element in a liquid. This requires the frequent evaluation of a specific temperature program for each indivudual material and always the application of peak area measurements.
- The possibility of introducing higher masses into the furnace for improved detection power or better representativity also requires altered temperature programs.
- Furthermore, the applicability to difficult samples, e.g. ores and dyes, which might have never been analyzed before by the classical approch and thus could cause in addition various problems that must be solved by careful program evaluation in using different m/V-preparations.

From this it is obvious that very often a reliable analytical approach is only possible if the program evaluation is verified by analyte additions and, if appropriate materials are available, by the careful use of certified reference materials (CRMs) in the method evaluation phase.

5.4.1
Matrix Modification

The usefulness of matrix modifiers in slurry-GF-AAS is controversially discussed in the literature. Some examples are given below, a more detailed overview is given in Table 5.6 at the end of this chapter.

Hinds and Jackson used matrix modifiers to overcome interferences due to organic carbon in soil slurries [513]. The same authors compared Pd nitrate and chloride as a chemical modifier for the determination of Pb in solutions and soil slurries and concluded that a mixture of Pd and Mg nitrates had the advantage of being applicable to a variety of analyte elements [557].

Lynch and Littlejohn found Pd modifier to be superior to ammonium dihydrogen phosphate for Pb determination in food slurries [524].

Haraldsen and Pougnet added Mg nitrate as a matrix modifier for the determination of Be in coal slurries to allow ashing temperatures of up to 1800 °C [559].

Ohta et al. determined Cd in various biological materials with a Mo tube electrothermal atomizer. Matrix modification was performed by adding 1 µl aliquots of a sulfur solution (10 mg/ml) in carbon disulfide [534].

Dolinsek et al. determined Cd and Pb in geological and plant materials. The authors reported that, for the analysis of plant materials, an ammonium dihydrogen phosphate modifier was necessary [137].

For the determination of As in different matrices, matrix modification was also employed. Garcia and Cordoba used Ni nitrate for Fe(III)oxide pigments [528], Pd-Mg nitrate was used by Mohl et al. for the analysis of citrus leaves [317] and by Bermejo-Barrera et al. [590] for mussels.

Bendicho and Sancho determined Se in wheat flour and reported the usefulness of a Pd/Mg matrix modifier and isothermal atomization [540].

In another paper, however, Hinds et al. using a fast temperature program for the determination of lead in soil-slurries stated that studies with mixed phosphate and Mg/Pd modifiers indicated that better results could be obtained without the use of these [538]. This study confirmed earlier work on the determination of several elements in coal and fly ash samples and Mn in biological material by fast sampling STPF procedures omitting the pyrolysis step and matrix modification [564, 530].

Miller-Ihli mentioned in a recent paper that often the use of matrix modifiers for slurries appeared to be unnecessary, but that in some cases, from their own experience, higher amounts of matrix modifiers than for aqueous solutions were required to achieve good results [502].

Krivan and coworkers found that for the analysis of technical materials only for a few elements, e.g. for Cr, Cu, Fe, and Mn in silicon nitride [238] and Al, Cd, and Si in zirconium dioxide based materials [526] was the use of matrix modifiers necessary.

Bendicho and De Loos Vollebregt reported the successful use of HF to remove silica as a volatile silicon fluoride by a stream of Ar from slurried glass and quartz samples [535].

López-Garcia et al. applied rapid furnace programs and also the addition of HF for the determination of Cr, Cu, and Pb in slurries of diatomaceous earth [568].

Hoenig and Cilissen showed that the use of rapid furnace programs for several metals in CRMs gave comparable results to conventional approaches using matrix modification, but that in all cases peak evaluation was mandatory [584].

5.4.2
Application of Oxidizing or Reducing Conditions

It has been well known for some time that introducing a reactive gas, either oxidizing or reducing, during the charring step can be successfully used for the minimization or even elimination of matrix interference, if very low metal contents in difficult materials have to be determined. This has been applied for the analysis of liquids by use of oxygen [591,592,593,594,595,596,597] but also hydrogen [598]. This approach was shown to be particularly useful to minimize background signals from mg-amounts of solid samples by use of an air or oxygen charring step [138,50,319,155]. Possible reactions of oxygen were also investigated in some detail [599,600].

These techniques were also applied in slurry GF-AAS in order to minimize matrix interference during atomization.

Ebdon and coworkers reported that an air ashing step for the determination of Mn, Cr, Co and Pb in biological materials and sewage sludge was similarly useful as the addition of modifiers [553].

Ebdon and Parry studied the determination of Se in coal using an AAS instrument with Smith-Hieftje background correction. The observed interference from Fe could be only partly overcome by the background correction system as well as by the use of matrix modifiers. By the introduction of air ashing in combination with a carefully optimized furnace program, the interference was completely eliminated [515].

Ebdon and coworkers used hydrogen in a graphite furnace to decrease the background signal arising from slurries of soil and sewage sludge samples for the determination of Cr, Mn and Pb. It was reported that the use of hydrogen was more effective for Cr in soil than the usual air or oxygen ashing step [546].

5.5
Methodological Features in Comparison

5.5.1
Slurry Sampling Versus Direct Solid Sampling

In Table 5.4 advantages and limitations of slurry sampling in comparison to direct solid sampling are listed.

In comparison to direct solid sampling, it is obvious from Table 5.4 that the most important advantages of slurry sampling are:

- Use of higher representative sample masses in between digest and direct solid sampling GF-AAS.
- Applicability to practically all GF-AAS systems, if manual injection is performed and the easier automated operation which has already led to a quite good, but still not perfect, commercially available automated system and simple, but working laboratory-made systems.
- Easy dilution and more convenient handling of the atomization process by matrix modification so that calibration might be simpler in many cases, leading to accurate results, however, within the precision limits of the method.
- Analysis of refractory elements can be relatively easily performed with the use of modern very small tubes.

Table 5.4 Comparison of typical properties of slurry versus solid sample analysis

slurry analysis	direct solid sample analysis
Effective injected mass typically 0.02 mg	Test sample mass typically 0.1-1 mg
Representative mass up to 50 mg (due to analyte extraction)	Representative mass up to 10 mg (number of replicates times mass of test samples)
Normal AAS instruments applicable	Special furnace-systems (tubes, boats) required
Automation commercially available, in laboratory made systems relatively simple to achieve	Complete automation difficult; at present a semi-automated device commercially available
Dilution easy	Dilution already senseless (equivalent to reduction of test sample mass)
Matrix modification easy	Matrix modification problematic (except graphite powder)
Thoroughly ground samples only (e.g. bits, lumps, pieces)	More types of samples possible
Elevated contamination risk due to the use of acids, other chemicals and matrix modifiers	No or very low contamination risk if no reagents are used
Mass of the laboratory sample at the mg-level required	Extremely low laboratory sample mass and single samples (organisms, parts of plants etc.) at the µg-level applicable
Homogeneity determination only advantageous with increased efforts	Homogeneity determinations easy with direct solid sampling
Limits of analysis comparable to these of digestion analysis, in some cases somewhat lower	Considerable lower limits of analysis attainable (for the real solid sample)

Limitations compared to direct solid sampling are:

- Increased risk of contamination by the use of reagents.
- Reduced "detection power" because of the dilution of the pulverized sample by the liquid medium.
- Very small and/or single samples such as small single particles, insects etc cannot be analyzed.
- Due to the higher sample mass requirements, analyte distributions in biological structures such as plants, technical and art objects cannot be studied in detail.
- For homogeneity studies there are advantages for the direct approach as far as simplicity and speed are considered. The higher intakes (representative masses) of 10–50 mg for homogeneity determinations by slurries, however, might be of advantage if practically complete extraction occurs for the studied analyte.

- From this it might be stated that the direct analysis of solid samples in principle is more versatile and unrivaled for a number of very valuable studies in biological and environmental research, but less suited for routine tasks for which the slurry approach is often more convenient. Thus, both approaches might ideally complement each other if available in the same laboratory.

5.5.2
Slurry Sampling Versus Sampling of Digested Samples

In Table 5.5 the properties of slurried samples in comparison to digested (liquid) samples are summarized.

From Table 5.5 it is obvious that, compared to classical decomposition-GF-AAS determination, the *advantages* of the slurry technique are:

- *Quick and comparatively inexpensive analysis* of unknown samples or of samples with limited amount of substance.
- Determination of trace elements in *difficult to decompose materials* like ores, silicates etc. and in materials with low elemental contents if digestion is prone to contamination

As far as *detection limits* in general are concerned, these are usually at the same order of magnitude as published for graphite furnace techniques in analyte solutions [601, 602, 23, 603, 604, 605] sometimes even better, due to the possibility of the introduction of relatively high sample masses for particular materials, thus permitting high m/V-ratios and the practical absence of contamination if only minimal reagent amounts are used and no additional grinding is necessary.

Limitations of the slurry technique are:

- The often necessary *vigorous grinding might introduce contamination* for particular trace metals or just problems with effective grinding if the material is not homogeneous in density and further physical differences (hardness) in sample constituents occur.
- Despite the fact that, in principle, the general application of all matrix modification and graphite furnace techniques already useful for liquids is possible, more complex atomization processes often require significant program alterations due to the fact that the *elements usually appear in both phases,* in the liquid (extracted) as well as – adsorbed or occluded – in the solid phase.
- *Technical limitations* because often still not optimized software and hardware might introduce problems also in the already commercially available automated systems, but particularly those made in the laboratory which require longer analysis time.In addition the still, for many GF-AAS-systems necessary, rather difficult manual slurry introduction makes the technique sometimes slow, prone to sedimentation errors and cumbersome.

Table 5.5 Comparison of typical properties of slurry versus digested samples analysis

slurry analysis	analysis of digested samples
Usually minimal requirements for chemicals and additional equipment	Laboratory equipment and stock of reagents considerable and expensive
Relatively small laboratory sample mass sufficient;	Higher laboratory sample mass for decomposition usually required
Limits of determination generally comparable, significantly lower at higher m/V-values and for difficult to digest materials	Worse limits of determination, if complete decomposition of samples cumbersome and often prone to contamination for low contents of ubiquituously distributed elements
Relatively fast analysis of unknown materials	Larger time consumption and running costs
Fine grinding (<100 μm) often necessary for optimal results	Fine grinding not necessary
Furnace program steps often more complex	Furnace programs generally more simple
Contamination possible if additional and vigorous grinding procedures are necessary	Contamination risk lower, because of no need for special grinding
Automated devices not for all systems and still not perfect	Automation has reached a high level in the entire process
"In-cup"-precision frequently >5% (a larger no. of injections sufficient)	"In-cup"-precision commonly <<5% (small no. of injections required)

5.6
Applications

In Table 5.6 examples for the determination of 30 elements are given in alphabetical order of elements and of groups of materials. The table includes, in addition to the abbreviation for the matrix group, a description of the investigated material, of the methodological approach (manual or automated) and instrumentation, on the concentration level or range and the reported standard deviation. The RSD is only an informative value and does not differentiate between in-slurry and between-slurry precision, because this is often not stated in the literature and because of very variable numbers of individual measurements. This is followed by information about the applied modifiers and calibration procedures, particular features of the method, and the references.

Table 5.6 Examples for the determination of elements (Al, As, Au, B, Be, Ca, Cd, Co, Cr, Cu, Fe, Ga, Hg, K, Li, Mg, Mn, Mo, Na, Ni, Pb, Pd, Se, Si, Sn, Ti, Tl, V, Zn) in different materials

Groups of considered materials with codes in the table in the following order: *biological* (**B**): plant materials, animal tissues, food, clinical samples; *environmental* (**E**): coal, ash, sludge, waste, sediment; *geological* (**G**): minerals and soils; *polymer* (**P**); *industrial* (**I**): Pharmaceutical, cement, rubber etc. The given RSD (either from only a few or a larger number of analyses) in many cases is worse than the accuracy compared with the values found for certified RMs (CRMs), see remarks. In this table a number of original papers will be cited under different elements and matrices. Thus, the first appearance in the table will contain detailed technical information. Only additional or absolutely necessary information will be given under the additional elements and matrices with reference to the first appearance in this table. Peak area (integrated absorbance) is always used for concentration evaluation. (Abbreviation: BC = background correction, AAM = analyte addition method)

Element	Matrix Code and Matrices	Instrumentation	Conc. (range) and typical RSD (range)	Slurry Approach and Matrix Modification (M:); Calibration (C:)	Special Features/Remarks	References (Literature, Analyte/ Matrixcode)
Al	**B:** Four NIST SRMs	Automated, ultrasonic, simultaneous multielement GF-AAS (SIMAAC) system	<10–10³ mg/kg RSD <25%	M: HNO₃ plus Triton X-100 C: Aqueous ref. solns.	First study with an automated ultrasonic system included in sample tray	[527]
	B: Vegetables	Manual, dried, ground samples with ethanol, P-E 1100B with HGA-400, wall atom.; D₂–BC	6.5–182 mg/kg RSD around 5%	M: H₂O₂-Nitric acid C: Aqueous ref. solns.	Results validated by CRMs and solution analysis. Fast furnace program with modified drying step	[606]
	E: Suspended matter from estuarine and sea water	Automated, ultrasonic, Varian SpectrAA-400 Z, GTA-96; programmable sample dispenser	3.75 mg/kg–6,5% RSD up to 4,4%	M: - C: Aqueous ref. solns.	Use of a less sensitive line and reduced atomization rate	[562]
	I: Silicon carbide	Automated, ultrasonic; magnetic stirring on sample tray; P-E 5000 Z; HGA-500 and AS-40	177–480 mg/kg RSD up to 10%	M: Mg nitrate C: AAM	Validation by analysis of decomposed samples with other methods	[607]

Table 5.6 Continued

Element	Matrix Code and Matrices	Instrumentation	Conc. (range) and typical RSD (range)	Slurry Approach and Matrix Modification (M:); Calibration (C:)	Special Features/Remarks	References (Literature, Analyte/Matrixcode)
	I: Silica gel	Automated, ultrasonic, Zeiss AAS-3, EA-3 and MPE autosampler; D_2-BC	up to 570 mg/kg RSD <13%	M: - C: AAM	Time between mixing and sampling 5 to 8s; values decrease after standing times of around 100 s	[517]
	I: Silicon nitride	Automated,ultrasonic, USS-100; P-E 4100 ZL with THGA, AS-70	11.6 + 380 mg/kg RSD 5.3 + 8.6%	M: - C: Aqueous ref. solns.	Comparison with other methods; the fine powder with particle size <1 μm required no grinding. Very low blanks for all investigated elements	[238]
	I: Zirconium dioxide based materials (ZrO_2 and ZrO_2 containing 3 mol% Y_2O_3)	Automated, ultrasonic (USS-100), P-E 4100 ZL with THGA, AS-70	6 and 25.6 mg/kg RSD 6.7 and 22%	M: H_3BO_3 in the presence of HF C: Aqueous ref. solns.	Comparison with other methods; very low blanks for all investigated elements. Detection limit for Al a factor of 7 lower compared to solution GF-AAS. Also radiotracer studies.	[526]
	I: High purity quartz	Automated, USS-100, P-E 4100 ZL, THGA and AS-70; work in a clean bench	3300 mg/kg RSD 12%	M: W carbide modified tubes C: AAM	Validation by independent methods; well suited method for ultratrace analysis of high purity quartz for microelectronic applications	[589]
	I: Three graphite powders	Automated, USS-100, P-E 4100 ZL, THGA and AS-70; no further grinding	5.0–30 mg/kg RSD 8–14%	M: 0,1 mol/l nitric acid and 0.008% Triton X-100 C: Aqueous ref. solns.	Collaborative study;good agreement with NAA and various other methods; median particle size of about 15 μm	[571]

As					
B: NIST SRM Citrus Leaves	Automated, USS-100; P-E Z5100, HGA-600, AS-60	3.1 mg/kg RSD -	M: Pd and Mg nitrates C: -	Comparative homogeneity study between slurry and direct solid sampling with Gruen SM 20	[567] [317]
B: Mussel tissue; BCR CRM 278 Mussel Tissue for QC	Automated, magnetic, P-E 1100B, HGA-700, AS-70; D_2- BC, grinding with zirconia beds to a particle size <10μm	2–9.3 mg/kg RSD <4–16%	M: Pd nitrate plus Mg nitrate C: AAM	Air ashing at 480°C and at 1200°C in argon. Detection limit of the final slurry solution 1 μg/L (1.3 mg/kg)	[590]
E: Selected coal samples including NIST SRM 1632a	Manual, P-E 460, HGA 76b; D_2- BC	1.2–96 mg/kg RSD around 5%	M: Ni and Mg nitrates C: Coal RMs	Detection limit <0.1 mg/kg	[520]
E: Three whole coal CRMs	Manual,Video 12 AAS, and IL-655 furnace, with D_2 and Smith-Hieftje BC	0.4–9.5 mg/kg RSD 7–25%,	M: Ni nitrate C: Aqueous ref. solns.	Study on elimination of interferences; only possible with Smith-Hieftje BC	[578]
E: NIST SRM River Sediment	Automated, ultrasonic (prototype), P-E Zeeman 5100 PC, HGA-600 and AS-60	21.7 mg/kg RSD <3%	M: Various modif. mixtures C: Aqueous ref. solns.	Test of different slurry prep. methods; variability increases with analyte extraction	[563]
E: NIST SRMs Coal Fly Ash and Trace elements in coal	Automated, ultrasonic, USS-100, P-E Zeeman 5100 PC, HGA-600 and AS-60	ca 10 and 140 mg/kg RSD <10%	M: - C: Aqueous ref. solns.	Fast sampling STPF procedures; charact. mass 14.8 pg/0.0044 A's; blank negligible	[564]
E: BCR CRMs Lake Sediment and River Sediment	Automated, ultrasonic, USS-100, P-E Zeeman 5100 PC, HGA-600 and AS-60	58.6 and 80.4 mg/kg RSD <5%	M: (Only for As) Pd/ Mg C: Aqueous ref. solns.	Ultrasonification system not powerful enough for these materials	[539]

Table 5.6 Continued

Element	Matrix Code and Matrices	Instrumentation	Conc. (range) and typical RSD (range)	Slurry Approach and Matrix Modification (M:); Calibration (C:)	Special Features/Remarks	References (Literature, Analyte/ Matrixcode)
	E: Contaminated soils and sediments	Automated, ultrasonic, USS-100, P-E 4100ZL and AS-70	11 µg/l RSD 13.6%	M: Pd/ Mg nitrates C: Ref. samples and ref. solns.	Validation with CRMs and by interlaboratory comparisons	[609]
	E: Marine sediment samples	Automated, vibromatic and magnetic agitation; P-E 1100 B; HGA-400 and AS-40	25.4–90.7 mg/kg RSD	M: Triton X-100, Pd- Mg nitrates C: AAM	Speciation study on total As and As(III). Detection limit for total As 44 µg/kg	[610]
	E: Ten marine sediments	Automated, vibromatic and magnetic agitation, P-E 1100B, HGA-400 and AS-40; samples ground to particle size <250 µm	1.2–10 mg/kg RSD around 3 %	M: Pd/ Mg nitrate C: Aqueous ref. solns.	Accuracy and precision tested with the sediment CRM, PACS-1	[611]
	I: Commercial Fe(III)-oxide pigments	Manual, magnetic, P-E 1100B and HGA-400; D_2-BC	8.4–20.2 mg/kg RSD around 10%	M: Ni and Triton X-100 C: Aqueous ref solns.	Detection limit of 0.2 mg/kg reduced to 0.05 mg/kg if preconcentration is used	[528]
Au	B: Bacterial cells	Automated, ultrasonic, Thermo Jarrell Ash SH-11; CTF-188; FASTAC nebulizer; S/H-BC	up to 2 ng/ml RSD up to 7%	M: - C: by bacterial samples	Experiments to accumulate gold by bacterial cells, tested with CRMs	[612]
B	B: Various Plant CRMs	Manual, mechanical agitation, Pye Unicam PU SP9, PU 9095 furnace; continuum source-BC	10–48 mg/kg RSD 5–20%	M: Ni, Mg and La nitrates C: Aqueous ref. solns. plus modif.	Samples were charred at 250 °C prior to slurry preparation	[551]

Element	Sample	Method	Result / RSD	M: / C:	Comments	Ref.
	B: Suspensions of cell cultures	Automated, magnetic on sample tray; P-E 4000, HGA-500 and AS-1; D_2-BC	up to 50 µg B/ml RSD <10%	M: - C: AAM	Larger tubes used for better passage of cell suspensions; det. limit 0.05 mg/L	[580] [613]
Be	B: Bacterial cells		up to 22 ng/ml RSD around 3%	M: - C: by bacterial samples	Experiments to accumulate Be on bacterial cells, checked with CRMs	612] see Au/B [614]
	E: Three South African coal CRMs	Automated, P-E 5000, HGA-500 and modified AS-40; D_2-BC	2.35-4.15 mg/kg RSD <7%	M: mg C: Aqueous ref. solns.	Method less tedious and more cost effective than ICP-AES	[559]
Ca	B: Four NIST SRMs		100 mg/kg to % RSD up to 2%	M: - C: Aqueous ref. solns.		[527] see Al/B
	I: Aluminium oxide	Manual, magnetic stirring; P-E 420 with HGA-74; D_2-BC		M: - C: Ref. samples and ref. solns.	Feasibility study	[615]
	I: High purity Mo trioxide	Automated, magnetic stirring on sample tray, P-E 5000 Zeeman, HGA-500 and AS-40	95 µg/kg RSD 15%	M: - C: AAM	Detection limit 2 µg/kg because of extremely low blanks	[541]
	I: Zirconium dioxide based materials		1 and 136 mg/kg RSD 15 and 20%	M: - C: Aqueous ref. solns.	Detection a factor of ten lower compared with solution GF-AAS.	[526] see Al/I
	I: Three graphite powders		6.5-98 mg/kg RSD 3-5%	M: - C: Aqueous ref. solns.		[571] see Al/I

Table 5.6 Continued

Element	Matrix Code and Matrices	Instrumentation	Conc. (range) and typical RSD (range)	Slurry Approach and Matrix Modification (M:); Calibration (C:)	Special Features/Remarks	References (Literature, Analyte/ Matrixcode)
Cd	**B:** Dried food samples; NIST SRM Bovine Liver for QC	Automated, magnetic; IL Video II, IL 655 graphite furnace; Fastac 254 autosampler	0.32–3.76 mg/kg RSD 3–30%	M: - C: Aqueous ref. solns.	In situ ashing; excellent recovery of Cd in bovine liver	[522]
	B: NIST plant SRMs, Bovine Liver and IAEA fish homogenate	Automated, Mini-Mix stirrer; Varian AA-1275 BD, GTA-95 graphite furnace; programmable sample dispenser; D₂-BC	28–300 µg/kg RSD up to 12%	M: glycerine, NH₄ phosphate and Mg nitrate C: Aqueous ref. solns.	Homogeneity satisfactory if a ratio of 1:1 glycerine: demineralised water is used	[576]
	B: BCR Milk Powder and NIST Rice Flour SRMs	Automated, ultrasonic (prototype), P-E Zeeman 3030, HGA-600 and AS-60	19 and 27 µg/kg RSD 5 and 15%	M: Pd- Mg C: Aqueous ref. solns.	Basic study; good agreement with certified contents	[529]
	B: Food samples and appropriate CRMs for QC	Automated, mechanically shaken; P-E 5000 Zeeman, HGA-400 and AS-40	20–300 µg/kg for CRMs RSD 10–25%	M: Pd C: Aqueous ref. solns.	Detection limit for a 50 mg/ml slurry 10 µg/kg	[577]

B:	Method	Conc. / RSD	M: / C:	Remarks	Ref.
B: NIST and NIES plant and liver CRMs	Manual, ultrasonic, metal tube atomizer (Mo); D$_2$-BC	0.1-6.7 mg/kg RSD <10-40%	M: Sulfur C: Aqueous ref. solns.	Absolute detection limit (0.31 pg) similar to or better than those of solution GF-AAS methods	[534]
B: NIST plant SRMs	Manual, ultrasonic; lab. assembled AAS system; D$_2$-BC	0.03-10 mg/kg RSD around 10% if finely ground	M: (NH$_4$)HPO$_4$ C: Aqueous ref. solns.	Comparative study: direct versus slurry sampling; ultrasonic homogenization preferred over vortex mixing.	[137]
B: 2 Tobacco Leaves candidate RMs	Automated, ultrasonic in autosampler vessels, Zeiss AAS-3, EA-3 furnace, MPE autosampler; D$_2$-BC	1.3 and 1.7 mg/kg RSD <10%, for slurry better than for direct sampling	M: Pd for Cd only C: AAM	Comparison of direct solid sampling, slurry sampling and determination after wet digestion.	[240]
B: NIST SRMs, OTL RMs	as above	0.08-1.28 mg/kg RSD <10-30%	M: Pd, for Cd only C: AAM	Good agreement with data from wet digestion; slurry stability and homogeneity study	[516]
B: Beech Leaves RM		0.1 mg/kg	M: Pd/ Mg C: -		[567] [317] see As/B
B: Biscuits, bread and cereal based products	Manual, dried and ground samples with magnetic stirring; P-E 1100B HGA-400, D$_2$-BC	<0.02 -0.03 mg/kg RSD 2.5-6.5%	M: addition of ethanol/H$_2$O$_2$, Pd and Cu plus NH$_4$OH C: Aqueous ref. solns.	Confirmation by analysis of 2 CRMs; due to the high solid content a conventional furnace program was used, detection limit 0.5 µg/kg	[617]
B: Vegetables	Manual, dried and ground samples,Ultra-Turrax and ultrasonification; P-E 1100B with HGA-400; D$_2$-BC	very low 0.01 to 0.1 mg/kg RSD 5-20%	M: addition of ethanol, H$_2$O$_2$ and (NH$_4$)H$_2$PO$_4$ C: Aqueous ref. solns	Confirmation by analysis of two CRMs; fast furnace program by replacing the drying and ashing step by a single modified drying step	[618]

Table 5.6 Continued

Element	Matrix Code and Matrices	Instrumentation	Conc. (range) and typical RSD (range)	Slurry Approach and Matrix Modification (M:); Calibration (C:)	Special Features/Remarks	References (Literature, Analyte/ Matrixcode)
	B: Six biological CRMs, frozen cervine liver and kidney	Automated, samples, blending with tissue mixer, flat valve homogenizer; Varian 300 GF- Zeeman AAS with autosampler; hot injection	0.3–16.5 mg/kg RSD 2.3–8% AAM	M: ethanol, water, tetramethyl ammonium hydroxide, $(NH_4)_2HPO_4$ C: AAM and external standards	First approach with this technique; acceptable agreement with certified values	[619]
	E: Various sewage sludges	Manual, Ultra-Turrax, P-E 305, HGA 72; D_2-BC	0.06–1.8 mg/L RSD 8%	M: - C: AAM	Comparison with wet digestion/flame AAS; "fast method" detection limit 0.1 µg/L	[560]
	E: Estuarine sediment		0.42 mg/kg RSD 7.2%	M: NH_4 phosp. and Mg nitrate C: Aqueous ref. solns.		[576] see /B
	E: Filter dust			M: Pd/ Mg C: Aqueous ref. solns.		[529] see /B
	E: Estuarine, Lake, and Marine Sediment, CRMs	Automated, mini-mix-stirrer; Varian SpectrAA-400, GTA-96 and programmable dispenser	0.28–9 mg/kg RSD 2.1–5%	M: glycerol, Pd/ Mg and 0.05% of matrix elements C: Aqueous ref. solns., glycerol and mixed modif.	Automated sequential multielement analysis feasible with this special modifier	[575]
	E: Airborne particulates and NIST SRM Urban particulate matter for QC	Automated, ultrasonic, Jarrel Ash 83000 with HGA-400, AS-40 and a lab. constructed line reversal BC system	filters: 1–9 ng/m^3 solids: 3.6 + 12.6 mg/kg NIST SRM 1648: 75 mg/kg RSD 2–4%	M: - C: Aqueous ref. solns.	Good agreement for the NIST SRM as well as between slurry and solution GF-AAS	[537]

Sample	Technique / Instrument	Concentration / RSD	Modifier / Calibration	Comments	Ref.
E: NIST SRMCoal Fly Ash and River Sediment		1.0–9.6 mg/kg RSD ≤10%	M: - C: Aqueous ref. solns.		[137] see /B
E: Real sediment samples and two NIST SRMs for QC	Automated, various mixing techniques, Varian SpectrAA-20, GTA-96, D$_2$-BC	RSD varied from 2.4 to 9%; error: 0 to 10.7%	M: Pd in dilute HCl C: Aqueous ref. solns. and CRM reference slurries	Comparative study of homogenization techniques; ultrasonification optimal	[239]
E: BCR sediment CRMs		0.48 + 1.68 mg/kg RSD 5.4 + 10.4%	M: - C: Aqueous ref. solns.		[539] see As/E
E: 14 diatomaceous earth samples; NIST SRM Coal Fly Ash	Manual, ultrasonic and magnetic; P-E 1100B, HGA-400; D$_2$-BC	0.02–0.2 µg/kg RSD <10%	M: addition of HF, C: Aqueous ref. solns.	Fast temperature program	[525]
E: Contaminated soils and sediments		0.75 µg/l RSD 2.21%	M: - C: Ref. samples and ref. solns.		[609] see As/E
E: Ten marine sediments		13–479 µ/kg around 3%	M: Pd/ Mg nitrate C: Aqueous ref. solns.		[611] see As/E
G: Sandy soils and clay soils	Manual, magnetic, IL 157 and P-E 2280 with electrothermal atomizers, D$_2$-BC	0.47–99 mg/kg RSD <4%	M: - C: Aqueous ref. solns.	Grinding to small particle size («30µm); good agreement of data with wet digested samples	[109]
G: IAEA CRM Soil-7		1.14 mg/kg RSD 1.5%	M: Mixed modifier C: Aqueous ref. solns. with modif.		[575] see /E

Table 5.6 Continued

Element	Matrix Code and Matrices	Instrumentation	Conc. (range) and typical RSD (range)	Slurry Approach and Matrix Modification (M:); Calibration (C:)	Special Features/Remarks	References (Literature, Analyte/ Matrixcode)
	G: IAEA CRM Soil-7		1.2 mg/kg RSD <10%	M: - C: Aqueous ref. soln.		[137] see /B
	I: Silicon carbide	Automated	<0.02 mg/kg	M: Mg nitrate C: AAM		[607] see Al/I
	I: Commercial silicagel samples		20 and 60 µg/kg RSD ≤10%	M: - C: AAM	Included slurry stability study showed an almost complete extraction of Cd	[517] see Al/I
	I: Zirconium dioxide based materials		<0.05 mg/kg	M: $(NH_4)_2$ HPO_4 plus $Mg(NO_3)_2$ C: Aqueous ref. solns.	Detection limit a factor of 5 higher compared with solution GF-AAS	[526] see Al/I
Co	B: Algae (*Chlorella*)	Automated, Varian AA-875, GTA 95 furnace and autosampler	typical recovery for Co: 73%	M: nitric acid suspension C: Aqueous ref. solns.	Study on the preconcentration of Co and Ni from seawater and riverine water by algae	[532]
	B: CRMs: Tomato Leaves, Human Hair, Mussel, Chlorella	Manual, milling with zirconia beds, Philips PU9095 and PU9390 furnace, D_2-BC	0.08–0.9 mg/kg RSD 3–38%	M: - C: Aqueous ref. solns.	Temp. program included air ashing < 600 °C, hair samples were ashed in a muffle furnace prior to analysis;	[553]
	B: 2 Tobacco Leaves candidate RMs		0.31 + 0.9 mg/kg RSD 6.4 + 9.7%,	M: - C: AAM		[240] see Cd,B

Sample	Instrument / Method	Concentration / RSD	Modifier / Calibration	Remarks	Ref.
B: Plant leave CRMs		0.08–0.79 mg/kg RSD 7.6–40%	M: - C: Aqueous ref. solns.	Co appeared to be somewhat heterogeneously distributed	[516] see Cd/B
E: Sediment CRMs:		10.7–22.3 mg/kg RSD 5.5–18.1%	M: Mixed modifier C: Aqueous ref. solns. plus modif.		[575] see Cd/E
E: BCR CRM Sewage sludge		8.63 mg/kg RSD 7.2%	M: - C: Aqueous ref. solns.	Good agreement with certified value	[553] see /B
G: Silicate rocks	Manual, rod stirring, Viscalex, P-E 306 and HGA-74	<10 -ca 110 mg/kg RSD up to 20%	M: Na-phosphate C: Aqueous ref. solns. if particle size is <25 μm	Comparison of flame, GF-AAS and ICP-AES; GF-AAS agreed quite good with certified values	[143]
G: IAEA Soil-7		9.0 mg/kg RSD 1%	M: Mixed modif. C: Aqueous ref. solns. plus modif.		[575] see Cd/E
I: NIST Glass SRMs and a quartz sample	Automated, P-E Zeeman 5000, HGA-500 and AS-40	5.5–36.1 mg/kg RSD 2.4–11%	M: glycerol or Ar stream C: Aqueous ref. solns.	Pyrolysis and matrix modification had no influence and thus were omitted	[497]
I: NIST Glass SRMs and a quartz sample	Automated, P-E Zeeman 5000, HGA-500 and AS-40	0.78–399 mg/kg RSD 2.8 -10.3%	M: HF HF and Ar stream combined C: Aqueous ref. solns.	HF treatment leads to almost complete extraction and improved precision	[535]
I: High purity Mo trioxide		<0.05 mg/kg	M:- C: AAM	Detection limit for integrated absorbance 0.05 mg/kg	[541] see Ca/I
I: Zirconium dioxide based materials		<0.6 mg/kg	M: - C: Aqueous ref. solns.		[526] see Al/I

Table 5.6 Continued

Element	Matrix Code and Matrices	Instrumentation	Conc. (range) and typical RSD (range)	Slurry Approach and Matrix Modification (M:); Calibration (C:)	Special Features/Remarks	References (Literature, Analyte/Matrixcode)
Cr	B: Ten pine needle samples and NIST Pine needle SRM	Manual, ultrasonic, Varian Techtron AA6 and Jarrel Ash 8300; HGA-2100 and CRA-90; D_2-BC	8.1–14 mg/kg RSD 2–6%	M: - C: NIST SRM slurries	Original samples were calcinated for 4 h up to 400 °C; good agreement with solution GF-AAS	[552]
	B: Various biological materials			M: - C: Aqueous ref. solns.	The negligible solubility of Cr did not favour the use of slurries	[529] see Cd/B
	B: Various biological CRMs; human hair:		0.56–4.4 mg/kg RSD ≤10%	M: - C: Aqueous ref. solns.	Air ashing; good agreement with certified values	[553] see Co/B
	B: Sweets and chewing gum	Manual, calcination, ultrasonification, mechanical stirring; P-E 1100B; HGA-400; D_2-BC	0.15–0.7 mg/kg RSD <8%	M: suspension: 8% ethanol, phosphate and H_2O_2, $(NH_4)H_2PO_4$ C: Aqueous ref. solns.	Results validated by analysis of several CRMs and solution analyses	[620]
	B: Vegetables		0.15–1.97 mg/kg RSD around 4%	M: H_2O_2-nitric acic		[606] see Al/B
	B: Seafood, vegetables and legumes	Manual, dried material ground, magnetic stirring; P-E 1100B; HGA-400, D_2-BC	0.04–0.64 mg/kg RSD up to 20%	M: Triton X-100, H_2O_2-nitric acid C: Aqueous ref. solns.	Validation by CRMs and conventional analysis. Fast furnace program, calcination of material prior to Co determination. Detection limit 4 µg/kg	[621]

Sample	Preparation	Conc./RSD	Modifier/Calibration	Comments	Ref.
E: Sewage sludges	Manual, Ultra-Turrax with Ti shaft; P-E 305, HGA 72; D_2-BC	0.5–6.0 mg/l RSD <10%	M: - C: AAM	Good agreement of digested samples with flame AAS:	[561]
E: Sediments		62–135 mg/kg RSD 3.3–21%	M: Mixed modif. C: Aqueous ref. solns. plus modif.	Cr possibly insoluble in the studied materials	[575] see Cd/E
E: 2 NIST Coal SRMs		2.5 + 33.6 mg/kg RSD 5 + 40%	M: Pd/ Mg nitrates C: Aqueous ref. solns.		[529] see Cd/B
E: Sewage sludge		121 mg/kg RSD 6.5%	M: - C: Aqueous ref. solns.		[553] see Co/B
E: BCR sediment CRMs	Automated	104 + 120 mg/kg RSD up to 25%	M: - C: Aqueous ref. solns.	Values in between certified aqua regia value and the total content	[539] see As/E
E: BCR 145 Sewage sludge	Manual, shaking with zirconia beds; Pye Unicam PU 9100x, ET atomiser PU9390x, D_2-BC	18 mg/kg RSD 11%	M: - C: Aqueous ref. solns.	Hydrogen (10% v/v) in the cool-down and autozero step effective for elimination of non-specific absorbance	[546]
E: Contaminated soils and sediments		13.3 µg/l RSD 4.5%	M: Mg nitrate C: Ref. samples		[609] see As/E
G: 5 rock samples	Manual, Viscalex; P-E 360, HGA 74		M: Na-phosphate C: Aqueous ref. solns. + Viscalex	First paper on the use of a thickening agent for slurry preparation	[519]
G: Silicate rocks		<5–ca 250 mg/kg RSD up to 20%	M: Na-phosphate C: Aqueous ref. solns. if particle size <25 µm		[143] see Co, G

Table 5.6 Continued

Element	Matrix Code and Matrices	Instrumentation	Conc. (range) and typical RSD (range)	Slurry Approach and Matrix Modification (M:); Calibration (C:)	Special Features/Remarks	References (Literature, Analyte/ Matrixcode)
	G: IAEA Soil-7	62.1 mg/kg RSD 4.7%		M: Mixed modifier C: Aqueous ref. solns.		[575] see Cd/E
	G: CANMET SO-2 soil	18 mg/kg RSD 11%		M: - C: Aqueous ref. solns.		[546] see /E
	G: 14 Diatomaceous earth and 6 silicate CRMs	Manual, ultrasonic, followed by magnetic; P-E 1100B; HGA-400, D_2-BC	2.1-325 mg/kg RSD 2-34%	M: HF added C: Aqueous ref. solns.	Drying step at higher than usual temperature, ashing and clean-up steps omitted	[568]
	I: Aluminium oxide			M: - C: Ref. samples and ref. solns.		[615] see Ca/I
	I: commercial glass materials and CRMs			M: - C: Aqueous ref. solns.		[497] see Co/I
	I: Silicon carbide		2.6-18.5 mg/kg RSD up to 10%	M: Mg nitrate C: AAM		[607] see Al/I
	I: High purity Mo trioxide		<0.05 mg/kg	M: - C: AAM	Detection limit 50 µg/kg.	[541] see Ca/I
	I: Silicon nitride		$0.85 + 3.8$ mg/kg RSD 2.6 + 4,7%	M: $(NH_4)H_2PO_4 + Mg(NO_3)_2$ C: Aqueous ref. solns.	Detection limit approx. 11 times lower compared with solution GF-AAS	[238] see Al/I

	Sample	Method/instrument	Concentration / RSD	M: / C:	Remarks	Ref.
	I: Zirconium dioxide based materials		<0.6 and 2.8 mg/kg	M: - C: Aqueous ref. solns.	Detection limit a factor of two lower compared with solution GF-AAS	[526] see Al/I
	I: High purity quartz		3.7 ± 0.6 mg/kg	M: W carbide modified tubes C: Aqueous ref. solns.		[589] see Al/I
	I: Three graphite powders		0.7–3.0 mg/kg RSD 3.3–14%	M: - C: Aqueous ref. solns.		[571] see Al/I
Cu	B: Aquatic plant and various biological CRMs	Manual, ultrasonic; IL Video 12 with IL 655 graphite furnace, Smith-Hieftje-BC	1.2–193 mg/kg RSD 4.9–20%	M: - C: Aqueous ref. solns.	Comparison between slurry and solid sampling; the latter was also used to determine the location of trace metals	[322]
	B: Seven NIST SRMs		<5–300 mg/kg RSD <10%	M: - C: Aqueous ref. solns.		[527] see Al/B
	B: Chaparro, water lettuce, tomato and tobacco; NIST SRMs	Automated, ultrasonic; Varian 875 GTA-95 furnace and PE AS-1 autosampler, D_2-BC	2.6–19.7 mg/kg RSD 1.9–6%	M: - C: CRMs	Good agreement between determination after wet digestion and slurry injection. The latter was more precise	[498]
	B: Pine needles		6.5–12.3 mg/kg RSD 3–5 %	M - C: NIST pine needle slurries	Calcination up to 400 °C	[552] see Cr/B
	B: Algae	Automated, Varian AA-875, GTA-95 and autosampler		M: nitric acid suspension C: Aqueous ref. solns	Study on the preconcentration of copper from seawater and riverine water by algae	[531]
	B: Food and food CRMs	Automated, ultrasonic, USS-100, SIMAAC and P-E Zeeman 5100 PC and autosampler	0.88–3 mg/kg RSD 2.5 –13%	M: - C: Aqueous ref. solns.	Evaluation of the slurry technique; basic data on physical parameters; foods homogenized prior to analysis with PTFE beds	[501]

Table 5.6 Continued

Element	Matrix Code and Matrices	Instrumentation	Conc. (range) and typical RSD (range)	Slurry Approach and Matrix Modification (M:); Calibration (C:)	Special Features/Remarks	References (Literature, Analyte/ Matrixcode)
	B: UPB Beech Leaves RM		9.1 mg/kg	M: Pd and Mg nitrates C: -		[567], [317] see As/B
	B: Sweets and chewing gum		0.2–2.3 mg/kg RSD <8%	M: $NH_4H_2PO_4$ C: Aqueous ref. solns.		[620] see Cr/B
	B: Biscuits and bread	Manual, ultrasonification; magnetic stirring; P-E 1100B; HGA-400; D_2-BC	1.06–2.96 mg/kg RSD below 10%	M: Ethanol-water and nitric acid, C: Aqueous ref. solns.	Fast program methodology with a modified drying stage. Good agreement with calcination/dissolution	[622]
	B: Biol. CRMs, cervine liver, kidney		3.94–83.3 mg/kg RSD 1.7–8.1%	M: NH_4NO_3 C: AAM and ext. stands.		[619] see Cd/B
	E: Various sewage sludges		3.93–16 mg/L RSD <10%	M: - C: AAM		[560] see Cd/E
	E: NIST Coal SRM		ca 20 mg/kg RSD <10%	M: - C: Aqueous ref. solns.		[527] see Al/B
	E: Various sediments		20–76 mg/kg RSD 2.8–6.8 %	M: Mixed modifier C: Aqueous ref. solns. plus modif.		[575] see Cd/E
	E: Airborne particulates		12–89 ng/m^3 RSD 1–3%	M: - C: Aqueous ref. solns.	Good agreement with wet decomposed material and a NIST SRM	[537] see Cd/E

Sample	Concentration / RSD	M / C	values between the certified aqua regia value and the total content	Ref.
E: BCR CRMs Lake and River Sediment	39,5–64.5 mg/kg RSD 4.7–19%	M: - C: Aqueous ref. solns.		[539] see As/E
E: Contaminated soils and sediments	31.1 µg/l RSD 2.88%	M: - C: Ref. samples and ref. solns.		[609] see As/E
G: Five rock samples	10–300 mg/kg RSD about 20%	M: Na-phosphate C: Aqueous ref. solns. + Viscalex		[519] see Cr, G
G: Silicate rocks	10–115 mg/kg RSD ca. 20%	M: Na-phosphate C: Aqueous ref. solns.		[143] see Co, G
G: IAEA Soil-7	13.5 mg/kg RSD 4%	M: Mixed modifier C: Aqueous ref. solns. plus modif.		[575] see Cd/E
G: Diatomaceous earth CRM	3.8–123 mg/kg RSD 1.6–11%	M: - C: Aqueous ref. solns.		[568] see Cr, F
I: Aluminium oxide		M: - C: Ref. samples and ref. solns.		[615] see Ca/I
I: Real glass samples and CRMs	2.3 + 39.7 mg/kg RSD 2.5 + 8.7 %	M: - C: Aqueous ref. solns.		[497] see Co/I
I: Real glass samples and CRMs	0.8–428 mg/kg RSD 1.4–7.7%	M: - C: Aqueous ref. solns.		[535] see Co/I
I: Silicon carbide	1,3–8.9 mg/kg RSD: up to 16%	M: Mg nitrate C: AAM		[607] see Al/I

Table 5.6 Continued

Element	Matrix Code and Matrices	Instrumentation	Conc. (range) and typical RSD (range)	Slurry Approach and Matrix Modification (M:); Calibration (C:)	Special Features/Remarks	References (Literature, Analyte/ Matrixcode)
	I: High purity Mo trioxide		0.08 mg/kg RSD 12.5 %	M: - C: AAM	Detection limit 20 µg/kg	[541] see Ca/I
	I: Silica gel samples		7.5 + 16,3 mg/kg RSD <8%	M: - C: AAM		[517] see Al/I
	I: Silicon nitride		0.6 + 1.56 mg/kg RSD 4.5 + 17%	M: $(NH_4)H_2PO_4 + Mg(NO_3)_2$ C: AAM		[238] see Al/I
	I: Zirconium dioxide based materials		<0.6 and 0.5 mg/kg	M: - C: Aqueous ref. solns.	Detection limit 200 µg/kg	[526] see Al/I
	I: High purity quartz		1.62 ± 0.06 mg/kg	M: W carbide modified tubes C: Aqueous ref. solns.		[589] see Al/I
	I: Three graphite powders		0.39–1.1 mg/kg RSD 5,1–18%	M: - C: Aqueous ref. solns.		[571] see Al/I
Fe	B: 6 NIST SRMs		$10–10^3$ mg/kg RSD <10%	M: - C: Aqueous ref. solns.		[527] see Al/B
	B: plant samples and NIST SRMs		98 mg/kg -9% RSD 0.6–11%	M: - C: CRMs		[498] see Cu/B
	B: Pine needles		37–220 mg/kg RSD 6-8%	M: - C: CRM	No calcination for Fe determination	[552] see Cr/B

Sample	Concentration	Modifier / Calibration	Notes	Reference
B: Food and food CRMs	3.5–45 mg/kg RSD 8.7–17%	M: - C: Aqueous ref. solns.		[501] see Cu/B
B: Sweets and chewing gum	1–200 mg/kg RSD <8%	M: $NH_4H_2HPO_4$ C: Aqueous ref. solns.		[620] see Cr/B
E: NIST Coal SRM	ca 2% RSD <10%	M: - C: Aqueous ref. solns.		[527] see Al/B
E: NIST River Sediment SRM	ca 4% RSD 4.2%	M: - C: Aqueous ref. solns.		[563] see As/E
I: Aluminium oxide		M: - C: Ref. samples and ref. solns.		615] see Ca/I
I: Real glass samples and CRMs	48.5 mg/kg RSD 3.7%	M: - C: Aqueous ref. solns.		[497] see Co/I
I: Real glass samples and CRMs	5.9–433 mg/kg RSD <6%	M: - C: Aqueous ref. solns.		[535] see Co/I
I: Silicon carbide	15.3–650 mg/kg RSD up to 8%	M: Mg nitrate C: AAM		[607] see Al/I
I: High purity Mo trioxide	0.40 mg/kg RSD 60%	M: - C: AAM	Detection limit 50 µg/kg	[541] see Ca/I
I: Silica gel samples	14.7 + 99.5 mg/kg RSD 8.8 + 11.6%	M: - C: AAM		[517] see Al/I
I: Silicon nitride	59 + 79 mg/kg RSD 2 + 7.6%	M: $(NH_4)H_2PO_4 + Mg(NO_3)_2$ C: Aqueous ref. solns.	Detection limit more than three times lower compared with solution GF-AAS	[238] see Al/I

Table 5.6 Continued

Element	Matrix Code and Matrices	Instrumentation	Conc. (range) and typical RSD (range)	Slurry Approach and Matrix Modification (M:); Calibration (C:)	Special Features/Remarks	References (Literature, Analyte/ Matrixcode)
	I: Zirconium dioxide based materials		6.2 + 117 mg/kg RSD 15 + 40%	M: - C: Aqueous ref. solns.	Detection limit 50 µg/kg; i.e. the same compared with solution GF-AAS	[526] see Al/I
	I: High purity quartz		360 ± 30 mg/kg	M: W carbide modified tubes C: Aqueous ref. solns.		[589] see Al/I
	I: Three graphite powders		35.8–490 mg/kg RSD 4–10%	M: - C: Aqueous ref. solns		[571] see Al/I
Ga	E: NIST SRMs Coal, Coal Fly Ash	Manual, mechanically stirred, P-E 3030, HGA-400, D$_2$-BC	1.05–59.7 mg/kg RSD 2.1–5.7%	M:. Ethanol-water plus 20 µg Ni as Ni(NO$_3$)$_2$ C: Aqueous ref. solns.	Good agreement with certified values	[650]
Hg	E: Marine sediments	Automated, mechanical stirring, P-E 1100 B, HGA-700 and AS-70, D$_2$_BC	2.2–3.2 mg/kg RSD around 2%	M: Triton X-100, Pd nitrate C: AAM	Detection limit 70 µg/kg, particle size <20 µm was sufficient to achieve total atomization of Hg	[623]
	E: Ten marine sediments		15–349 µg/kg RSD around 3%	M: Pd/ Mg nitrate C: AAM		[611] see As/E
K	I: High purity Mo trioxide		0.061 mg/kg RSD 11.5%	M: - C: AAM	Detection limit 1 µg/kg; i.e far below solution GF-AAS	[541] see Ca/I

Element	Sample	Conditions	Concentration	Modifier / Calibration	Notes	Ref.
	I: Silicon nitride		0.75 mg/kg RSD ca 10%	M:- C: Aqueous ref. solns.	Detection limit approx. 40 times lower compared with solution GF-AAS	[238] see Al/I
	I: Zirconium dioxide based materials		0.05 and 9 mg/kg RSD 5.5 and 20%	M: - C: Aqueous ref. solns.	Detection limit a factor of 30 lower compared with solution GF-AAS	[526] see Al/I
	I: High purity quartz		665 ± 40 mg/kg	M: W carbide modified tubes C: Aqueous ref. solns.		[589] see Al/I
	I: Three graphite powders		15.9–42.3 mg/kg RSD 2–8%	M: - C: Aqueous ref. solns.		[571] see Al/I
Li	I: High purity Mo trioxide		<0.04 mg/kg	M: - C: AAM	Detection limit 40 µg/kg	[541] see Ca/I
	I: High purity Mo trioxide	Automated, magnetic stirring on sample tray (AS-40); P-E 5000, WETA 82 W tube; D₂-BC	<0.002 mg/kg	M: - C: AAM	Detection limit 2 µg/kg, Det. limits for Cr, Mn and Ni similar to reported values by solution GF-AAS [541]	[542]
	I: Zirconium dioxide based materials		0.14 mg/kg RSD 2.2%	M:- C: Aqueous ref. solns.	Detection limit a factor of 6 lower compared with solution GF-AAS	[526] see Al/I
	I: High purity quartz		1,3 ± 0.2 mg/kg	M: W carbide modified tubes C: Aqueous ref. solns.		[589] see Al/I
Mg	B: Different NIST SRMs		– RSD 2% (Spinach)	M: - C: Aqueous ref. solns.		[527] see Al/B
	I: High purity Mo trioxide		0.033 mg/kg RSD 34%	M. - C: AAM	Detection limit 0.5 µg/kg, far below that of solution GF-AAS	[541] see Ca/I

Table 5.6 Continued

Element	Matrix Code and Matrices	Instrumentation	Conc. (range) and typical RSD (range)	Slurry Approach and Matrix Modification (M:); Calibration (C:)	Special Features/Remarks	References (Literature, Analyte/ Matrixcode)
	I: Silicon carbide		4.0–30.8 mg/kg RSD up to 20%	M: Mg nitrate C: AAM		[607] see Al/I
	I: Silicon nitride		55 and 3 mg/kg RSD ca 7%	M: - C: Aqueous ref. solns.	Detection limit a factor of 9 lower compared with solution GF-AAS	[238] see Al/I
	I: Zirconium dioxide based materials		0.55 and 17 mg/kg RSD 5.5 and 13%	M: C: Aqueous ref. solns.	Detection limit a factor of 6 lower compared with solution GF-AAS	[526] see Al/I
	I: High purity quartz		130 ± 30 mg/kg	M: W carbide modified tubes C: Aqueous ref. solns.		[589] see Al/I
	I: Three graphite powders		1.3–10.7 mg/kg RSD 8.6–15%	M: Pd C: Aqueous ref. solns.		[571] see Al/I
Mn	B: Six NIST SRMs		ca 10–ca 10^3 mg/kg RSD <10%	M: - C: Aqueous ref. solns.		[527] see Al/B
	B: Plant samples		23.5 -750 mg/kg RSD 0.8-4.3 %	M: - C: NIST SRM slurries		[498] see Cu/B
	B: NIST SRMs r	Automated, USS-100; P-E Zeeman 5100 PC, HGA-600 and AS-60	9.1-95 mg/kg RSD 2-8.1 %	M: - C: aqueous ref. solns.	Manual sonication of samples(10 mg) prior to analysis. No pyrolysis step.	[530]
	B: Various biological CRMs		61-260 mg/kg RSD 2-10%	M: - C: aqueous ref. solns.		[553] see Co/B

B: Food materials		0.30–8.60 mg/kg RSD 4.7-11%	M: - C: Aqueous ref. solns.		[501] see Cu/B
E: NIST Coal SRM		ca 60 mg/kg RSD <10%	M: - C: Aqueous ref. solns.		[527] see Ca/B
E: NIST River Sediment SRM		560 mg/kg RSD 3.6%	M: Various C: Aqueous ref. solns.		[563] see As/E
E: Sewage sludge CRM		244 mg/kg RSD 6.2%	M: - C: Aqueous ref. solns.	Air ashing	[551] see Co/B
E: BCR Sewage sludge		241 mg/kg RSD 5%	M: Hydrogen C: Aqueous ref. solns.		[546] see Cr/E
G: Sandstone fractions	Automated, USS-100; P-E 3030 Zeeman, HGA-600 and AS-60	40–420 mg/kg RSD <10%	M: - C: Aqueous ref. solns.	Accuracy usually <15%; also other elements studied; short summary paper	[565]
G: Diatomaceous earth and CRMs		17–175 mg/kg RSD 2–10 %	M: - C: Aqueous ref. solns.		[525] see Cd/G
I: NIST Glass SRMs and a quartz sample		29 and 38 mg/kg RSD 4 and 8.3%	M: - C: Aqueous ref. solns.		[497] see Co/I
I: NIST Glass SRMs and a quartz sample		1.5–480 mg/kg RSD 1.3–13%	M: C: Aqueous ref. solns.		[535] see Co/I
I: Silicon nitride		2 and 0.23 RSD ca 5%	M: $(NH_4)H_2PO_4 + Mg(NO_3)_2$ C: AAM	Detection limit 6 µg/kg; i.e the same compared to solution GF-AAS	[238] see Al/I

Table 5.6 Continued

Element	Matrix Code and Matrices	Instrumentation	Conc. (range) and typical RSD (range)	Slurry Approach and Matrix Modification (M.); Calibration (C:)	Special Features/Remarks	References (Literature, Analyte/ Matrixcode)
	I: Silicon carbide		0.2–6,9 mg/kg RSD up to 40%	M: Mg nitrate C: AAM		[607] see Al/I
	I: Zirconium dioxide based materials		0.08 and 1.1 mg/kg RSD 12.5 and 27%	M: - C: Aqueous ref. solns.	Detection limit a factor of three lower compared to solution GF-AAS	[526] see Al/I
	I: High purity quartz		17.4 ± 4 mg/kg	M: W carbide modified tubes C: Aqueous ref. solns.		[589] see Al/I
Mo	B: NIST SRM Bovine Liver		ca 6 mg/kg RSD <10%	M: - C: Aqueous ref. solns.		[527] see Al/B
Na	I: Aluminium oxide			M: - C: Ref. samples and ref. solns.	Feasibility study; but problems with Na because of contamination	[615] see Ca/I
	I: High purity Mo trioxide		0.059 mg/kg RSD 20%	M: - C: AAM	Detection limit 1 µg/kg; far below that of solution-GF-AAS	[541] see Ca/I
	I: Silicium nitride		8.2 and 6.5 mg/kg RSD ca 6%	M: - C: Aqueous ref. solns.	Detection limit a factor of 100 lower compared with solution GF-AAS	[238] see Al/I
	I: Zirconium dioxide based materials		30 and 168 mg/kg RSD 6.5 and 17%	M: - C: Aqeous ref. solns.	Detection limit a factor of 20 lower compared to solution GF-AAS	[526] see Al/I

	Sample	Concentration	Method	Notes	Ref.
Ni	I: High purity quartz	69 ± 9 mg/kg	M: W carbide modified tubes C: Aqueous ref. solns.		[589] see Al/I
	I: Three graphite powders	1.9–19.8 mg/kg RSD 4–5.3%	M: - C: Aqueous ref. solns.		[571] see Al/I
	B: Algae		M– C: Aqueous ref solns.		[532] see Co/B
	B: Tobacco Leaves candidate RMs	1.6 and 5.6 mg/kg RSD 7.2 and 9.2%	M: - C: AAM		[240] see Cd/B
	B: Plant samples and NIST Orchard Leaves	0.53–5.6 mg/kg RSD 6.4–17%	M: - C: Aqueous ref. solns.		[516] see Cd/B
	B: Seafood and plant materials	0.98–3.58 mg/kg RSD up to 20%	M: H_2O_2-nitric acid C: Aqueous ref. solns.	Detection limit for Ni: 36 µg/kg	[621] see Co/B
	E: Four sewage sludges	0.4–3.2 mg/kg RSD <10%	M: - C: Aqueous ref. solns.		[561] see Cr/E
	E: Various sediments	29–40 mg/kg RSD 3–17%	M: Mixed modifier C: Aqueous ref. solns. plus modif.		[575] see Cd/E
	E: Airborne particulates and SRM	1.8–8 ng/m³ RSD 1–3%	M: - C: Aqueous ref. solns.		[537] see Cd/E
	E: IAEA Soil-7	29.5 mg/kg RSD 6.2%	M: Mixed modifier C: Aqueous ref. solns. plus modif.		[575] see Cd/E
	E: BCR sediment CRMs	53–66 mg/kg RSD 9–24%	M: - C: Aqueous ref. solns.		[539] see As/E

Table 5.6 Continued

Element	Matrix Code and Matrices	Instrumentation	Conc. (range) and typical RSD (range)	Slurry Approach and Matrix Modification (M:); Calibration (C:)	Special Features/Remarks	References (Literature, Analyte/Matrixcode)
	E: Contaminated soils and sediments		16.7 µg/L RSD 3.92%	M: - C: Reference samples, ref. solns.		[609] see As/E
	I: Glass samples and CRMs		5.4 and 37.6 mg/kg RSD 5.3 and 9.3 %	M: - C: Aqueous ref. solns.		[497] see Co/I
	I: Glass samples and CRMs		1.1–410 mg/kg RSD <5%	M: - C: Aqueous ref. solns.		[535] see Co/I
	I: Silicon carbide		3.0–21.4 mg/kg RSD up to 15%	M: Mg nitrate C: AAM		[607] see Al/I
	I: High purity Mo trioxide		<0.05 mg/kg	M: - C: AAM	Detection limit 50 µg/kg	[541] see Ca/E
	I: Zirconium dioxide based materials		0.3 and 1.9 mg/kg RSD 5.3 and 33%	M: - C: Aqueous ref. solns.		[526] see Al/I
	I Three graphite powders		1.1–5.5 mg/kg RSD 7–9%			[571] see Al/I

Pb	Sample / matrix	Instrument / preparation	Conc. / RSD	Modifier / Calibration	Remarks	Ref.
	B: NIST Orchard Leaves	Manual, dental mill, Vortex mixer; P-E 303 with HGA-2000; D_2-BC	50 mg/kg RSD around 10%	M:- C: Aqueous ref. solns.	Acceptable agreement with the certified value of 45 mg/kg	[545]
	B: Spinach	Automated, Agate ball mill, Viscalex; P-E 3030, HGA-500 and AS-40; D_2-BC	ca 1.5 mg/kg RSD <10%	M: - C: Aqueous ref. solns and analyte. add.	O_2-ashing; acceptable agreement with solution GF-AAS; up to 10% m/V can be tolerated in the suspension	[544]
	B: vegetables and foods; NIST SRM Pine needles for QC	Automated, Viscalex; P-E 3030, HGA-500 and AS-40; D_2-BC		M: - C: Aqueous ref. solns.	O_2-ashing; detection limit of approx. 5 µg/kg in fresh material	[521]
	B: Various NIST CRMs		0.135–45 mg/kg RSD up to 10%	M: - C: AAM	Feasibility study; Pb detection limit 0.87 mg/l	[616] see Cd/B
	B: Various biological CRMs		0.38–11.4 mg/kg RSD <1%–11%	M: NH_4 phosphate + Mg nitrate C: Aqueous ref. solns.		[576] see Cd/B
	B: Pine needles		0.75–15 mg/kg RSD 4–8%	M: - C: SRM Pine needles slurries:		[552] see Cr/B
	B: Milk powder and pine needles		1.06 + 10.2 mg/kg RSD 3 + 9%	M: Pd/ Mg C: Aqueous ref. solns.		[529] see Cd/B
	B: Various food and other biological samples	Manual and automated, magnetic; P-E Zeeman 5000, HGA-400 and AS-40	0.37–64 mg/kg RSD 10–55%	M: Pd nitrate C: Aqueous ref. solns.	Slurries up to 4% m/V could be analysed; detailed investigation of matrix modification with Pd nitrate	[524]
	B: Various biological CRMs		1,2–6.1 mg/kg RSD 10–15%	M: - C: Aqueous ref. solns.		[553] see Co/B

Table 5.6 Continued

Element	Matrix Code and Matrices	Instrumentation	Conc. (range) and typical RSD (range)	Slurry Approach and Matrix Modification (M:); Calibration (C:)	Special Features/Remarks	References (Literature, Analyte/ Matrixcode)
	B: Various NIST biological SRMs	Manual, Viscalex; P-E 3030, HGA-400; D_2-BC	6.1–62.3 mg/kg RSD 3.2–8%	M: $(NH_4)H_2PO_4$ C: Aqueous ref. solns.		[536]
	B: Commercial paprika samples	Manual, P-E 1100B, HGA-400; D_2-BC	1.5–9.5 mg/kg RSD 3.2–7%	M: calcination, suspension; $(NH_4)H_2PO_4$ C: Aqueous ref. solns.	The results for 7 paprika samples agreed with those by solution GF-AAS	[554]
	B: NIST biological SRMs		6.5–52.7 mg/kg RSD 4–10%	M: $(NH_4)_2 HPO_4$ C: Aqueous ref. solns.	Precision improved if finely ground	[137] see Cd/B
	B: Two tobacco leave candidate RMs		3.88 + 23.4 mg/kg RSD 13 + 17%	M: - C: Aqueous ref. solns.		[240] see Cd/B
	B: Various plant samples and SRM		0.85–47.3 mg/kg RSD 8.5–15%	M - C: AAM		[516] see Cd/B
	B: UPB Beech Leaves RM		4.9 mg/kg	M: Pd and Mg nitrates C: -	homogeneity study	[567], [317] see As/B
	B: Sweets and chewing gum		0.07–0.26 mg/kg RSD <8%	M: $(NH_4)H_2PO_4$ C: Aqueous ref. solns.		[620] see Cr/B
	B: Mussels	Automated, milling with zirconia beads; magnetic stirring; P-E 1100B, HGA-700; AS-70, D_2-BC	4.15–6.08 mg/kg RSD up to 10%	M: Triton X-100, Pd- Mg nitrate C: AAM	Study on the effects of different slurry stabilization agents, also air ashing applied	[553]

		M / C		Ref.
B: biscuits, bread and cereal based products	0.01–0.091 mg/kg RSD 4.5-14 %	M: Phosphate C: Aqueous ref. solns.	Detection limit 8 µg/kg	[617] see Cd/B
B: Vegetables	0.1–13.4 mg/kg RSD up to 12%	M: Phosphate C: Aqueous ref. solns	Fast furnace program	[618] see Cd/B
B: CRMs, cervine liver and kidney	0.036–0.345 mg/kg RSD 4.4–12%	M: Pd nitrate C: AAM/external stands.		[619] see Cd/B
E: Various sewage sludges	3.9–25 mg/kg RSD <10%	M: - C: AAM		[560] see Cd/E
E: NIST SRM Estuarine Sediment	29.2 mg/kg RSD 5.5%	M: NH_4-phosphate + Mg nitrate C: Aqueous ref. solns.		[576] see Cd/B
E: NIST SRM River Sediment	165 mg/kg RSD 7%	M: Various C: Aqueous ref. solns.		[563] see As/E
E: NIST Coal Fly Ash; Trace elements in coal	9–73 mg/kg RSD<5%	M: - C: Aqueous ref. solns.		[564] see As/E
E: NIST Coal SRMs	3.38 + 12.8 mg/kg RSD 7 + 11%	M: Pd/ Mg C: Aqueous ref. solns. plus modif.		[529] see Cd/B
E: Various sediment CRMs	28–120 mg/kg RSD 1.8–8.6	M: Mixed modif. C: Aqueous ref. soln. plus modif.		[575] see Cd/E

Table 5.6 Continued

Element	Matrix Code and Matrices	Instrumentation	Conc. (range) and typical RSD (range)	Slurry Approach and Matrix Modification (M); Calibration (C:)	Special Features/Remarks	References (Literature, Analyte/Matrixcode)
	E: Sewage sludge CRM		365 µg/kg RSD 6%	M: - C: Aqueous ref. solns.	Air ashing	[553] see Co/B
	E: NIST Coal Fly Ash and River Sediment		67.4–774 mg/kg RSD 6–14%	M: - C: Aqueous ref. solns.	Influence of particle size observed	[137] see Cd/B
	E: BCR sediment CRMs		52.3 and 105 mg/kg RSD ca 10%	M: - C: Aqueous ref. solns.	Pb values in SRMs systematically about 20% too high; reason unknown	[539] see As/E
	E: Sediment samples and sediment SRMs		27.2–30.1 mg/kg RSD 2.7–7.2%	M: Pd in HCl C: Slurries/aqueous ref. solns.		[239] see Cd/E
	E: Ten sediment samples		9.8–28.6 mg/kg RSD about 3%	M: Pd/ Mg nitrate C: Aqueous ref. solns.		[611] see As/I
	E: Contaminated soils and sediments11		26.5 µg/l RSD 4.24%	M: Pd/ Mg nitrates C: Ref. samples and ref. solns.		[609] see As/E
	G: Soil samples		50–156 mg/kg RSD ca 10%	M: - C: Aqueous ref. solns.		[142] see /E
	G: Sandy soils/ soils with high clay content		11–156 mg/kg RSD <10%	M: - C: Aqueous ref. solns.		[109] see Cd/G

Sample	Instrument/Method	Results	Modifier/Calibration	Remarks	Ref.
G: Model and real soils	Manual, magnetic; P-E 2280, HGA-500, optical pyrometer and micro computer etc.; D_2-BC	no detailed data; basic study	M: Mg nitrate and NH_4 phosphate C: -	Study on the effects of matrix components on the absorbance versus time profile for lead in soil slurries	[513]
G: Model and real soil samples	as above	14–21 mg/kg RSD up to 17%	M: Mg/Pd C: Aqueous ref. solns.	Study with various modifiers and modifier mixtures, O_2-ashing, wall and platform atomisation etc.	[579]
G: Canadian soil CRMs from three regions	as above	14.2–21.2 mg/kg RSD around 10%	M: Pd/Mg C: Aqueous ref. solns. plus modif.	Study on Pd plus Mg as a matrix modifier for Pb in solutions and slurries	[523]
G: IAEA Soil-7		62 mg/kg RSD 1.8%	M: Mixed modifier C: Aqueous ref. solns. plus modif.		[575] see Cd/E
G: Soil CRM: SO-4	Manual, magnetic; P-E 2280, HGA-500, optical pyrometer and microcomputer etc., D_2-BC	15.2 mg/kg RSD 6.6%	M: Pd/Mg C: Aqueous ref. solns. plus modifier	Study on Pd nitrate and chloride as chemical modifiers	[557]
G: Iron rich soil CRMs (Canada)	see above [557], [579]	16.7–21 mg/kg RSD 7.1–18%	M: - C: Probably aqueous ref. solns.	Study to show the usefulness of vortex mixing for soil samples	[533]
G: 3 Canadian soil CRMs	see above [533]	11.5–19.3 mg/kg RSD <1–18%	M: - C: Aqueous ref. solns.	Fast temperature program; better results without charring and modifiers	[538]
G: IAEA Soil-7		57.6 mg/kg RSD <10%	M: - C: Aqueous ref. solns.		[137] see Cd/B
G: CANMET SO-2 soil		22 mg/kg RSD 18%	M: - C: Aqueous ref. solns.	Hydrogen in Ar during cool down and autozero steps prior to atomization	[546] see Cr/E

Table 5.6 Continued

Element	Matrix Code and Matrices	Instrumentation	Conc. (range) and typical RSD (range)	Slurry Approach and Matrix Modification (M); Calibration (C)	Special Features/Remarks	References (Literature, Analyte/Matrixcode)
	G: Diatomaceous earth and CRMs		1.6–71.3 mg/kg RSD 4.2–13%	M: - C: Aqueous ref. solns.	Fast temperature program	[525] see Cr/G
	I: Aluminium oxide			M: - C: Ref. samples and ref. solns.	Feasibility study	[615] see Ca/I
	I: Different Al matrices (adsorbed and occluded Pb)	Manual, magnetic; P-E 2280, HGA-500, microcomputer; D_2-BC with fast response		M: Mg and phosphate modifiers C: Aqueous ref. solns.	Comparative study of slurry vs solution GF-AAS	[558]
	I: 4 different Al matrices; see above	see above		M: NH_4 phosphate and Mg nitrate C: Aqueous ref. solns.	Adsorbed Pb could be determined by use of modifiers; difficulties, however, with occluded Pb	[514]
	I: Commercial iron oxide pigments	Manual, mechanical stirring; P-E 1100B with HGA-400; D_2-BC	8.4–41.2 mg/kg RSD 2-5%	M: Triton X-100, NH_4 phosphate C: Aqueous ref. solns.	Accuracy verified by hydride AAS; particle sizes of the samples <100 μm; larger particles discarded by sieving	[626]
	I: Silica gel samples		1.4 and 3.1 mg/kg RSD ca 10%	M: - C: AAM		[517] see Cd/I
	I: Three graphite powders		0.11–1.0 RSD 10%	M: - C: Aqueous ref. solns.		[571] see Al/I

	Sample	Method	Concentration / RSD	M / C	Comments	Ref.
Pd	P: Polymer-supported catalysts in organic synthesis	Manual, suspension of finely ground (agate mill) material in triethyleneglycol (TEG) P-E 400, HGA-2100; D_2-BC	0.17–2.17% RSD 1.3–8.3%	M: - C: Homogeneous Pd suspension in TEG	Good agreement between calculated and experimentally determined values;	[627]
Se	B: 5 commercial wheat flour samples and NIST SRM Wheat Flour	Automated, 100 USS-100, P-E 4100 ZL, THGA and AS-70	0.88–1.05 mg/kg (SRM) RSD <10%	M: nitric acid/Triton X, Pd/ Mg C: Aqueous ref. solns. plus modif.	Good agreement with certified value; and characteristic mass (37 pg) with manufacturers value (45 pg)	[540]
	B: Seafoods	Manual, freeze dried, ground material; P-E 1100B and HGA-400; D_2-BC	0.41–18.2 mg/kg RSD around 5%	M: water-ethanol, conc. H_2O_2/Ni nitrate C: Aqueous ref. solns.	Good agreement with a CRM and with determinations by hydride AAS. Fast heating program with no ashing or cleaning step	[628]
	E: Five NIST Coal CRMs	Manual, magnetic, particle size <10 μm; IL Video 12, IL655; S/H and D_2-BC	0.6–2.1 mg/kg RSD 7.9–57%	M: Ni and Mg nitrates C: Aqueous ref. solns.	S/H-BC and in situ air ashing preferred; acceptable agreement with certified values; detection limit 0.05 mg/kg for a 15% m/v slurry	[515]
	E: NIST SRM Fly Ash		9.8 mg/kg RSD 15%	M: Pd C: Aqueous ref. solns.	Se was more difficult than the other elements	[564] see As/E
Si	I: Boron nitride	Automated, USS-100; P-E Zeeman 4100 ZL, THGA graphite furnace and AS-70	21.6 mg/kg RSD 8.3 %	M: Mg nitrate C: AAM	Good agreement with XRF values; detection limit 3 mg/kg	[518]
	I: Titanium dioxide and zirconium dioxide	Automated, USS-100; P-E Zeeman 4100 ZL, THGA and AS-90	25–244 mg/kg RSD 5.3–30%	M. Ca nitrate C: AAM	Good agreement with other methods; but lower blanks and less time needed	[543]

Table 5.6 Continued

Element	Matrix Code and Matrices	Instrumentation	Conc. (range) and typical RSD (range)	Slurry Approach and Matrix Modification (M:); Calibration (C:)	Special Features/Remarks	References (Literature, Analyte/Matrixcode)
	I: Zirconium dioxide based materials		25 and 436 mg/kg RSD 15 and 16%	M: Ca nitrate C: AAM	Detection limit 2 mg/kg	[526] see Al/I
	I: Three graphite powders		41 and 138 mg/kg RSD approx. 12%	M: Pd and Mg nitrates C: Aqueous ref. solns.		[571] see Al/I
Sn	E: Ten marine sediments		1.6–13.8 µ/kg RSD about 3%	M: Pd/ Mg nitrate C: Analyte addition		[611] see As/E
	I: Three graphite powders		< 0.2 mg/kg	M: - C: Aqueous ref solns.		[571] see Al/I
Ti	B: Paprika and pepper samples, pea, spinach and lettuce	Manual, ashing at 350 °C, magnetic; P-E 1100B, HGA-400; D_2-BC	0.32–21.8 mg/kg RSD 4.2–5%	M: Hexametaphosphate C: Aqueous ref. solns.	Good agreement with wet chemical procedure	[556]
	I: Silicon carbide		40–107 mg/kg RSD: up to 10%	M: Mg nitrate C: AAM		[607] see Al/I
	I: Silica gel samples		830–1300 mg/kg RSD ca 11%	M - C: AAM		[517] see Al/I
Tl	E: NIST Fly Ash SRM		5.6–6.7 mg/kg RSD ca 10%	M - C: Aqueous ref. solns.		[564] see As/E

	Sample	Preparation	Concentration/RSD	Modifier/Calibration	Remarks	Ref.
V	I: Cement	Manual, ultrasonic treatment then magnetically stirred; P-E 1100B with HGA-400; D$_2$-BC	0.1–0.5 mg/kg RSD ca. 25%	M: 10% ethanol, 1% nitric and hydrofluoric acids C: Aqueous ref. solns./AAM	Good agreement with different calibrating methods and CRM (Fly Ash)	[569]
	E: Airborne particulates and NIST SRM for QC		SRM: RSD 5% Particulates: 8.6–48 ng/m^3	M: - C: Aqueous ref. solns.		[537] see Cd/E
	G: Various silicate rocks		<10– ca 400 mg/kg	M: Na-phosphate C: Aqueous ref. solns.		[143] see Co/G
	I: Silicon carbide		9–34 mg/kg RSD: up to 4%	M: Mg nitrate C: AAM		[607] see Al/I
Zn	B: Various plant samples		20.6–127 mg/kg RSD 3.9–6%	M - C: NIST SRM slurries		[498] see Cu/B
	B: 4 NIST biological SRMs		10- to ca 100 mg/kg RSD <10%	M- C: Aqueous ref. solns.		[527] see Ca/B
	B: Sweets and chewing gum		RSD <8%	M: (NH$_4$)H$_2$PO$_4$ C: Aqueous ref. solns.		[620] see Cr/B
	E: 4 sewage sludges		19–120 mg/L RSD <10%	M: - C: AAM	Detection limit 1 µg/l;	[561] see C/E
	G: Diatomaceous earth and		5.6–226 mg/kg RSD <2–9%	M: (NH$_4$)H$_2$ HPO$_4$ C: Aqueous ref. solns.		[525] see Cd/G
	I: Silicon carbide		0.18–0.68 mg/kg RSD: up to 10%	M: Mg nitrate C: AAM		[607] see Al/I

Table 5.6 Continued

Element	Matrix Code and Matrices	Instrumentation	Conc. (range) and typical RSD (range)	Slurry Approach and Matrix Modification (M:); Calibration (C:)	Special Features/Remarks	References (Literature, Analyte/ Matrixcode)
	I: Silicon nitride		0.13 + 0.34 mg/kg RSD 6 and 15%	M: - C: Aqueous ref. solns.	Detection limit 1 µg/kg; i.e. a factor of 9 lower compared to solution GF-AAS	[238] see Al/I
	I: Zirconium dioxide based materials		0.09 and 1.3 mg/kg RSD 22 and 23%	M: - C: Aqueous ref. solns.	Detection limit a factor of 30 lower compared with solution GF-AAS	[526] see Al/I
	I: Three graphite powders		0.5–3.2 mg/kg RSD 2–10%	M: - C: Aqueous ref. solns.		[571] see Al/I

CHAPTER 6

Advantageous Fields of Application for Solid Sampling Analysis

JÜRGEN FLECKENSTEIN, MARKUS STOEPPLER AND ULRICH KURFÜRST

6.1 Introduction

The analysis of samples in a graphite furnace directly from the solid state can be attractive for a number of reasons already discussed in Chapter 1 in detail. Langmyhr [629] pointed out what the solid sampling technique of atomic spectrometry can do:

> "*As compared to other methods for the quantification of trace elements the direct analysis of solids by atomic absorption spectrometry offers some definite advantages: being direct the method involves only three steps, viz. the sampling, the sample preparation and measurement, the time-consuming decomposition step can be omitted, and the analysis can be carried out without addition of reagents and without any separation and concentration steps; the risks of introducing contaminants and of losing the elements to be determined are thus considerably reduced.*"

Moreover, in comparison to analysis after chemical matrix decomposition, hazards associated with acid digestion are eliminated, the dissolution of the sample that may involve excessive dilution is avoided, and the selective analyses of solid samples for traces of elements is facilitated.

The discussion about "applications" of a method in the scientific analytical community generally only covers the purely analytical viewpoints e.g. accuracy, limit of detection, calibration techniques, required sample preparation, interferences due to concomitances ("matrix-effects"), as described in this book for the different techniques of solid sampling, e.g. in Sects. 3.4, 4.7, 5.6. The original problem, that is to clarify by the analytical results is only seldom discussed in analytical papers. Actually, the decision to introduce or apply a method for a certain task is often made not by considering the analytical quality parameters alone but also by other features, such as speed of the analysis (in what period are the data available?) or the costs (investment in instruments, consumables, personnel deployment). These viewpoints are discussed in general in Chapter 1 for the solid sampling techniques.

Frequently, a lack of mutual appreciation is apparent between "researchers" and "practitioners" when advantages and disadvantages of a particular analytical method are discussed and it is not necessarily conducive that the transfer of scientific results into practice is carried out mainly by instrument manufactories. However, with the marketing of the solid sampling AAS-instrument "SM1" from Grün-Optik, the situation was somewhat different: A majority of scientists raised some serious objections against the broad application of the method, so that the instruments could only be sold if the "interested party" was convinced individually that the particular tasks in the laboratory could be solved with the instrument and the method. As a consequence, very close cooperation between this small company and its customers is growing with the development of practical applications. The first events of the Colloquium *"Solid Sampling with Atomic Spectrometric Methods"* in 1984 and 1986 have basically been users conferences at which numerous practical applications of the Grün instrument from the first customers, from scientific, administrative and industrial laboratories were reported.

With this process and with the further development of the method in the last decade using other instruments and other techniques, various fields of analytical tasks were found and described that are particularly advantageous for direct solid sample analysis. These are outlined in the following sections for the completion of the answer "What can the method do?" the question that has already been discussed in Chapter 1 under the pure analytical viewpoint.

6.2
Determination of a few Elements in Many Samples

Solid sampling is, under certain circumstances, a method that can provide analytical results faster than any other method, as discussed in Sect. 1.2.1. As long as solid sampling is not fully automated, the deciding factor for using this feature is a low number of trace elements to be determined, preferably only one. Actually there are some characteristic analytical tasks for which frequently only one to three trace elements are of interest. Examples for investigations that are carried out by direct solid sampling with low expenditures in time, costs and personnel are the study described by Pesch et al. [630] about the content of toxic metals in cigarette tobaccos in East and West Germany in 1987 and 1991 (100 samples) and the studies for 37 and 150 wheat samples from ecological cultivation [631, 632] (see Fig. 6.1) for which the analyses for these studies were carried out in only some days by graduate students.

Typical assignments for solid sample analyses are stated in the headings of the following sections in which practical examples are given:

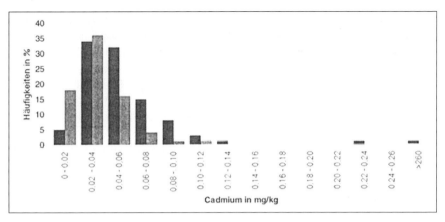

Fig. 6.1. Histogram of the content of cadmium in wheat from ecological cultivation in Germany 1989 (*gray columns*, n=151). From the comparison with results of a nationwide monitoring of conventionally cultivated wheat (from 1993, *black columns*, n=477) two significant differences can be read: For biological cultivated wheat the mode, the median and the average are at lower contents average (average 0.034 mg/kg compared to 0.056 mg/kg). 2. Higher contents are more rare, e.g. for biological cultivated wheat only 2% of the values are higher than 0.08 mg/kg compared to 15%) (from [632], with permission)

6.2.1
Bio-Monitoring and Investigations of Environmental Burden

Whenever a basic question in the bio- or ecosphere arises or a serious problem in the environment emerges and these are connected with a certain trace element a large number of samples often of the same kind have to be analyzed. Generally, measures for protection from toxic effects of pollutants are a combination of political regulations and economic constraints. Examples of such analytical tasks are the huge number of samples that had to be analyzed for Tl, when it was realized in 1979 that a Westphalian cement plant had heavily polluted the surrounding environment with the toxic element (for the reason for this pollution see below in Sect. 6.3.1) or the enormous number of samples that had to be analyzed for Hg as a consequence of the fire in a chemical factory in Basel in 1986 when the river Rhine was heavily polluted by mercury compounds. In both cases, direct solid sampling has shown that it offers the highest analytical speed. It can be assumed that direct solid sampling has also played a significant analytical role in similar incidents, although in such cases the data and the methods applied have mostly not been widely published.

Within the scope of a inquiry for the extension of a hazardous waste incineration plant at Bibesheim/Rhine, Germany, Steubing and Grobecker [633, 708] used direct solid sampling GF-AAS for their study of the pollution burden of Pb and Cd in the surrounding urban and agricultural ecosystem. The investigations were carried out by active monitoring with grass cultures and by passive monitoring with various plants growing in the region. In a monitoring area, wild or cultivated plants will often be found unequally distributed and in small numbers. Therefore exposed standardized grass cultures (*Lolium multiflorum*) are a preferred indicator for heavy metal uptake caused by air pollution or by the burden in the soil. Over a period of two years, 74 soil samples, 180 samples of turnip-leaves and 1088 samples of defined grass cultures exposed at different sites were taken. The authors presented the huge amount of data as "maps of burden", an example is given in Figure 6.2.

Active and passive biomonitoring was applied by Grobecker [708] for an investigation of an aquatic ecosystem (River Toce, Lago Maggiore, Italy). The Hg contamination of the sediments (50 mg/kg d.m.) of the Toce had been caused in the past by gold mining using mercury for amalgamation and waste water of industrial chlorine alkali electrolysis. For active monitoring at 10 sampling stations along the river and an accumulation experiment, an aquatic moss (*Platyhypnid-*

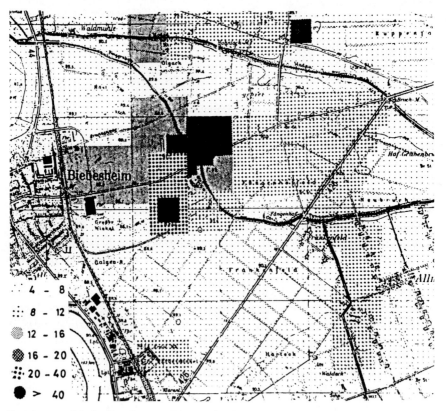

Fig. 6.2. Map of the lead burden in in the area of Biebesheim, Germany, based on lead contents (mg/kg) of standardized grass cultures exposed for 1 month. A suspected source for the emission of lead is situated on the road crossing in the south-west edge of the higher polluted areas, that corresponds with the predominant direction of the wind (from [633], with permission)

ium riparioides) was used that was native to the river. These plants were bred under controlled conditions and then exposed in special containers in the water. The large number of naturally grown plants, standardized plants and sediment samples were analyzed directly using the Hadeishi-furnace (see Sect. 3.3.3). For analyses of the water from the River Toce and Lago Maggiore with this technique, samples were reduced to a tenth of their volume by evaporating at 40 °C and analyzed. The mean Hg content of the water samples was found to be in the range of 0.25 µg/l, that is, higher than comparable running waters but below legal values. In 1988, the sediments had an Hg content that reflects more the geological background level (0.16–0.41 µg/g), while in the periods 1962–1985 sediment contents were found that were one to two orders higher. This improvement was probably achieved by a changeover of the chlorine alkali electrolysis from amalgam to diaphragm processing in the years 1982–1985 in Rumianca in the upper course of the river.

Lehmann et al. [323] carried out an investigation of the "Heavy metal status of the Plözensee in Berlin" for the elements Pb ,Cd and Cu. They sampled a large number of soils, sediments and leaves around this lake near Berlin city. Direct SS-GF-AAS offered the chance, for carrying out the voluminous sampling and analysis in the framework of an University diploma. Beside the environmental study an elaborate methodological study was performed, showing that the precision and the trueness of solid sampling results are fully comparable to that of digested samples.

In a study about the heavy metal burden of mushrooms [Fleckenstein, unpublished] direct solid sampling has proven to be an excellent and a suitable method for the determination of Cd, Pb, Hg, Cu and Zn on meanwhile more than 900 samples. Wild living mushrooms show a species specific and element selective behavior to heavy metals. The wild mushroom (*Langermania gigantea*) a member of the puffball family, sampled at many sites accumulates about 80 mg/kg d.m. Hg, otherwise *Agaricus* species accumulate Cd up to 100 mg/kg d.m. The analytical investigation of wild mushrooms is of interest, even their occurrence and collection is by chance. The mushrooms have a widespread and long-standing hyphal system. On fructification, an osmotic regulated stream of the stored ions flows into the fruiting body. The contents in the fruiting body may reflect the state and the occurrence of the heavy metal burden.

Generally, mushrooms have a high water content (90–95%), therefore the amounts of samples available are often not sufficient for digestion procedures. After oven drying or freeze-drying, mushrooms samples can be ground to a homogenous powder without any problem. It does not matter which analytical method is used, the sample preparation and pretreatment are a crucial step. Because of the small amounts (0.5 to 10 mg) very homogenous material is necessary for SS-GF-AAS. If for homogeneity, samples are ground in an agate mortar, Hg losses on sample preparation are assumed because very high local friction temperatures may be generated.

The uptake of heavy metals by the cultivated mushroom *Agaricus bisporus* can be investigated by growing samples on contaminated substrates [Fleckenstein, unpublished] – the behavior of the bound metals depending on the preparation steps during the analyses could be ascertained. The mushrooms were cultivated on horse manure and horse manure with additional sewage sludge waste compost. The mushroom *Agaricus bisporus* grow in four fleshes during the cultivation. About 30% of the organic substrate will be decomposed. The harvested fruit bodies of the mushroom were homogenized with bi-distilled water in the ratio 1 to 2 in a mixer. Afterwards these mixtures were freeze dried for preserving and for the analyses. The availability of the heavy metals during the decomposition of the substrate changed. Increasing content of Hg from 3 mg/kg d.m. in the first flesh until 27 mg/kg d.m. in the fourth flesh occurred in the harvested mushrooms, which were cultivated on horse manure mixed with sewage sludge waste compost. The pure horse manure compost resulted in an equal content between 3 to 4 mg/kg Hg d.m. in the mushrooms over all flushes. Twelve freeze dried samples from the experiment were measured by SS-GF-AAS in the nickel furnace. In order to check any volatilization effects of Hg according to different pretreatments such as grinding and drying, the measuring was carried out three times with various methods of sample preparation – direct input into the furnace as small crumbled freeze dried material, oven drying of the freeze dried sample once at 60 °C and twice at 105 °C, measuring without and with intensive grinding in an agate mortar:

The lowest standard deviation of the values were found when the samples were ground intensively in the agate mortar. Deciduous losses of mercury can not be confirmed in dependence with both oven drying and grinding. The subsumption of the values shows the availability of mercury significantly in the different types of substrates and the accuracy of these Hg measurements is more than sufficient.

Gellert and Wittassek [634] investigated the copper content of sediments and macrozoobenthos in a devastation region of the river Sieg below a cable metal factory. Gradients in sediments (up to 291 mg/kg Cu) and organisms showed a Cu enrichment originated from the cable metal factory. The larvae of *Ephemerella ignita* (949 mg/kg), *Asellus aquaticus* (479 mg/kg) and caddis fly *Hydropsyche sp.* (384 mg/kg) accumulate Cu to a great extent. Mayflies *Baetis sp.* (275 mg/kg) seems to regulate the Cu uptake, whereas the swimming beetle *Stictotarsus 12-pustulatus* (93 mg/kg) has a restricted Cu concentration. It was shown, how in singular organisms the uptake of copper is dependent on the concentration of the sediment as well on their habit.

In another case, direct SS-GF-AAS was applied for a collaborative study of the state of heavy metal contamination in a river ecosystem in the Harz mountain area [286]. The sediments of the river Oker have a heavy metal burden caused by the mining and smelteries in the Harz mountains in former times, whereas the tributaries Ecker and Schunter are not polluted. The samples for the analysis of Cu, Cd, Pb and Zn ranged from species of aquatic insects, exposed aquatic fungi, algae, to organs and tissues of ducks and water rails. In microbiological studies the number of samples, which have to be analyzed, increased rapidly with the variations and the statistical proof.

Evidently species-specific differences exist in the behavior to heavy metal uptake by the insects. The chironomids and Mayfly larvae have a high intake of copper in the contaminated area. The aquatic fungus *Mucorflavus*, which was cultivated in the laboratory as a test organism for water quality monitoring had a content of Cd 0.04 and Zn 30 µg/g after an exposure time of 3 days at the sampling point of the river Oker and about 200 µg/g for Cd and 3520 µg/g for Zn. The natural range of the element content in all the samples is essentially higher than the difference between replicates of the measurements.

Bioindication procedures with the aid of organisms are intended to indicate environmental conditions over a wide area and at reasonable expense. The necessary prerequisite for using this approach is that a clearly recognizable and unambiguous relationship exists beetween the environmental feature to be investigated and the reaction of the organism studied and that this can be quantitatively documented. In acquiring information on pollutants, animals - in contrast to plants that due to their sessile mode of life only integrate pollutants over time locally - can provide additional information on pollution burden over an area. The reference area for the pollutant integration is given by the sphere of action of the species in question. Long-term studies in the sampling areas of the German Environmental Specimen Bank and other European regions with several bird species have shown that standardized feathers from goshawks (*Accipiter gentilis*) and magpies (*Pica pica*) but also from other species (e.g. whimbrel, pigeon, eagle owl, peregrine falcon) are suitable acceptors for particle-borne air pollution in order to cover atmospheric pollution by Cd, Cu and Pb [635, 636, 637, 638]. As far as sampling of feathers is concerned this is feasible without impairing the respective individuals and thus can be recommended from the aspect of animal protection and nature conservation. Moreover, as a dead tissue, feathers are easy to store. Sampling by collecting moulted feathers is simple and no sophisticated preservation techniques are required for their storage. Since mercury was analyzed in feather vanes of a number of selected species (e.g. tawny owl, eagle owl, common buzzard, wood pigeon, common raven, herring gull, magpie, osprey, pheasant, and white-tailed eagle) it could be proven that there are no mercury losses from the matrix since mercury occurs by far predominatly as methylmercury which is firmly bound to the keratin structure of the feather vanes.

Significant progress was achieved by Hahn et al. when after thorough rinsing of the feathers with acetone to remove adhering dirt, the analysis of microsamples cut from defined parts of the feathers was performed by direct SS-GF-AAS [637, 638]. Depending on the element to be investigated and the local concentration sample weights varied between 50 and 1000 µg. In the course of the studies performed the method has proved to be an invaluable aid for the analysis of the distribution of cadmium, lead and mercury over the feather vanes. The evaluation of the pattern for each element together with some basic considerations of flight aerodynamics [636] showed very clearly that, due to typical patterns for lead and cadmium on the one hand and mercury on the other the incorporation mechanisms are completely different. A typical increase of Pb and Cd contents occurs from the feather base to the tip and also from the areas close to the quill towards the edge . Thus, as is demonstrated in Fig 6.3 a, the highest contents are observed in those parts of the feathers most greatly exposed to atmospheric influences. The contents in feathers depend on several factors. These are i) the age of the feathers (i.e. the time of exposure to atmospheric influences), ii) the proportion of the feather surface and the intensity with which it is exposed to these impacts and iii) the local heavy metal pollution in the sphere of action of the individual bird. This clearly shows that Pb and Cd in feathers originate from *exogenous* deposition.

In contrast to this, the mercury contents in the individual parts of the vane of the feathers studied are approximately equally high as is shown in Fig. 6.3 b, i.e. the concentration gradients observed for Pb and Cd did not occur for Hg. The reason for this behavior is that Hg enters into the feathers by completely different mechanisms. The homogeneous distribution can be explained by the *endoge-*

Fig. 6.3. Distribution of trace metals in bird feathers (from [638] with permission):
a Distribution of lead in the 3rd primary of an eagle owl (*Bubo bubo*) from the North Eifel, Germany and cadmium in the 5th primary of a magpie (*Pica pica*) from a polluted region around Stolberg/ Rhineland, Germany.
b Distribution of mercury in primaries of a herring gull (*Larus argentatus*) from the North Siberian Taimyr Peninsula, Russia and a goshawk (*Accipiter gentilis*) from the Saarbrücken area, Germany

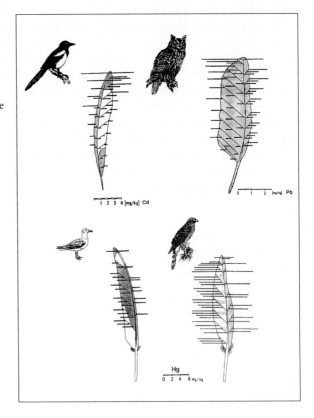

nous incorporation of Hg from the bird's food. Thus it is obvious that the highest Hg contents were measured in feathers from sea eagles and ospreys. Both these species are predators at the top of the food chain mainly feeding on fish, bird and mammals. Within these studies it was also observed that the feathers of ospreys and white-tailed eagles from Mecklenburg and Brandenburg in Eastern Germany had high Hg contents which probably can be attributed to the high mercury pollution still present in these regions.

6.2.2
Element Distribution Analysis

An impressive proof of the potential of direct solid sampling for the elucidation of trace element distribution was given from Pieczonka and Rosopulo [113]. Different morphologically defined fractions of wheat grains were prepared and analyzed separately using the platform technique. The content of Cd, Cu, Zn in the caryopsis of wheat were quantitatively determined in the whole grain, the germ, the aleurone layer, the outer pericarp, and the endosperm. The investigated wheat was grown on fields fertilized with sewage sludge of industrial origin, and therefore heavily burdened with heavy metals. Metal content markedly differed among the tissues investigated as can be seen in Fig. 6.4. (see also Sect. 2.3.1.1, Table 2.4). In comparison with flame AAS both methods produced almost identical heavy metal content, however, the authors pointed out that direct solid sampling needed considerably smaller sample amounts, thus the method is specially advantageous when only a few samples are available and sample production is time consuming. A consequence of these results is that consumers of whole meal products must be secure that the wheat is not burdened, while for white flour the fractions that are particularly burdened with Cd are separated. For wheat from ecological cultivation the situation is favorable also concerning this toxic element [631, 632] (see Fig. 6.2).

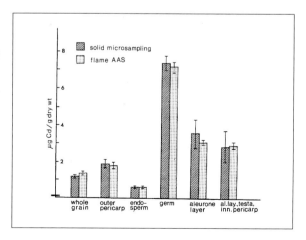

Fig. 6.4. Content of cadmium in morphologically defined fractions of wheat grains. The results from direct SS-GF-AAS are in very good agreement with the results from Flame-AAS (from [113], with permission)

Ebdon and Evans [322] determined Cu in biological microsamples. Direct solid sampling graphite furnace AAS with deuterium correction was applied to portions of caddis fly larvae and aquatic plants from a copper-contaminated stream. The Cu was determined in the heads (3930; 1660; 800 mg/kg), thoraxes (e.g.1130, 590; <50 mg/kg) abdomens (1690; 770; 380 mg/kg) of individual larvae by direct platform atomization in a microboat with amounts of 50 µg. However, the sample size was too small to ascertain the significance in the different concentration found. However, the authors conclude: " The application of the technique to both animal and botanical samples shows them to offer a speedy analytical method. Such a technique provides a powerful tool for the elucidation of the location of trace metals in such samples of wide environmental and clinical applications."

Klüßendorf et al. [639] studied the distribution of Pb and Cd in porcine livers in order to ascertain the sampling procedure for the analysis of the fresh organ for legal control. Figure 6.5.a shows the sampling scheme for the 4 different livers. These subsamples (totally 162) were analyzed with 5–10 test samples each. It was shown by ANOVA that a sufficiently homogeneity is necessary for the required accuracy, see Fig. 6.5.b. Similar results were achieved for bovine livers. Over 300 microsamples for the investigation of the distribution of Pb and Cd in equine liver were analyzed by Lücker et al. [640] and it was confirmed that the distribution of Pb and Cd in fresh equine livers is virtually homogenous. Distinct heterogeneity is not to be expected on the anatomical level. Nevertheless, it is recommended that at least 6 microsamples from different parts of the liver are taken in order to reduce the influence of minor heterogeneity.

A comprehensive study using direct solid sampling for trace element distribution in fresh renal tissues was carried out by Lücker et. al. [641]: They focused on the "duality of kidney, the division into cortex and different portions of medulla, as well as into submacroscopically distinct areas with varying connective tissue content" that "increase the possible source of variation in this organ so that a point is reached where an analysis based on minute sample size must appear futile". Figure 6.6 shows the sampling scheme for 3 bovine kidney pairs in their study. They found that there are very distinct inhomogeneities. "In the renal cortex, as was to be expected, especially for Cd, the highest Pb and Cd concentrations can be found; the lowest are in the Zona basalis of the renal medulla." – "The distribution of Pb and Cd in the kidneys analyzed proved to be sufficiently homogeneous in the renal cortex only." For the method of legal inspection they conclude that the "standard recommendation values existing in the FRG relate to the entire organ – The method of sampling in conventional procedures is also a source of errors that must not be underestimated. – Therefore, in addition to the paramount importance of the cortical tissue for the retention of heavy metals and the morphological proportions within the kidney, the cortex can be used for a uniform standardization of sampling as well. – A future provision of standard recommendation and maximum values for Pb and Cd concentrations in kidneys must rest exclusively upon the element concentration within the renal cortex."

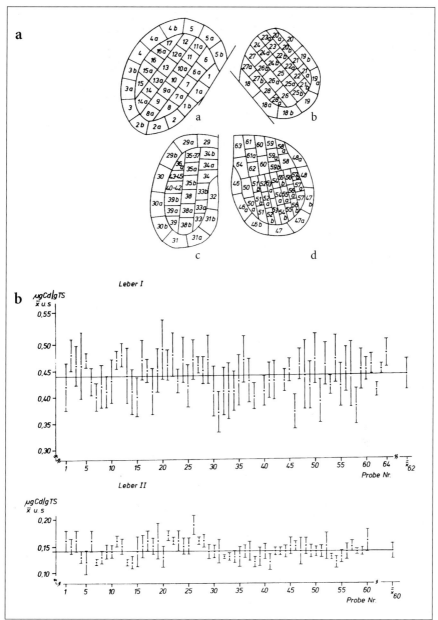

Fig. 6.5. Investigation about the homogeneity of lead distribution in pig-livers (from [639], with permission).
a Sampling scheme of about 60 subsamples for the different parts of the liver (*a* Lobus dester lateralis, *b* Lobus sinister lateralis, *c* Lobus dexter medialis, *d* Lobus sinister medialis)
b Cadmium content (μg/g d.m.) in the freeze-dried subsamples of pig-livers I and II and the mean value and the standard deviation of the total organ (on the right). Each subsample is analysed with 5-10 test samples (replicates)

Fig. 6.6. Sampling scheme for the examination of homogeneity of pig kidneys. The *black-bordered rectangle* represents a kidney. It depicts the division into cortex, as well as Zona intermedia (Medulla I) and Zona basalis (Medulla II). The kidney is dissected into three separate "localizations" (*A* anterior part, *B* central part, *C* posterior part); each tissue from each locality is then dissected tenfold and from each area a sample is taken, the resulting 270 samples from 3 kidneys were measured six times each with SS-GF-AAS (from [641], with permission)

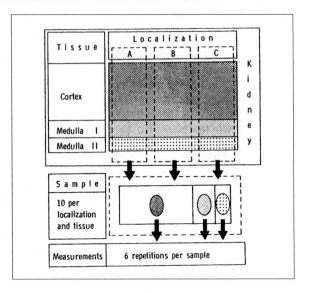

Analyzing the distribution of Pb and Cd in 121 livers of mallards [642], the SS-GF-AAS turned out to be a fast and easy applicable method, extremely robust concerning contamination and furthermore nearly independent of sample sizes.

By means of SS-GF-AAS Lücker and Thorius-Ehrler [643] determined the endogenous Pb contamination in muscle tissue caused by calcification of *Cystcercus bovis*, the larval stage of *Taenia saginata*. In Figure 6.7 the enormous differences in the Pb content are visualized that occur with the progress of calcification of the larval capsules. This serious case of sample heterogeneity and its consequence for solid sampling analysis is described and discussed in Sect. 2.3.2. However, the authors show that larger sample sizes in sample decomposition methods may also give rise to serious sampling errors when faced with the problem of this kind of endogenous contamination. Worst-case calculations using the experimental data show that calcified cysts can increase the original Pb content of muscle tissue to 150 times and above if only 1 mg sample mass is used. Even if the sample mass for decomposition is 100 g the analytical result can be 2.5 times higher than the original Pb content of the muscle tissue. This may explain some of the extremely high levels reported in literature for bovine muscle tissue. 79 samples of cysts and the respective muscle tissue were analyzed by SS-GF-AAS. The maximum Pb content of calcified cysts of *Cystcercus bovis* was 3.0 mg/kg, that of muscle tissue 0.02 mg/kg.

Fig. 6.7. The results of solid sampling Pb determination of bovine muscle tissues and the respective calcified cysts (*lower plot*) as well as the ratio between the content of the cysts and the respective muscle tissue (*upper plot*) (note: log-scale on the ordinate). The material was taken from 70 cattle during official meat inspection; 79 cysts were found, in 5 animals more than one (from [643], with permission)

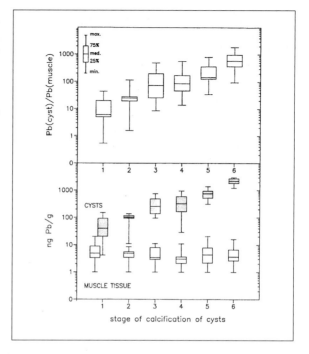

6.2.3
Transfer Chain of Metals in the Environment

While, for the above described studies, the samples are more or less of the same (matrix-) type, for the study of transfer chains for trace elements, different sample types must be analyzed in a project (e.g. soil, fertilizer, plants as animal feed, animal organs). In these cases, the independence of the matrix composition that is given for SS-GF-AAS to a great extent is an advantage (direct or slurry analyses based on reference solutions and confirmed with suitable CRMs).

The aim of the studies cited in Sect. 6.2.1 is connected with finding out the burdening of the environment, biosphere and food with toxic trace metals, here the aim is the transfer and distribution in the various trophic steps.

The efficiency of earthworms in producing humus in cultural soils is well-known, as well as for the bioconversion of residues and municipal wastes, and the hygienic destruction of sewage sludge [644]. They are also usable as animal feed. However, earthworms have immediate contact with all polluted materials in their surrounding and they are the pathway for the introduction of undesired and hazardous pollutants, e.g. heavy metals into the food chains. In food chains the uptake of heavy metals by the earthworm *Eisenia foetida*, which were kept in waste compost and sewage sludge, was investigated. The collected earthworms were fed with wet cellulose powder for evacuation of the intestine, afterwards killed in liquid nitrogen, freeze-dried and powdered in an agate mortar. The analyses of Zn, Cd, Pb, Cu and Hg were carried out by direct solid sampling GF-AAS (SM1, Grün-Optik). The availability of the appropriate equipment for direct solid sample analyses has supported and facilitated such experiments.

A species-specific behavior of earthworms to heavy metals and a considerable transfer and accumulation of heavy metals was found. In a pot experiment with a substrate prepared from 30 g straw, 4 g peat and 4 quartz sand and 100 g water the earthworms *Eisenia foetida* and *Eudrilus eugeniea* (3 animals in each pot) were fed with different contaminated cow faeces (total amount: 6 g d.m.) over 72 days.

The Cd content 3.32 mg/kg in the cow faeces (control: Cd 0.88 mg/kg d.m.) resulted to an increase in the Cd content in the substratum from 0.92 to 1.60 mg/kg d.m. The Cd content of the earthworm *Eisenia foetida* increased from 13 µg/g to 25 µg/g and of the earthworm *Eudrilus eugeniae* from 2.9 µg/g to 8.2 µg d.m.. In a further test, the earthworm *Eisenia foetida* was fed with powdered fruit bodies of the mushroom *Agaricus bisporus* (Cd content: 7.8 mg/kg; control: 0.90 mg/kg) , which were cultivated on variously burdened substrates. A food amount of 6 g d.m. led to an increase in the Cd content in the substrate (total amount 38 g d.m.) from 0.2 mg/kg to 1.1 mg/kg Cd (d.m.). The effect was an increase in the Cd content of the earthworm *Eisenia foetida* from 1.01 µg/g to 3.25 µg/g (d.m.). The input of matter containing only a single contaminant (nearly 4% of the prepared substrate) into the substrates or cultural soils effects a high transfer of heavy metals into the earthworms.

Kjer et al.[645] investigated the content of Cd, Zn, and Cu in hops and barley from ecological (here, in this chapter, the term is used in the German sense i.e. using only natural plant protection measures and fertilizers) and conventional cultivation and their relation to the brewery products. The aim of the analysis was to compare the quality terms regarding these raw materials and their side- and end products. The raw material of these two very different methods of cultivation varied considerately regarding the heavy metals. It became evident that the use of copper-containing sprays in hop cultivation presented a major problem. A total of 236 samples were examined. These can be divided into different fractions. 35 barley samples were taken from both ecological and conventional cultivation and then 35 samples of malt from both ecological and conventional cultivation. These samples were examined for content of Cd and Zn. In addition, 48 hop samples were examined for the content of Cd and for Cu. To a certain extent the amounts of Cd, Cu and Zn could be determined in relation to soil types. Finally the content of Cd and Cu in the end products (spent grain and beer) were analyzed. Conventionally cultivated barley contains a significantly higher amount of Cd (18.8 µg/kg d.m.) than ecologically cultivated barley (9.9 µg/kg d.m.). Furthermore, even the malt from conventionally cultivated barley contains significantly more cadmium 15.2 µg/kg than ecological malt 8.9 µg/kg. Ecological barley tends to contain more Zn than conventional barley. The Zn content of ecological malt are even significantly higher. Corresponding values of Zn contained in the soil and plants show that raising the amount of Cd prevents Zn absorption in the plants. The Cd content of hops (25 µg/kg) is evidently not influenced by the method of cultivation. However, hops contain large amounts of Cu

(260 mg/kg) which far exceed the copper content of the soil. Plants absorb Cu primarily through Cu-containing sprays. The heavy metal content from the raw materials can also be seen in the end products, spent grain and beer. The Cd content of ecological spent grain has been shown to be significantly less than those from conventional spent grain whereas the Cu content (273 mg/kg) in all the spent grain is so high that it considerably exceeds the permissible amounts for animal food. Even conventional beer contains more Cd (0.19 µg/l) than ecological beer (0.09 µg/l).

On composting of the different organic wastes, the question of the state and behavior of the toxic elements is an important point. Fuchshofen et. al. [646] investigated the influence of the fertilization with biological waste compost, cow manure compost and mineral fertilizer on different crop rotation. Pb, Cd, Cu, Ni, and Zn analyses by means of direct solid sampling GF-AAS are presented for celeriac and cabbage. The results from a 5 year study covering a huge number of samples showed that intensive fertilization with biological waste compost does not increase the heavy metals in soils and plants. In the variants with mineral fertilization, increased Cd content was found in the soils and plants.

A successful application of direct SS-GF-AAS for the determination of Cd, Pb and Cu in water-borne suspended matter was tested by Van Son and Muntau [42]. Due to the low concentration of suspended matter in freshwater bodies, the determination in water-borne particles represents a problem of trace analysis on the microchemistry scale. Concentrations of the suspended matter are common in the range from 5 mg/l to 0.1 mg/l. 250 up to 1000 ml of the collected water samples were filtered by means of 0.45 µm filters. The filter were dried by room temperature in an exsiccator and from each of the filters, which had an effective filtration surface of 1000 mm^2, disks with a surface area of 24 mm^2 were taken using a perforator. Each disc was analyzed separately. The blank values were acceptably low compared to the lowest amount of metals found on the disk (31 pg Cd; 630 pg Pb, 750 pg Cu). The concentration of the metals in the suspended matter in the lake water ranged from 1.15 up to 39 ng/l Cd, from 45 up to 1440 ng/l Pb and from 35 up to 2100 ng/l Cu.

6.3
Control Analysis

6.3.1
Analytical Control in Industrial Production

The development of the direct solid sample method with GF-AAS was closely related to its application in laboratories of the *steel industry*. In Sect. 3.3.1., the analytical details of metal analysis are discussed. At a very early stage, this method was used in routine analysis of metallurgical and related samples. As an example, the investigations of Sommer and Ohls [107] and Bäckmann and Karlsson [22] should be mentioned. Both made use of the first commercial sample carrier system (microboat, see Table 2.1 and Fig. 2.5) for the introduction of solid test samples into the graphite furnace. The latter authors emphasized in the conclusion of their basic study: " During the time the technique has been in routine use, clear non-homogeneity has been established in only about ten out of a total of about some 8000 samples." It can be supposed that – in addition to other techniques of solid sample analysis - solid sampling with GF-AAS is also widely established in this field, without further reflection of this application in scientific papers.

With the invention of the advanced background correction systems, routine use for other materials became possible in industry. In 1987, Esser gave an overview of the applications of direct solid sampling with the "SM1" (Grün-Optik) in German industry [647]. In the introductory section, he discussed the "Requirements on analytical information for industrial practice" as follows: "It is a well-known fact but nevertheless worthwhile discussing that analytical practice in industry especially in product control is often very different from that what happens in a laboratory of a university or other research institutions. While research deals with the elaboration of new techniques, the lowering of detection limits and similar subjects, in industry, routine applications are the norm." ... "But what should be emphasized is that accuracy and precision are not features of the highest priority in a lot of cases when an application in industry is to be discussed. This statement may appear provocative to an analytical chemist and to prevent misunderstanding, it has to be formulated more precisely. What is obtained by an analytical procedure are data of the analyzed substances, in our case, usually data of metal concentrations. These data must be a sound basis for the decisions which have to be taken in the industrial process and, in this sense, they do indeed have to be accurate and precise. However, for a wide range of industrial applications the information required for the decision process is comparatively simple if one considers what a technique such as AAS could do. A brilliant example is the monitoring of concentration limits as is the case in Germany for Cd in plastics. The information "the Cd concentration is below 75 ppm" is easy to obtain with solid sampling AAS in most cases but this is much more inaccurate than this technique is capable of reaching. Although, for an analytical chemist who wants to come nearer to the "analytical truth", this information is clearly insufficient, in industrial practice it is useful."

Concerning the introduction of direct SS-GF-AAS in the *cement* company Esser explained [179]: "Up to the late seventies there seemed to be no reason for the cement industries to deal with the determination of trace elements. In 1979, however, the thallium emission from a Westphalian cement plant showed the necessity of analyzing materals even in the ppm-region. An iron carrier with a low thallium content needed for a special product had been fed into a rotary kiln where thallium was accumulated in a heating-cooling cycle via volatile thallium chloride. In connection with this incident, an interlaboratory study on thallium analysis of inorganic material had been conducted under the direction of the Forschungsinstitut der Zementindustrie (then Research Institute of the Cement Industry). The scattering of the results, covering two orders of magnitude, pointed to a considerable lack of knowledge of thallium analysis at low concentrations. Obviously a lot of work had to be done to obtain adequate results. With this in mind, in 1982, the

"Anneliese Zementwerke AG" decided to buy an atomic absorption spectrometer mainly for monitoring the kiln in- and output. Direct analysis of solids had become a more and more valuable technique for trace determination even in geological and related inorganic materials."

After the presentation of methodological investigations Esser concluded that: "For two years direct ZAAS has been used in the laboratory of the "Anneliese Zementwerke" to determine trace elements in a variety of inorganic matrices. Participation in interlaboratory studies as well as the measurement of standard reference materials indicate the usefulness of this method for a lot of elements. The time-saving sample preparation makes it especially recommendable for routine analysis of well-known matrices."

The excellent results of direct solid sampling with GF-AAS in several interlaboratory collaborative studies in the German cement industry and research institutes, documented in [648, 295], caused other cement factories to establish direct solid sample analysis in their own laboratories. In a case study, the accumulation of various trace metals in the cement production process was investigated using direct solid sample analysis by Desmorieux [335]. Figure 6.8 a, b, c shows the kiln in- and output balance of thallium in comparison to that of zinc. This work was carried out

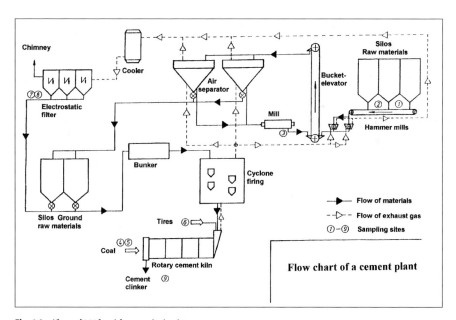

Fig. 6.8. (from [335], with permission).
a Flow chart of a cement plant: The production of cement is connected with grinding and milling of great amounts of raw materials (lime stone, silicates, oxides of aluminium and iron) and an high input of fuels (coal, waste tyres). The firing in the rotary cement kiln and in the cyclone causes a lot of exhaust gas and dust, which will be separated by electrostatic filter. The filter dust is lead back into the production process. On sampling at different sites (1-9) the amounts of different elements (Zn, Pb, Tl) are monitored and balanced.

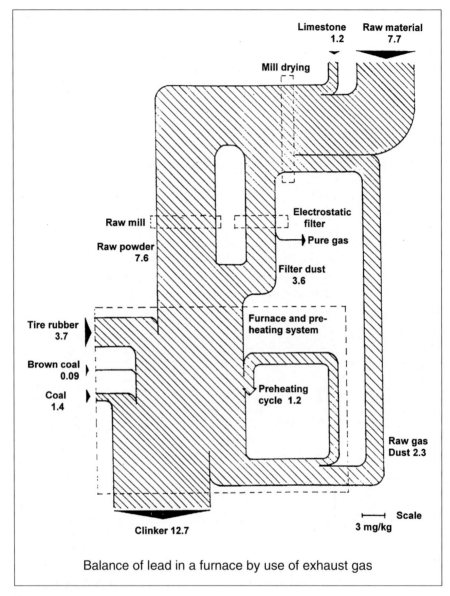

Balance of lead in a furnace by use of exhaust gas

Fig. 6.8. continued.
b Balance of Pb: In a scale the contents of lead are shown according to the flow chart of the cement production. The balance demonstrates, that the input of lead is mainly coming from the raw materials and the tire rubber. Nearly all the lead is bound in the final product (clinker).

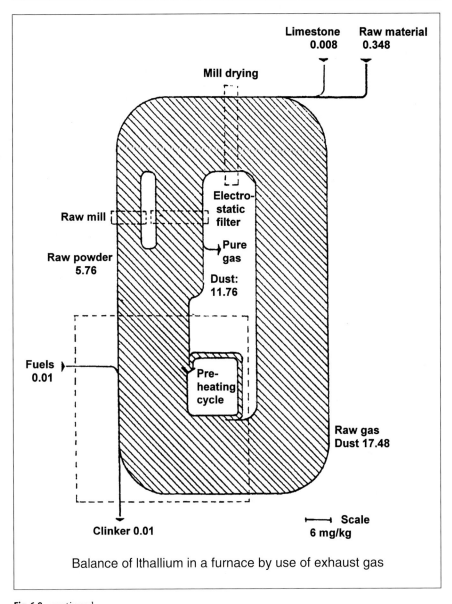

Limestone 0.008 Raw material 0.348

Mill drying

Raw mill

Raw powder 5.76

Electro-static filter

Pure gas

Dust: 11.76

Fuels 0.01

Pre-heating cycle

Raw gas Dust 17.48

Clinker 0.01

Scale
6 mg/kg

Balance of lthallium in a furnace by use of exhaust gas

Fig. 6.8. continued.
c Balance of Tl: In a scale the contents of thallium is shown according to the flow chart of the cement production. The flow chart presents very low inputs of thallium with raw materials and fuels, but a significant enrichment of thallium due to the raw gas dust cycle. In case of breakdown of the electrostatic filter, dust carrying high amounts of thallium may be released

for a thesis by only one student. With the classical approach of digestion analysis, it would not have been possible to analyze such a large number of samples in a period of only three months.

López-Garcia et al. [569] documented the analysis of cement slurries for Cd, Pb and Tl in an ethanol, nitric/hydrofluoric acid medium. They pointed out that the procedures are rapid and reliable, thus may be used for routine control purposes. The above mentioned analysis for cadmium in *plastics* was carried out at the Adam Opel Aktiengesellschaft (General Motors) on a very large sample scale, where direct SS-GF-AAS has become a high ranking control method. Rühl described the problem and the introduction of a reliable control process in [649]:

"Starting point of these activities was a Swedish law for environmental protection. According to this law, the Swedish Government will gradually prohibit the import of products and components which contain intentionally added cadmium; 75µ g/g (ppm) are stated as the permissible contamination. The effects of the German automobile industry's plan to eliminate cadmium from their products as far as possible will have a great effect on the various branches of the supplier industry and is a great challenge to the manufacturers of raw materials and the processing industries. ..."

"From the extent of the entire project, at the Adam Opel Aktiengesellschaft, as one of the major German automobile manufactures, the requirement to install a testing capacity adequate for supervising the limiting value for cadmium (75 ppm) on approx. 20,000 parts p.a. was established. The test method in question had to fulfill the requirements of large series, both with respect to cost and time, and also had to be sufficiently accurate to avoid test-related controversies with suppliers." ...

"The electrothemal atomic absorption method with background correction seems to be suitable for the direct use of solid samples if certain requirements of the instrumentation are fulfilled. One of these requirements is the use of a relatively large graphite oven capable of burning such a large number of samples which can be quickly and accurately weighed in a normal industrial laboratory without any particular provisions." ...

"It was found that most samples can be tested with an already available wave length-dispersive -X-ray-fluorescence spectrometer. This applies to samples with a cadmium content of >>75 ppm or <<75 ppm, which fit into the sample support. Individual small pieces and all samples having a cadmium content near the limiting value of 75 ppm to be tested cannot be accurately evaluated by X-ray-fluorescence analysis without extensive calibration work. For these samples, an appropriate analysis method was sought, preferably one that does not require chemical decomposition. According to more recent literature, a graphite oven AAS unit with Zeeman background correction, particularly designed for the direct analysis for solid samples should be suitable for the analytical problem. For this reason, a comparative experiment with material representative of those to be tested was conducted in which the cadmium was determined by conventional methods and compared with the respective results of the solid sample analysis (ZAAS). ... "

"Series testing of plastics used in automotive engineering has been done for about one year now, according to the scheme shown." (see Fig. 6.9). "By September 1984, approx. 16,000 parts were tested in this way. The experience so far gathered with the analysis of solid samples shows that approx. 80 samples/8 h of work on the average can be evaluated …."

"The continuous analysis of parts for cadmium shows that the change-over to cadmium-free materials from the technical point-of view, is meanwhile far advanced in automotive engineering. While at the beginning of the activities approx. 50% of the parts submitted to laboratory testing were found to have more than 75 ppm cadmium, today this is true for only about 5% of the parts, with a further downward tendency."

The analysis of polymers is of course not restricted to the determination of Cd. Other trace metals deliberately added for different reasons or contaminants e.g. from catalytic processes are of interest. Anderson and Skelly Frame [36] presented a methodological study on the determination of Pb, Cr, Co in polymers. The conditions of routine analysis of PVC for Pb are the background for the methodological study from Belarra et al. [20]. Additional respective analytical studies were performed, see Sect. 3.3 and Table 3.5.

Another example for useful applications of direct solid sampling in industry is given by Erb from the tesa-Werke Offenburg GmbH [710]. All raw materials, processing steps, and products for adhesive tapes are controlled for the content of Cu, Mn, Fe, Cd, and Pb because these metals act as "rubber poison" by catalytic effects if these are present even as traces. The introduction of direct SS-GF-AAS as an important measure of *internal quality control* "has led not only to considerable savings in working time, but also to increase the availability of the control data. For the control laboratory, which, in the course of this work, is permanently under time pressure, this was a remarkable alleviation." (Translated from the German by the author.)

Fig. 6.9. Flow chart of series of plastic parts for cadmium in product control in the automobile industry (58% of samples were analysed with SS-GF-ZAAS) (from [649], with permission)

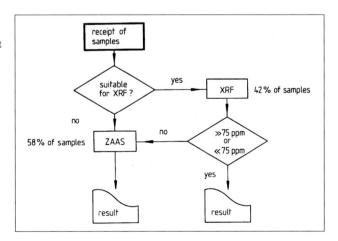

The special conditions for the analysis of high purity materials in industry and research are discussed below in Section 6.4.2.

Finally, the control for toxic metals of *fuels and refuse materials* in industrial production should be mentioned. Especially, the fast analysis of coals have been reported repeatedly as an advantageous application of direct solid sampling [153, 179, 292, 51, 280] and slurry sampling [520, 650, 651].

6.3.2
Analysis of Foodstuffs

For the analysis of process control and the official inspection of foodstuffs, a combination of features of solid sample analysis are particularly advantageous, namely (1) the fast availability of the results, (2) the low limit of analysis that is achieved (3) no danger of secondary contamination or analyte loss in the analytical process. The speed of analysis has already been pointed out in the discussion of industrial control analysis above. If only a few contaminants are to be determined (e.g. Cd, Pb, Zn), solid sampling (direct or slurry) is faster than every other analytical technique. The content of trace metals in (uncontaminated) food is generally low, mostly in the lower ng/g-range, and often below that of ubiquitous materials like smoke, dust, air particulate matter.

Hence, direct solid sampling was more and more applied in food industry and food control laboratories after this method was facilitated by modern GF-AAS instruments. Esser reported in [647] on several applications e.g. in the production of children's food for the control of raw materials (e.g. wheat meal) and products (e.g. baby cereals) and in the contamination control of soy and herb products. From the authors knowledge direct solid sample analysis has proven to be so beneficial that recently – 10 years after the first introduction in the laboratory - two German food companies have bought a second instrument system from Grün-Optik for the same purpose. It is really a pity that companies working on the "sensitive" dietary market are particularly reserved about announcing their methods of quality control (this is also characteristic of the pharmaceutical industry). As a consequence, no concrete procedures and data are available for this field. However, scientific papers have been published that give an impression about favorable applications of solid sampling in food analysis.

The application of direct SS-GF-AAS in a official food control laboratory was reported by Fecher and Malcherek [132]. They analyzed a great variety of foodstuffs such as milk powder, flour, cocoa, chocolate, meat and seeds. Although the graphite furnace system used (GBC, Carbon Rod 1000) impeded the introduction of test samples, the method was successfully used in routine analysis and complementary tasks like contamination control or distribution analysis. For example, monitoring Cd in milk (freeze dried) from the region is documented and this showed a median of 14.5 ng/g from 62 samples but single values were up to 65 ng/g.

The very low content in beer was determined by the solid sample analysis of the *dry mass residues* [645]. With freeze drying of the original beer a significant enrichment factor can be achieved. While the original concentration of the trace metals is in the lower ppb-range (Cd: approx. ng/ml), the respective content in the dry mass is in the ppm-range (Cd: approx. µg/g). This technique was occasionally applied to the analysis of other liquid foodstuffs, e.g. for the analysis of wine and orange juice and is also used for the analysis of *milk* [8, 318].

The applicability of direct solid sampling for the analysis of *animal tissues* was thoroughly investigated in the "Institute for veterinary foodstuffs" at the University of Giessen, based on a concept from Rosopulo for the legal inspection of fresh animal organs and muscles. In contrast to scientific investigations, where the analyte content of samples is related mostly to the dry weight for reaching a defined state, legal maximum values of toxic element content in foodstuffs are mostly related to the fresh weight. Consequently it seems to be advantageous to analyze the samples without a prior drying step. If homogeneizing can also be avoided the analysis is notably facilitated. The latter precondition was verified for the specific organs in a number of studies (compare above, Sect. 6.2.2).

Lücker and Schuierer [652] studied the error due to the weight loss during the sampling period of fresh tissues. They conclude that "weight determination is only a minor source of error in SS-ZAAS analysis of trace elements in matrices with a high water content. However, precautions have to be taken such as rapid microsampling, adequate setting of microbalance, low sample- and normal room temperature."

Grobecker and Klüßendorf [294] analyzed Cd, Pb and Hg in seafood of various origins (e.g. canned shrimps, mussles, tuna) directly but without any pretreatment and for comparison with the samples that had been dried previously. Considering the loss of water, the analysis of the dry and fresh material of the same sample leads to corresponding results.

The determination of Hg in seafood has gained a lot of attention in the last two decades. Muntau [653] developed a concept for meeting the analytical conditions of legal control (e.g. European Directive 93/351/CEE) by direct determination using the fresh material with the Hadeishi-furnace (the development of direct solid sample analysis with AAS is closly connected to this problem, compare Sect. 3.3.3.2.). As no tolerance range has been fixed by law, the accuracy of results and the validation of the data are fundamental in the decision process of whether or not a fish can be accepted for human consumption. The aim of the study described in [653] "was to check whether the sample of different part of dorsal muscle of the same fish could show different levels of mercury, considering the fact that in the various sections of this muscle there are different concentrations of fat and therefore different blood flows. Although the samples analysed were relatively few considering the two wet methods of mineralisation, no significant differences were shown in mercury concentration between the various parts of the same fish. ... With the Zeeman method of solid-sampling, data were obtained showing a mercury content in the tail of 19% higher than in the other parts of the dorsal muscle. ... All these problems related to the validity of the data, show the difficulty of obtaining accurate (i.e. true and precise) data for the analysis of samples of fresh fish and stress the need for thorough investigations and optimization of techniques for improving the state of the art for this type of chemical analysis."

Krebs [341] gave an overview of the problem of nickel contamination by the catalytic production process in order to convert liquid vegetable oils into a solid spreadable product. The content of Ni should be below 0.2 mg/kg in margarine because oral nickel uptake can cause allergic reactions if people are highly sensitized ("nickel dermatitis"). With her direct solid sample analysis of 25 margarines she found 2 with 0.2 mg/kg, 2 with 0.3 mg/kg and one with 1.8 mg/kg. Undoubtedly for the production control of Ni in margarine, direct solid sampling offers benefits with respect to rapidity and simplicity of the analytical procedure. Van Dalen and De Galan [573] tested the ultrasonic slurry sampler for the analysis of edible oils and fats and found that with this technique, the contribution of particulate Fe and Ni to the analytical result is improved.

In 1988, the research project "German Food Contamination Monitoring Programme" started in the Federal Republic of Germany for a term of 5 years. It was supported by the Federal Ministry for Research and Technology (BMFT) and the Federal Ministry for Health (BGA). Schauenburg and Weigert [29] reported in 1991: "The aim of this project is to establish conditions for continuous determination and description of the contamination of food with hazardous or unwanted substances in the Federal Republic of Germany. Several pesticides and other environmental contaminants as well as various heavy metals have to be determined in selected foods. The chemical analysis is to be carried out by laboratories of the official food control of the federal "Länder" (states). ... At present, 37 laboratories of the eleven "Länder" are participating in this research project. Thus, extensive analytical quality assurance is indispensable and a number of corresponding measures have been developed. The most important one is an especially designed collaborative study carried out twice a year using suitable certified biological reference materials currently available or candidate reference materials and materials of similar quality. The participants receive materials with known concentrations of analytes as well as materials with unknown concentrations of analytes. In general, the matrix of the reference material is the same or similar to that of the food being examined in the research project. ... Specific methods of analysis are not prescribed, because all investigations should be done according to the normal laboratory routines examining real food samples."

After the presentation of results from the collaborative studies (see Chapter 1, Figs. 1.3 a and b) the authors conclude: "In laboratories of official food control, numerous samples have to be analyzed every day. This work is usually done using well known spectroscopic and electroanalytical methods with a lot of necessary sample pretreatment steps. The establishment of solid sampling with Zeeman GF-AAS in the laboratories of official food control may be very helpful, because it seems to be suitable for routine analysis as well as the other methods mentioned. The avoidance of chemical sample treatment is a sound advantage in minimizing errors caused by contamination and also time saving." In the internal milestone report, they pointed to the need of further automation that would make direct solid sampling even more attractive.

The analysis of foodstuffs is also a preferred application for *slurry sampling*. A large number of studies are performed for the slurry analysis of this sample type that range from the determination of As in the ashed residues of evaporated beer [654] to the determination of Al in chewing gum [655] (see Table 5.6). The methodology for the slurry method using ultrasonic agitation was investigated thoroughly by Miller-Ihli at the Food Composition Laboratory, Beltsville Human Nutrition Research Centre (US Department of Agriculture) with the analysis of a multitude of food samples [e.g. 527, 157, 501]. López-García et al. from the University of Murcia, in particular, investigated the suitability of this method for the analysis of food samples, as there are biscuits and bread (Co) [622], paprika (Pb) [554] and (Ti) [556], cereal-based products (Cd, Pb) [617], sweets and chewing gum (Co, Pb, Zn, Fe, Cr) [620], vegetables (Cd, Pb) [618] and (Al, Cr) [606], seafood (Co, Ni) [621] and (Se) [628]. For the particular elements, the different analytical conditions for the determinations are worked out in these studies.

6.4
Determination of Very Low Content Levels

Direct solid sampling offers the two conditions that are mandatory for the determination of very low analyte amounts in solid materials: (1) Very low limits of analysis, i.e. the limit of detection of the method can be fully utilized for the analysis of the solid sample material. For the definitions and the dependence of limit of analysis and limit of detection see Sects. 1.2.4, 3.2.1 and 4.5.1.4 and (2) the method must allow one to reduce all blanks to a very low level, i.e. the transfer of the analyte from reagents, surfaces and procedures to the test sample must be avoided. The last condition is also of importance if higher contents are to be analyzed, in case the analyte occurs ubiquitously at high content (e.g. Zn and Al).

6.4.1
Analysis of Biological Materials with Low Contamination

In several biological materials - specially those which are often used for food production e.g. milk powder, muscles tissues, flours - the content of toxic metals is very low, but must be analyzed in routine and legal analysis. For these tasks solid sampling can be the method of choice. The danger of contamination, large with the conventional methods for such samples, is considerably diminished with direct solid sampling, explaining the preference for this method in the food industry (see examples in Sect. 6.3.2).

Akatsuka and Atsuya [312] have carried out an investigation into the determination of vitamin B_{12} as cobalt in solid pharmaceutical samples (mecobalamin) with direct solid sampling. They reached a limit of analysis <0.03 ng/mg of cobalt, and a "realistic" limit of 0.15 ng/mg, i.e. 4 ng/mg of vitamin B_{12}. This value can only be reached with other methods if extensive chemical pretreatment is performed.

For some elements like Co, Ni, Mn, Pb even the low limit of analysis of direct solid sampling GF-AAS is often insufficient for the determination in non-contaminated biological materials. Atsuya et al. [87, 185, 186] have developed the *preashing technique* that concentrates the analyte by the mass reduction of the matrix ("concentration factor" from 2 to 50). However, the inorganic matrix components (Na, K, Mg, Ca) are also concentrated, which may cause strong matrix effects (see Sect. 2.5.3).

6.4.2
Analysis of High Purity Materials for Technological Use

Direct solid sampling with graphite furnace and AAS detection can be an excellent analytical method for the examination of high purity materials, offering limits of analysis that can hardly be achieved with other methods. The reason is that no dilution is performed and no analyte blank occurs in the analytical procedure of the solid sample.

Hadeishi and Kimura demonstrated that at an early stage of the present method with the analysis of *GaAs crystals* [55]: "The analysis of impurities in GaAs is of particular interest because of their possible role in the mechanism of thermal conversion and resistivity degradation. These effects presumably involve impurities and their interaction with the point defects such as vacancies." An observed photoluminescence peak from annealed Cr-doped GaAs are attributed to manganese on Ga sites, however, the presence of Mn was not detectable by mass spectrographic analysis which show (at that time) a limit of analysis of 3×10^{15} cm^{-3}. The researchers applied solid sampling by injecting crystal test samples through a removable injection funnel directly into the preheated graphite furnace. This introduction technique corresponds to the probe technique and thus excellent isothermal conditions and limits of detection are achieved (see Sects. 3.1.2.1 and 3.1.2.2). They found the Mn content in two different samples to be 42 ± 4 ng/g (2.4×10^{15} cm^{-3}) and 26 ng/g, respectively, these findings are consistent with the photoluminescence observations. The limit of analysis for Mn was estimated with this technique to be $\sim 1 \times 10^{14}$ cm^{-3}. The determination of Cr, Cu and Ag has revealed similar impurities in the GaAs samples.

GaAs samples were analyzed directly for different elements by a group of scientists at Sheffield University [98]. Mn and Ag content was found that are similar to those from the analysis described above, however, the content of Bi, In, and Pb were even below the respective limits of analysis which are found to be in the lower ng/g-range. Because the semiconductor material was chromium-doped, the determination of Cr shows a higher content (approx. 500 ng/g). Thus the AAS-analysis of dissolved samples was possible for this element. Comparable results for GaAs samples for both methods were achieved. In order to minimize undesirable contamination of the furnace with the matrix elements, the samples were pretreated in a separate furnace in a stream of argon to remove arsenic (for the determination of Cr) and of argon and chlorine to remove both gallium and arsenic. In a separate study [332], it was shown that "a calibration based on aqueous synthetic standard solutions may be used for the direct determination of Cr in gallium arsenide". This group of researchers also analyzed *semiconductor silicon*. They summarized their results as follows [97]: "Antimony as a dopant at a level of ca. 35 atom/10^6 atoms (ppm, atomic) and ultra-trace concentrations of lead and manganese (<0.02 ppm, atomic) are determined in semiconductor silicon by atomic absorption spectrometry after introduction of milligram samples of silicon into a pyrolytically-coated graphite furnace. Calibration was done with standard aqueous solutions. Iron, silver, zinc and cadmium were investigated but were at concentrations below the limits of detection. The graphite microboats used for sample introduction were useful for only 3–10 samples because of silicon carbide formation."

In a basic methodological study performed at the VHG Labs (Manchester NH, USA) [89] matrix effects, interferences, calibration methods, homogeneity with the analysis of semiconducting materials by direct solid sampling using the platform boat technique were investigated in order to find out the suitability of this method for *semiconductor* technology. The researchers stated in the technical abstract of their final report: " A new method for determining three trace elements in II - VI semiconducting materials was investigated. Three elements (Cu, Fe and In) were chosen due to their deleterious impact on the electro-optical characteristics of the semiconductor, as well as the inability of currently available analytical techniques to determine these elements at critical levels (sub ppm by weight). Recoveries and detection limit studies in CdTe material show that the technique of solid sampling graphite furnace atomic absorption is superior to mass spectrometric (SS, ICP and GD) and optical emission (ICP and DCP) techniques. Moreover, results obtained to date indicate that further reduction of detection limits can be achieved."

In an early study on the analysis of Cr in GaAs Guenais et al. [333] pointed out the fact that solid sampling allows the detection of heterogeneous distribution of the dopant element to be made.

Hiltenkamp and Jackwerth [83] determined traces of Bi, Cd, Hg, Pb and Tl of *gallium* in the lower µg/g range by direct SS-GF-AAS in samples of about 5 mg of the metallic material. Electrolytically spiked samples of high-purity gallium were used for calibration (see Sect. 2.2.1.3). For the analysis of gallium (7–8 N) the limit of analysis for Bi. Hg, and Pb were improved by 2–3 orders of magnitude by previous partial dissolution of larger weighed test samples in acid.

High purity tin is used as the starting material for making certain semiconductors or superconductors. Takada and Hirokawa [274] summarized their methodological study of direct solid sample analysis of this material using Zeeman-AAS and a graphite-cup cuvette as follows: "Lead at the μg/g level and cadmium at ng/g–μ g/g levels in high-purity tin have been determined by polarized Zeeman atomic-absorption spectrometry with direct atomization of the solid sample. Pieces of high-purity tin weighing up to 5 mg for lead and 20 mg for cadmium were analyzed. Calibration graphs were constructed by use of standard solutions of lead and cadmium in the presence of pure tin having lead and cadmium content below the detection limit. The tin matrix remained in the graphite-cup cuvette after atomization and did not adhere to the wall of the cuvette, so it could be easily removed and the same cuvette repeatedly used."

The difficulties of the determination of zinc and bismuth in this material were described by Takada and Shoji [275]: "Usually, in order to determine trace elements in tin, a sample of over 1 g was dissolved, matrix tin was removed as stannic bromide and the analytes were concentrated. However, traces of zinc and bismuth in tin could not be determined by these treatments, because zinc was contaminated by the reagent used and bismuth was lost as bromide with stannic bromide. Sodium-DDTC-MIBK extraction-flame atomic absorption spectrometry could not be applied to the determination of zinc and bismuth in high purity tin because of the low sensitivity." They concluded their experiment with direct atomization of the solid samples: "Traces of zinc (0.002 ~ 0.05 ppm) and bismuth (0.06 ~ 1.3 ppm) in high purity tin were determined by polarized Zeeman atomic absorption spectrometry. Standard solutions of zinc and bismuth were used for standardization. The relations between the peak areas of the absorption signals for zinc and bismuth and the amounts of zinc (0 ~ 0.6 ng) and bismuth (0 ~ 3 ng) were linear. The peak areas for zinc and bismuth were not influenced together with tin up to 25 mg for zinc and up to 10 mg for bismuth. … Time required for one sample was about 5 min."

Slurry and direct sampling in the graphite furnace for the analysis of high purity materials used in semiconductors and other electronic component parts were investigated at the ´Sektion Analytik und Höchstreinigung´ (University of Ulm, department of ´Analysis and Ultrapurification´) in a series of recent studies which will be presented below. Krivan et al. focussed their attention on the determination of relevant trace impurities, especially of alkali and alkali earth elements; these show an "extraordinarily high contamination risk" because of the ubiquitous occurrence of these elements, i.e. there are often high blanks of reagents and vessel materials and surfaces. Accordingly the low limit of detection that the GF-AAS technique offers in principle can often not be realized (compare Eq. 4.5). The problem was clearly described in [180]: "Solution procedures involving sample digestion and analyte/matrix separation are commonly used for the determination of trace impurities in … materials in modern microelectronic technology. However, the conventional digestion and chemical separation procedures represent a considerable source of contamination, especially for the ubiquitous alkali and alkaline-earth elements. Furthermore, the medium resulting from the decomposition has to be changed often to another one suitable for the separation. The step of the medium change involves the removal of the remaining decomposition acid, usually by evaporation and addition of new reagents. For example, the medium of a reported ion-exchange chromatographic separation must be free of hydrofluoric acid which, however, is required for the decomposition of molybdenum silicide. Thus, the limits of detection usually achieved for the contamination-risk elements Ca, K, Mg and Na have been at the 100 ng/g level. Therefore, methods involving sample digestion and separation are not suitable for the determination of these and some other very important elements at the lower ng/g level."

Docekal and Krivan [541] illustrated the problem of blanks impressively for the analysis of *molybdenum trioxide* after a standard dissolution procedure. Figure 6.10 a shows the signals for Ca of dissolved samples in comparison to the

Fig. 6.10. **a** Results of three parallel determinations of Ca in molydenium trioxide by solution GF-AAS and corresponding values for five parallel determinations of blanks obtained by using a dissolution procedure (mean and standard deviations of replicate measurements). The water blank is indicated as *W* (from [541], with permission). **b** Fluctuation of the Mn blank in four series of dissolution experiments (in each: n=5) for the determination of magnesium in ultra high purity molybdenium powder. It can be recognized, that the estimated limit of detection (*black column*) is higher than the content of the sample (from [180], with permission)

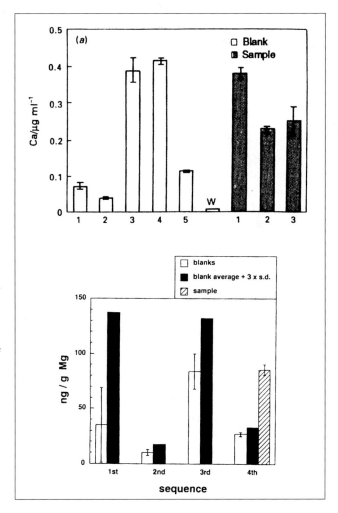

blank signals; this occurs in a similar way for the determination of Mg, K, and Na. Even though they were able to reduce the blank level of the dissolution process by simplifications of sample handling, use of selected vessels made of an ultra-pure material and the avoidance of contact of the sample solution with the neck of the flask, further reductions were achieved by the "*blank controlled slurry sampling technique* …(that) allows an essential reduction of the contamination risk and , consequently, a decrease in limits of detection to be achieved."

One of the main impurities in many high-purity materials is Si because of its prevalence in the lithospere. Hauptkorn et al. [543] found for the analysis of *ceramic materials* that "slurry sampling GF-AAS has proved to be an advantageous method for the determination of silicon in powdered titanium dioxide and zirconium dioxide samples, when calcium nitrate is used as a chemical modifier to overcome otherwise strong matrix interferences. Compared with methods that involve sample decomposition and separation of the silicon, this method leads to lower blanks, it is less time consuming and no hazardous acids are needed." Similar conclusions were drawn for the analysis of *silicon carbide* by Docekal and Krivan [607] and *boron nitride* by Hauptkorn and Krivan [518], however, the latter authors stated that for these samples "a drawback of this technique can be seen in the relatively short lifetime of the graphite tube (about 100 atomizations) …". A slurry sampling GF-AAS method for the determination of Al, Cr, Cu, Fe, K, Mn, Na, Zn in *silicon nitride* powders was developed by Friese and Krivan [238]. Presenting the results that are achieved with different methods (sample decompositon-GF-AAS and ICP-MS, NAA) they concluded: "… the slurry sampling technique provides the best limits of detection of all available nonradioactive methods for a number of elements, and for K and Zn, even including RNAA. At the same time, it is well suited for routine analysis."

Nakamura et al. [194] analyzed the *ceramic materials* that are used as heat-resistant, high-strength materials for turbine blades and ceramic engines by direct atomization. They studied the influence of the particle size, mixing with graphite powder and the size of the test sample on the accuracy of the analysis. Unfavorable effects were observed e.g. double peaks and bent calibration curves, however, their investigation and methodological developments showed that "the proposed method was successfully applied for the determination of Cu and Pb in several Si_3N_4 and SiC ceramic powders".

Similar conclusions were drawn after a thorough study of the analysis of *zirconium dioxide based materials* with the slurry technique by Schneider and Krivan [526]. They carried out investigations into the behavior of the zirconium matrix in the atomizer by electron microscopy, energy dispersive X-ray spectrometry and radiotracer technique. Systematic studies of the analysis of zirconium dioxide applying direct solid sampling were documented by l´vov [3] with the development of graphite tube AAS.

Schäffer and Krivan [571] determined a large variety of elements in high purity *graphite powders* using the slurry technique. They described the difficulties with the analysis of this material by other methods (e.g. incomplete decomposition, insufficient limits of analysis, no suitable calibration materials) and conclude that this method seems well-suited for routine application, because it provides extremely low limits of analysis, low contamination risk, the possibility of standardization using aqueous reference solutions, short analytical procedure, and simple handling. Results of direct solid sampling analysis of graphite powders were documented by l'vov [3] and Belysev et al. [657] at a very early stage of this method. Kowalska et al. [345] determined lead in graphite powder and found much better limits of analysis compared to emission spectrometry. Koshino and Narakawa [184] found good agreement of their solid sampling results of graphite powder to those from analysis after decomposition for a variety of elements (Li, Na, K, Mg, Al, Fe, Ni, Cr).

In order to achieve a further reduction of blank values, Docekal and Krivan [180] used direct solid sampling for the analysis of *molybdenum metal* and *molybdenum silicide* powder. The high density of these materials prevents the preparation of a sufficiently stable slurry [336], see Sect. 5.2.2.3; additional problems occur with the slurry technique with the accumulation of interfering refractory residues on the tubewall or the platform. With the method developed the blanks were reduced even more effectively than with the slurry technique. In a preliminary study [336], a basic radiotracer investigation was carried out, making in situ a simulation of the production process of the molybdenum metal. They were able to prove that the analyte elements can be directly and effectively separated from these materials by "electrothermal extraction". The method was optimized with regard to possible blank fluctuation and detection power, proper standardization and correction for background attenuation. Figure 6.10 b shows the fluctuation of blanks of dissolution experiments. It can be seen that the blank values and thus the limits of detection are in the range of or even higher than the content of the analyte element Mg, which was determined by direct atomization of solid test samples. Two different solid sample introduction techniques and graphite furnace systems were compared (Varian rod cup atomizer and PE-HGA72 tube and removable platform boat, see Fig. 2.5 and Table 2.2) . The researchers found that with the cup technique, the potentially contaminated surface is limited to the inner wall of the cup, whereas, in the platform technique, the whole surface of the platform can contribute to the blank, moreover it is well known that the contaminated cold ends of the tube may release a low amount of analyte during platform introduction will produce blanks. Consequently the cup technique provided lower limits of detection than the platform technique. Despite this observation, the authors concluded generally: "The direct solid sampling technique for GF-AAS represents a simple, powerful and fast method for the determination of several detrimental ultra-trace impurities such as alkali elements and magnesium, in molybdenum-based materials for microelectronic applications. This technique improves the detection limits of the commonly used solution GF-AAS by 1–2 orders of magnitude. Comparable detection limits can be only achieved by highly instrumental methods such as RNAA and GDMS. The method described seems to be applicable also to a number of related high purity materials with similar physico-chemical properties."

Friese et al. [181] used direct solid sampling for the analysis of *high purity tantalum* powders used in advanced capacitor technology, for very/ultra large scale integration devices (V/ULSI), and as a gate material in MOS-components. The authors describe the problems that are connected with the use of different analytical methods (e.g. the extremely high matrix activity limits the performance of radioactive NAA, because of the line-richness of the matrix emission spectrum, a complete matrix-analyte separation is required for the application of ICP-AES). The conclusion of their methodological study (see also Sect. 3.3.1 and Fig. 3.12) may be regarded as representative for the application of solid sample atomization for the analysis of high purity materials: "Solid sampling GF-AAS using an automated platform sampler proved to be a rapid and reliable method … Its advantageous features include excellent limits of detection, especially for elements prone to contamination in digestion processes, low matrix interferences and avoidance of the toxic reagent hydrofluoric acid. Considering the performance obtained, this method represents a promising approach to the analysis of the high-purity refractory materials."

6.5
Analysis of Filter Media

Several attempts have been reported on the direct analysis of particulate airborne or water suspended matter on filters. In addition to these techniques sampling techniques have been developed that take advantageous of the features of direct solid sampling in the graphite furnace.

6.5.1
Direct Analysis of Loaded Filter Media

Generally the direct analysis of organic filter materials (cellulose or paper) raise no particular problem. The calibration is carried out mostly with reference solutions. If the reference solution is pipetted on a blank filter that is used as the test sample, a kind of synthetic solid reference material is produced [37, 24].

If the filters are ground prior to the analysis, the normal sampling procedure for pulverized materials can be carried out. However, the obvious sampling procedure is the analysis of "cuts" from the filter disc. Such test samples were produced by the use of "cutters" e.g. made from Perspex bars [246], by a steel circular cutter [344], a desk-top paper punch [42, 37, 41] or by a scalpel [42, 41]. A prerequisite for the use of this sampling technique is a sufficiently homogeneous distribution of the particulate matter and thus presumably also for the analyte on the filter surface. Moura et al. [344] performed an ANOVA study of the distribution of Pb on cellulose ester filters and found that no difference was found at the 5% significance level, however, the dispersion of values was larger at the periphery than at the center of the filter disc. The relative standard deviations of the analyte amounts on the test sample cuts are reported e.g. for Cd to be between 8–15% [41, 42, 37, 246], however, may reach up to 30–50% [42]. Van Son and Muntau [42] showed that a large range for the precision may occur even in a series in which the experimental conditions are kept identical. The precision of the analysis of a large number of filters (~150) for water-borne suspended matter showed a distribution that was skewed considerably to the right for Cd, Cu, and Pb; while the mode was generally at a RSD between 5–15% (for ~60% of test results), the tail reached up to 40–70%.

Schothorst et al. [37] compared the direct introduction of filter cuts with that of ground filters. They found that in most cases the variations in the results of the grinding technique were lower, indicating a more homogeneous sample. Thus, "the grinding technique has to be preferred ... especially when analyzing filters with relatively thick layers of dust, which can be expected from air sampling in industrial environments ...". However, the mean content achieved with both methods shows good agreement. Slight differences might be "due to the locations of the sampling head (of the air sampling device) that were not exactly the same."

A systematic methodological study for the accuracy of filter analysis was performed by Almeida et al. [246]. They determined Cd, Cr, Cu, Ni, and Pb in industrial atmospheric particulate matter by GF-AAS using solid samples directly from the filters used for trapping and compared the results with these from analysis after chemical decomposition of the filter. They concluded:" The results obtained ... for solid sampling analysis show that it is a viable alternative to the conventional procedure. The proposed procedure is simple, yields results acceptable on the basis of currently established statistical evaluation criteria and eliminates the sample preparation steps". Grobecker and Muntau [41] documented a comparison of the Hg determination of 35 air filters using direct solid sampling GF-AAS and NAA. The results are in very good agreement (the solid sampling results lie mostly between these of two different NAA analyses).

Some workers pointed to the blank problems associated with the analysis of filters. For large analyte quantities collected on the filter, these values may be neglected [44, 42], however, blank values must generally be inspected, a simple task using direct solid sampling. Rettberg and Holcombe showed that for a paper filter no blank for Ag could be observed but for Mn the blank content created an absorbance of 0.15 [24].

6.5.2
Special Sampling Techniques for Particulate Matter

In addition to investigations of analytical characteristics some publications report on the studies of the sampling process and special filter techniques are proposed that take advantage of the features of direct analysis.

Siemer and Woodriff [40] developed a carbon rod atomizer, using the *tubes directly as filters* for airborne particulates. They found a filtration efficiency for lead particulates from 97–99%. The limits of analysis are from 20 ng/m^3 (Hg, Se) to 0.1 ng/m^3 (Cd) based on a 10 l air sample. The chief reason for the very low detection levels is that all of the collected material is atomized and measured in one instant. An advantage of the technique lies in the fact that essentially the same apparatus and procedures are applicable for any element which can be determined by atomic absorption. The authors pointed out: "Other approaches usually require dissolution of a filter material with subsequent analysis of a fraction of the resulting solution. The high reagent or filter blanks unavoidable by some techniques drastically limit the detection limit capabilities of these methods. For example, a single square centimeter of common organic membrane filters contains an average of 8 ng of lead. This ´blank´ amount of lead is 800 times the sensitivity of the direct approach. The ´blank´ associated with handling the porous tubes, contact with the filter adaptor, storage, etc. is usually less than the sensitivity of the final analysis; i.e. for lead this blank is on the order of 10^{-11} g or less." They described a practical experiment as follows: "The ambient level of lead and mercury was determined in a laboratory in the chemistry building at Montana State University to assess the degree of reproducibility possible at fairly low levels. ... Values of 0.144 µg/m^3 with a relative standard deviation of 13 per cent were achieved for 9 lead determinations and 0.23 µg/m^3 and an RSD of 8 per cent for four mercury measurements."

A method for direct analysis of solid airborne particulate matter with GF-AAS was developed by Chakrabarti et al. [43] by using a *porous graphite probe as a filter* for air particles. The probe is then subjected to atomization by being inserted into a graphite furnace which has been pre-heated. After their preliminary study of the collection system, the filtration efficiency, the atomization conditions and the precision occurring in practical experiments, the authors considered: "Important advantages of the method include high relative sensitivity and short analysis time. It is relatively free of contamination from sample handling, reagents, and from matrix interferences. ... A very attractive feature of the method is that, by combining a highly efficient filtering system with the extreme sensitivity of solid sampling, it offers the potential of monitoring sources of air particulate emissions with sufficient temporal and spatial resolution for unequivocal identification of the pollution source, at least on a local scale."

A combination of air filtering on cellulose nitrate membranes and *dry deposition on graphite platforms* and the direct GF-AAS analysis of both samples was used by Low and Hsu [44] in order to determine the dry deposition velocities of lead under various meteorological conditions. Figure 6.11 a shows the experimental set up for both sampling schemes, Figure 6.11 b shows the micro-deposition unit using 16 platform boats. "The micro-depositon samplers were exposed to the air for a certain period of time ... After sampling, the sample holder was coverd, sealed and double-packed ... The analysis of the micro-deposition samples was comparativly easy, as no further preparation was required. Each platform boat was inserted into the graphite furnace and analysed directly." With the known surface area exposed to deposition (i.e. surface of the platform boats),

Fig. 6.11. **a** Experimental set up for low volume air sampling, micro-deposition and mass particle-size sampling: *1* Anderson cascade impactor, *2* vacuum pump, *3* and *4* low volume air samplers, *5* and *6* sampling heads or filter holders, *7* and *8* aspiration tubes, *9* and *10* micro-deposition samplers, *11* polystyrene board. All sampling systems were run concurrently. **b** Sampling matrix for the micro-deposition in detail (from [44], with permission).

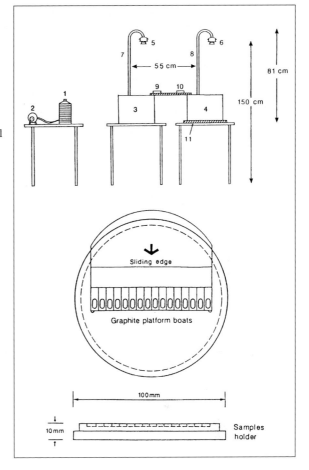

it is possible to calculate the dry deposition flux. Thus, together with the known concentration of lead as measured simultaneously it is possible to estimate the dry deposition velocity V_d. The authors concluded: "Using dry deposition flux as a yardstick, the micro-deposition sampling provides a useful means of assessing the spatial variation in lead pollution and its effects in any area, particularly in highly polluted areas where sampling time can be less than 1 h. … In particular, the simple micro-deposition technique can potentially become a useful tool for routine monitoring of lead pollution and its effects on the environment. It may also be possible to extend the technique to other collection surfaces (including natural surfaces) so that the dry deposition velocities of lead and other trace metals to these surfaces can be estimated."

Results obtained from a collection system for solid airborne particulate matter based on *direct impaction in the graphite tube* was developed and described by Sneddon [658]. The atmospheric sample is drawn by vacuum through a jet, with the output stream directed against the inner wall of a graphite tube that acts as the impaction surface. An almost identical area of deposition for the analyte amount in the tube was achieved by the use of an aerosol deposition system (Fastac II, [659]). With this technique, nearly real-time monitoring of the ambient air can be attained. A collection technique and prompt analysis that was originally proposed by John M. Ottaway was described by Sneddon [337, 338]. He developed theory and principles, design and construction and evaluated the system on the *electrostatic precipitation* of the analyte (Pb, Mn, Hg) in aerosols on a tungsten collecting electrode, followed by injection of this rod into an electrothermal atomizer for AAS determination. The practical application showed: "The precision of the analysis was difficult to obtain as the homogeneity of the air sampled will contribute significantly, e.g., the short-term precision for lead obtained by sampling disturbed air for 1 min at a flow rate of 10 l/min⁻¹ was 10.8% for four measurements. However, ... 50 measurements ... (gave) precision in excess of 100%." No definite statement about the trueness of the analysis could be given, because "the task of producing a standard aerosol of known, variable and reproducible particle size is extremely challenging." However, the author pointed out the obvious advantage of such a system: "If the levels (of metals in the air) are ca. 20 ng/m³ and higher then ... it is possible to obtain the concentration in ca. 3 min."

6.6
Homogeneity Testing

The increase in the *sampling error* in solid sample analysis due to *heterogeneous distribution of the analyte elements* in the laboratory sample is frequently considered in the literature as a disadvantage or a limitation of this method. However, as documented extensively in this book, it has been proven in practice and it is theoretically confirmed that solid sampling results are as "certain" as results achieved with the established methods - with the condition that all effects, i.e. sources of uncertainty in the analytical processes under consideration are evaluated [660].

Actually, the contribution of analyte heterogeneity in the laboratory sample on the precision of direct solid sampling results using test samples in the mass range of 0.1–1 mg are generally dominating. It is a distinguishing mark of the direct solid sampling method that it is "under statistical control", i.e. all other significant random effects can be quantified or can even be neglected (see the discussion in Sect. 2.4.1.2). Thus, the *relative sampling error* S_H occurring for a *test sample mass m* can be estimated directly from the overall *experimental relative standard deviation RSD* or can easily be evaluated considering the other random errors in the analytical process. As the consequence a *homogeneity constant* H_E (for the analyte element H) that is a property of the laboratory sample can be determined according to (see Sect. 2.3.1.2)

$$H_E = S_H \sqrt{m}$$
$$\leq RSD \sqrt{m}$$

Eq. 6.1

This characteristic makes direct solid sampling highly suitable for detection and quantification of sample homogeneity. For other methods of trace element determination, either the process is not entirely controlled for random effects (e.g. in the decomposition process) or the required test sample is too large for revealing micro-heterogeneity effects, e.g. for XRF; (for the suitability of the slurry method for homogeneity test see below, Sect. 6.6.3)

With the development of the graphite furnace technique, L`vov pointed to the potential of direct solid sample analysis for the determination of heterogeneity [3]: "It is interesting to note that it is even possible to establish a quantitative correlation between the coefficient of variation w, which represents the reproducibility of the results within a series of measurements of a particular element in samples, and the mean size P of the samples used for analysis." Experiments performed by Katskov and L´vov [3] with the determination of numerous trace elements in zirconium dioxide confirmed the expected relation between test sample mass and analytical imprecision. From the graphical documentation of their results of a series with different test sample masses homogeneity constants in the range of $H_{Pb} \sim 5$ mg$^{1/2}$ and $H_{Sb} \sim 13$ mg$^{1/2}$ can be estimated.

In another early approach to homogeneity testing using direct solid sample GF-AAS, Lord et al. [46] analyzed freshwater mussel aggregate samples (twelve animals of the same species freeze-dried and ground). With the relative large number of 30 test samples they gained a reliable characterization for the homogeneity of copper that was typically 10–15% RSD using test samples of 1–5 mg. From the observation that for metallurgical samples the analytical precision of direct solid sampling in the mg-range is also strongly influenced by analyte heterogeneity Bäckman and Karlsson proposed "that the method can also be used for studies of heterogeneity" [22].

The concept for the determination of the homogeneity constant as a measure for sample heterogeneity using direct solid sampling was proposed by Kurfürst et al. and discussed in a series of papers [12, 117, 141]. The theoretical basis and some examples are provided in Sect. 2.3.1. This concept has already found applications for the characterization of reference materials (see below). Frech and Baxter [156] gave examples for the determination of Al in biological reference materials. They found homogeneity constants of about 16 mg$^{1/2}$ using different ranges of test sample mass. With the development of a programm package "SOLIDS" for the evaluation of direct solid sampling data, Berglund and Baxter [95] included the continuous calculation of the homogeneity constant of the analyte element in the analyzed sample. This information may also be important for routine analysis, giving the operator "on-line"-information about the degree of homogeneity that is independent of the test sample mass actually used.

Undoubtedly, homogeneity control by solid sampling using the graphite furnace will find further acceptance and will be a preferred application of this method in the future.

6.6.1
Production Control of Reference Materials

The producers of reference materials have made great efforts in the past to achieve reliable characteristic numbers for the homogeneity property of CRMs. However, for biological, geological and environmental materials up till now only roughly estimated minimum masses (e.g. 100–250 mg) are given in the certification reports, for which the certified content for the trace elements are suggested as being not (significantly) influenced by a heterogeneous distribution of the analyte. However, as the results of direct solid sampling show, the conjecture that acceptable results cannot be achieved using test samples below the recommended minimum weight are not justified on the basis of data achieved with the methods applied. The homogeneity data that are given nowadays by the producers of CRMs characterize sufficiently the "between-bottle" heterogeneity but are, up to now, not suitable for the characterization of the "in-bottle" (micro-) heterogeneity that may also be different for the various analyte elements.

Advanced studies on the assessment of microhomogeneity were performed at the *IRMM (Institute for Reference Materials and Measurements, Geel, Belgium)*. Pauwels et al. have laid down three criteria that a method should meet [661]: "The microhomogeneity test should preferably be carried out using a method that does not require a sample pretreatment, so that both analyte loss and contamination can be avoided; (that) allows a sufficiently precise analysis of small samples, preferably <1 mg, but which are representative of the bulk material; (and that) is fast, so that a large number of analyses can be carried out in a short time." Obviously direct SS-GF-AAS fulfills these criteria. Because of the proven "usefulness of SS-ZAAS for the microhomogeneity control of CRMs" [661] the method was so far applied during the production of several CRMs at the IRMM which are described below.

In order to assess the homogeneity of silver for the candidate reference material EC-NRM 522, "Copper" (for neutron dosimetry) 9 samples were taken at each of 6 different locations [84]. The results of the solid sample analysis are displayed in Fig. 6.12. The statistical evaluation reveals that no significant differences in concentration exist between the six locations and from the experimental standard deviation the "sampling standard deviation" is estimated with a 0.45 mg sample size to be 0.11 µg/g (11%) and at the 4 mg level 0.046 µg/g (4.8%).

On behalf of the German automobile industry federation (VDA) the IRMM produced a set of four plastic materials certified for Cd in the range of 40–400 mg/kg (see also Sect. 6.3.1). Prior to the certification [624, 326] a carefully performed homogeneity study was performed using direct solid sampling GF-ZAAS [123, 661]. 60 test samples out of different bottles of each material were analyzed using test sample masses between 60–250 µg. For the different materials the following homogeneity constants were determined: VDA-OO1 "Sicolen Yellow" H_{cd}=3.1 mg$^{1/2}$, VDA-OO2 "Sicolen Orange" H_{cd}=2.5 mg$^{1/2}$, VDA-OO3 "Sicolen Red" H_{Cd}=1.7 mg$^{1/2}$. For the sample VDA-OO1 "Sicolen Bordeaux" a sig-

Fig. 6.12. Results of homogeneity control of candidate reference material Copper EC-NRM 522: Ag content fo fifty four 450 µg test samples taken at six different locations in the batch (from [84], with permission)

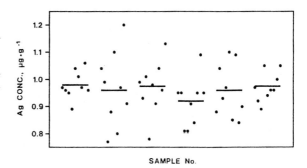

nificantly larger spread of test results were found (RSD ~ 35%) that were not even normally distributed. For this sample, a heterogeneous distribution of the cadmium was confirmed by visual observation of darker spots yielding effectively higher results. For the use as a reference material in digestion analysis, in this case, a minimum sample mass of 1000 mg should be used, while for the other three materials only 27 mg, 18 mg and 13 mg, respectively, are required (see below "minimum sample mass").

At the request of BCR at the IRMM, fresh seafish muscles were prepared as a candidate reference material to be certified for some trace elements [134]. Direct SS-GF-AAS was already used for the control of the production process. As shown in Fig. 2.21, an accentuated decrease in the distribution width of test samples resulted for mercury after each separate production step could be observed. Similar observations were made for Fe, Cd and Zn. The final product (BCR CRM-422 Cod Muscle) was extensively tested for homogeneity in 85 to 138 test samples for five elements: Pb, Cd, Hg, Fe and Zn [127]. The samples were collected from one single bottle (50% of the analyses) and from a mixture of 20 different bottles (also 50%). As no differences were observed between both series, the results were pooled to draw conclusions about the microhomogeneity of the entire material. While for Zn, Fe and Hg the distributions of test results could be considered to be normal, for Cd and Pb skewed distributions were achieved, probably due to the "nugget effect". As an example Figure 6.13 a, b shows the time sequence and the distribution of the test sample results for Fe and Figure 6.14. a, b the respective diagramms for Cd. The evaluation of the homogeneity constant for the normally distributed elements are performed by using the relation given by Eq. 6.1, yielding H_{Hg} <=26 mg$^{1/2}$, H_{Fe} <=21 mg$^{1/2}$, H_{Zn} <=5.3 mg$^{1/2}$. Because it was assumed that the experimental RSD is dominated by the sampling error, these values may be overestimated. (A correction for other random errors could be performed using Eq. 2.16.) The not normally distributed data sets were evaluated using the "nugget-model", i.e. considered as a Poisson distribution. In Figure 6.14 b, the designated fractions of results containing different numbers of "Cd-nuggets" (x=0, 1, 2, 3) are indicated. The histogram of the nugget distribution is plotted in Figure 6.14 c a fitted with of a Poisson probability function,

Fig. 6.13. Microhomogeneity study of Fe in BCR CRM 433, Cod Muscle (n=85, m=1.71 mg, s=15.9%, mean content normalized to 1). **a** Time series of the test sample measurements. **b** Frequency histogram and normal distribution fit (from [127], with permission)

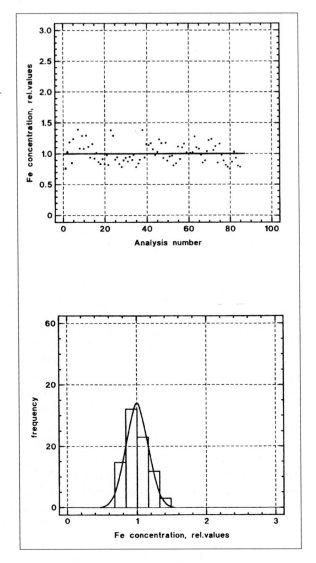

yielding z=0.546 (m=1.18 mg). Using Eqs. 2.20 and 2.16, a relative homogeneity constant of H_{Cd}=28 mg$^{1/2}$ can be estimated. The respective evaluation of the experimental data for Pb yielded H_{Pb} =27 mg$^{1/2}$. If the "nugget-model" is applied, other random effects are not considered, thus the estimated value is actually based only on the sampling error (compare with the discussion in Sect. 2.3.2).

Basic investigations on the homogeneity of reference materials using direct SS-GF-AAS are performed in the framework of the *Environmental Speciment Bank (ESB)* at the Research Centre Jülich, Germany. This work was initiated by Stoeppler et al. [117]. With the revelation of the occurrence of "outliers" which

Fig. 6.14. Microhomogeneity study of Cd in BCR CRM 433, Cod Muscle (n=108, m=1.18 mg, s=23.9%, mean content normalized to 1). **a** Time series of the test sample measurements **b** Frequency histogram for the content and normal distribution fit, subdivision of the results in the number of nuggets (x=0, 1, 2, 3) **c** Frequency histogram for the fractions with different numbers of nuggets and Poisson distribution fit (compare Figs. 2.18 a, b, c.) (from [127], with permission)

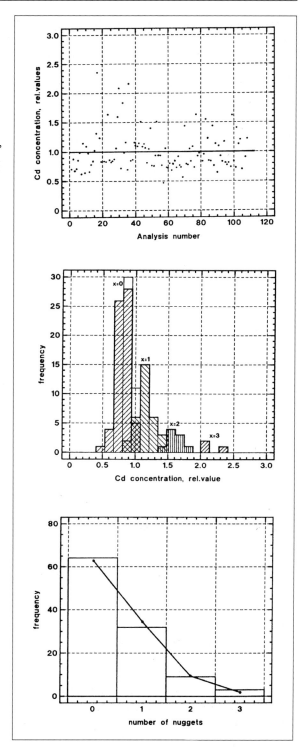

represent real analyte amounts in solid sampling results by Mohl et al. [124] the investigation of the "nugget-effect" had begun. In the following years a number of materials have been characterized for homogeneity of different elements, some of which are described below.

Bagschik et al. [662] documented a series of solid sample analyses in a variety of reference materials (pine needles, spruce shoots) and bar diagrams for the ranges of test sample results (spinach, tomato leaves, tea leaves) for Pb, Cd, Cu and Ni. In the case of asymmetric distributions, they propose the difference between mean and median for a preliminary characterization of heterogeneity. Series of the analyses of wheat gluten, corn bran, wheat flour and whole egg powder samples for Cu, presented by Ihnat and Stoeppler [663] indicated the good homogeneity of these materials (for the interpretation of the Pb results, see below 6.6.4).

Rossbach, Sonntag et al. [664, 136] performed notable investigations about the homogeneity of IAEA and ESB-own reference materials. Various direct solid sampling analyses were carried out in different ranges of test sample mass. The data were evaluated for the homogeneity constant of the respective trace elements and the validity of the relationship between test sample mass and sampling error. The documented plots "relative sampling error over test sample mass" for Pb in single cell algae [664] and for Pb and Cu in IAEA-393 Algae [136] correspond clearly with the correlation given in Eq. 6.1, confirming that the concept of the homogeneity constant is suitable for the characterization of the heterogeneity effect. Figure 6.15 a, b show two time series of results (of four in all) using different test sample masses for the determination of Cd in a mussel material from the ESB and in Figure 6.15 c the respective plot for the relative standard deviations over test sample mass. From the curve fitting, a homogeneity constant of $H_{Cd}=4.3$ mg$^{1/2}$ can be evaluated. Applying direct solid sampling GF-AAS, this group has characterized a large number of reference materials for different elements (mussel, tobacco, single cell algaea, urban dust) for Cu, Cd, Hg, Pb and Tl. All investigated materials show homogeneity constants <10 mg$^{1/2}$, thus can be considered suitable as reference materials even for micro-analysis.

At the *Institute of Veterinary Food Hygiene* (Giessen University, Germany) a large number of the laboratory's own reference materials (RM) from animal tissues for the use in analytical quality control have been produced. Lücker et al. [128] gave the following reasons for the great effort they invested in this work: A lab-internal reference material can be used in addition to and even as a substitute for a CRM. This saves money, especially if these materials are used for methods with prior sample decompositon. Using typical test sample masses of 0.1–1 g a bottle of a CRM is used up in a short period. Secondary, the matrix composition of the sample and the reference material should be as similar as possible. However, in many cases a CRM which is, in matrix composition, equal or similar to the sample is not available. The same can be said for the analyte content of reference materials. A further reason for the production of RMs is its positive effect on the laboratory's analytical quality assurance. It will ultimately be nec-

Fig. 6.15.
Microhomogeneity
study of Cd in a mussle
reference material from
ESB. **a** Time series of the
test sample measure-
ments with a test sample
mass m=0.04 mg **b** Time
series of the test sample
measurements with a
test sample mass
m=0.51 mg **c** Plot of the
relative standard devia-
tion vs the test sample
mass and a regression
with the theoretical
function (from [136],
with permission)

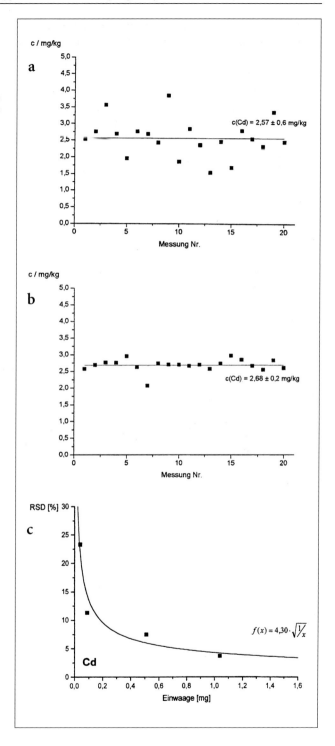

essary to verify the content of the element of interest in the laboratory's own reference material. This involves the use of a certified reference material, the application of standard methods, or even the development of additional and complimentary methods and an exchange of reference materials and results with other laboratories.

RMs produced at this institute comprised the following materials: bovine liver [120, 128], bovine teeth, bone, muscle, blood, equine renal cortex, and porcine kidney [128]. Different techniques of direct solid sampling were applied to the production and homogeneity control, but were also used for the determination of the final reference value. Figure 6.16 shows the scheme for the lab-internal production of bovine liver reference material.

In connection with this work, Lücker et al. revealed the endogenous origin of strong heterogeneity effects in pulverized bovine muscle samples that lay in the occurrence of calcified cysts of *Cysticercus bovis* (larval stage of the "ox tapeworm") that may be in the tissue. This effect was carefully investigated and documented in [128, 643]. In Sect. 2.3.2 such a sample material is used for the clarification of the "nugget effect".

Fig. 6.16. Scheme for the internal laboratory production of Bovine Liver reference material at the Institute of Veterinary Food Hygiene (Giessen University, Germany)(from [128], with permission)

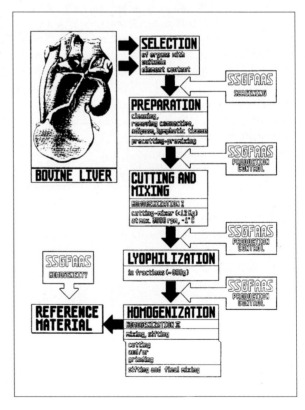

In order to assess the homogeneity of synthetically produced reference materials, Akatsuka and Atsuya [86,312] (see Sect. 2.2.1.3) calculated the relative standard deviation for a variety of trace elements on the basis of 7–9 test samples each. From the low values between 1–5% for sample amounts between 0.1–1.0 mg homogeneity constants of 1–3 $mg^{1/2}$ (!) can be estimated, so that the conclusion drawn by the authors that the analytes are homogeneously distributed is fully justified. Moreover, it can be supposed that even these low values are influenced by random errors of the method and the instrumentation (sampling procedure, baseline scatter etc.).

6.6.2
The Concept of a Minimum Representative Test Sample Mass

In the certificate of a solid CRM usually only the certified content and its uncertainty is given. The recommendation for the use of a minimum test sample mass is generally not based on a rigorous quantitative determination of the sampling error, but rather on a conservative evaluation based on the sample sizes effectively used during the characterization of the material in the certification campaign. As a consequence, the use of CRMs for calibration or control of "microtechniques" (such as solid sampling GF-AAS) is hindered.

For overcoming this lack of information Pauwels and Vandecasteele [118] propose that additionally a minimum test sample mass should be given that is based on a homogeneity study performed on solid sampling results for a definite separation of the sampling error from other random effects. The authors emphasise that the test samples used should be as low as possible, because "possible overestimation decreases when extrapolation from smaller samples to larger samples are made ... When, on the contrary, large samples are used to assess the homogeneity - as is the case for most of the existing CRMs - extrapolation to smaller samples is always hazardous".

The demanded value should give the *minimum representative sample mass M* for which the certified uncertainty interval can be expected to be met by the users analysis of the CRM. This can assumed to be true if the *tolerance interval* Δ_m (=$k's_m$) for results of *test sample mass m* is equal the certified *uncertainty interval U(c_r)* of the CRM. Considering the fact that the relation of M/m is given by the inverse relation of the respective variances (compare Eq.2.17), one can get the expression for M [118]

$$M = \left(\frac{k's_m}{U(c_r)} \right)^2 m \qquad \text{Eq. 6.2}$$

where
s_m=is the sampling error of the CRM for test samples of mass m expressed as a standard deviation, k'=factor for two sided tolerance limits.

The tolerance interval can be interpreted as the range in which future values can be expected. The *factor k´* is chosen for at least a *proportion p* of results according to a *level of confidence* and the *number of test samples n*. Tables for k`-values can be found in standard statistical text books, e.g. [665, 126].

Example

From the homogeneity study for Ag of the reference material EC-NRM 522 "Copper" (see Fig. 6.12) the following values are determined [118]: m=0.45 mg, s_m=9.5% . The (relative) uncertainty is certified for this material to be $U(c_r)$=4.2%. The k`-value for a proportion of 95% and 95% confidence (n=54) is given by 2.358. According to the above-derived expression from these values, a minimum representative sample mass of M=13 mg can be estimated. Using test samples larger than M, the certified uncertainty is valid for the actual analysis. If the user analyses smaller samples of the CRM than that, he can estimate the larger interval in which his results can be expected. In this example e.g. using 1 mg test samples the interval is increased from the certified value of 0.04 mg/kg to 0.14 mg/kg.

In [118] it is pointed that Eq. 6.2 is only strictly valid if $U(c_r)$ comprises only random ("statistical") components. As long as (unknown) systematic effects between different analytical methods play a significant role in the certification, these also contribute to the certified uncertainty. Consequently, the minimum mass will be overestimated to a certain extent.

This concept was applied in various case studies, e.g. on the plastic reference materials [123] and on the Cod Muscle BCR CRM-422 [127] (see above). From these studies summarized in [661] it can clearly be deduced that the test sample mass of the second and third generation of CRMs are mostly significantly smaller than those that have been recommended so far. The value of M would give important information in the case when only a few replicate measurements can be performed as is the case for the classical analytical methods based on decomposed and slurry samples, where in routine analyses the number of prepared test samples is mostly only 1 (!) to 3. For direct solid sample analysis, the declaration of a homogeneity constant would be more convenient than the mimimum test sample mass, because the decision, e.g. whether a CRM is suitable for calibration purpose can be made more directly. Basically, both values must be determined by a carefully performed homogeneity study.

Sonntag and Rossbach [136] summarizes the usefulness of reliable homogeneity characterization of CRMs based on their extensive homogeneity`studies and experiences in the production of reference materials as follows:

"As the relative homogeneity factors H_E are determined in an accurate way for individual elements in a given material, they can be used to assign the uncertainty of a certified concentration which is due only to material heterogeneity. Today the uncertainty given in certified reference materials is a composite of several uncertainties – i) the systematic error of the analytical technique, ii) bias from different standardizations of all the techniques used for certification, and iii) the material's inherent heterogeneity. By assigning a certified value with the homogeneity factor related to the sample mass consumed for analysis it is possible to calculate a mass dependent uncertainty which is derived strictly from the quality of the material itself. Hence systematic errors and bias of analytical techniques will be easier to recognize and the analytical process will be more transparent."

"The accurate determination of element specific homogeneity factors for certified reference materials is a must for all CRMs to be used for quality control in micro-analytical trace element determinations. Suitable techniques such as INAA and SS-ZAAS with no sample digestion and accurate control over the total sample mass analyzed are available and should be applied regularly in the course of certification of CRMs for the quantification of trace element distribution in natural matrix materials."

"The "true value" of an element in a matrix is not a very useful parameter unless there is reliable information on the probability of regaining exactly the same value in repeated measurements (reproducibility) of the same material. This probability is clearly dependent on the number of particles with potentially different concentrations in an analytical aliquot and hence on the sample mass used for analysis. Hitherto certified values are given with a rather conservative estimate (amounting to up to 50% in some cases) of the overall uncertainty. The empirical statement "a minimum sample weight of 250 mg of the dried material (…) is necessary for any certified value … to be valid within the stated uncertainty" is misleading and may only have a commercial background. Precisely determined individual homogeneity factors for each element would therefore help the CRM users to: i) save precious material, ii) determine the systematic error of the analytical technique applied, and iii) test the reliability of the sample preparation techniques more accurately and should therefore also be endorsed by CRM producers."

6.6.3
Homogeneity Testing Using Slurry Sampling

In principle slurry sampling can also be used for the determination of homogeneity constants. However, in the discussion and assessment one must distinguish between "in-slurry" sampling, i.e. test samples are out of one slurry preparation (one cup) and "between-slurry" sampling for different slurry preparations (various sample weights in different cups).

As described in Sect. 5.2.2.1, in a slurry dispersion, the analyte can be partly or even nearly totally extracted into the liquid phase. Accordingly, the representativity of results from an analysis out of one cup could be referred to a larger sample mass than that corresponding to the actually introduced solid material into the furnace. This effect is an advantage for the determination of the mean content, but leads to a limitation for homogeneity testing. In a simplified model, the *in-slurry precision* would be smaller due to the extraction compared to the solid heterogeneity. However, it should be pointed out that if the extraction varies for sample fractions due to differences of particle size or particle composition, the sampling error might also be increased. There is still a lack of knowledge in this field, a study by Hinds et al. [109] about the extraction of analytes from soil particles, throws a spotlight on this problem.

Consequently, a reliable value for the degree of homogeneity can only be achieved from in-slurry precision if the assumption of *no analyte extraction* is ensured (or only an extraction of a small portion). In this case, the very low effective test sample mass of slurry pipettings can help to reveal low levels of heterogeneity. In [501], Miller-Ihli gives an example of a Cr determination of a coal sample, where the extraction is only 2%. The sampling precision was found to be 3.5% (n_{ss}=10) using an effective sample mass of only 30 µg, hence a sampling constant of H_{Cr}=0.6 mg$^{1/2}$ (!) can be estimated. This value, however, can only be regarded as being representative if the number of replicate measurements is large, e.g. n_{ss}>100, so that rare particles are "caught" with sufficient probability [125]. Moreover, if only one slurry is prepared and analyzed, the precision of the (pipetted) test samples may be determined by a combination of several effects, e.g. volumetric effects (see Sect. 5.2.2.2) and instrumentel effects so that the precison is not only due to sample heterogeneity (see discussion below in Sect. 6.6.4). In [501] it is demonstrated how the instrumental error in slurry analysis can be determined and deconvoluted (corresponding to Eq. 2.16).

In contrast to this, it can be assumed that the *between-slurry precision* is dominated by sample heterogeneity, comparable to the situation with direct solid sampling. Thus, this value can be used for homogeneity characterization if a sufficiently large number of preparations are analyzed (e.g. n_s>20), however, each value must be based on a sufficient number of pipettings (e.g. n_{ss}=10) in order to keep the in-slurry uncertainty small. As Bradshaw and Slavin [564] pointed out: "Differences between the averages of replicate determinations in each cup that exceed the statistical limits of the replicate results will indicate heterogeneity of the sample taken". Under these conditions a homogeneity constant can be calculated from slurry sampling data using Eq. 6.1.

Sandoval et al. [239] documented extended studies of ultrasonically suspended slurry analysis. From the experimental design (n_s=14, n_{ss}=12) a first homogeneity characterization can be expected. For Pb in Estuarine Sediment (NIST SRM 1646) the between-slurry RSD of 4% and the mean sample mass for the slurry preparations of m_s=4.6 mg gives an estimate for the homogeneity constant of H_{Pb}=8.5 mg$^{1/2}$. For Cd in River Sediment (NIST SRM 2704) the calculation leads to H_{Cd}=7.3 mg$^{1/2}$ (4.8%, 2.3 mg).

Because the sample mass m_s used for a slurry preparation is usually larger than for one test sample in direct solid sampling, the corresponding small sampling precision may be "hidden" in the overall RSD or only more distinct heterogeneity effects will come out, respectively. Mohl et al. [317] presented comparative results using direct and slurry sampling from homogeneity testing of biological materials. These measurements were performed for a diploma thesis [567] at the Research Centre Jülich and included a large number of experiments for different samples and various elements. The pattern of the slurry results are often obviously statistically not consistent (e.g. see Fig. 6.17 b). Steps and drift effects are superimposing the scatter due to sample heterogeneity to such a degree, so that the calculated values for the homogeneity constants (50–100 mg$^{1/2}$) do not reflect real sample heterogeneity and are significantly larger than these achieved with direct solid sampling (Fig. 6.17 a). From the discussion above, it follows that the proposal of the authors [317] to use the effective sample mass instead of the sample mass for one slurry preparation for the calculation cannot be followed.

However, the extensive data sets presented in [567] show qualitatively that slurry sampling also has the potential for homogeneity testing, if interfering effects are controlled or corrected (see discussion below). Finally, it should be pointed out that the fact that a significant larger total sample mass usually will be used with slurry sampling (if a comparable number of preparations are analyzed). For example, in the case of the example given in Fig. 6.17 a, b, the total laboratory sample mass analyzed with direct solid sampling was 7 mg while with slurry sampling it was 260 mg. Although the difference will not in practice be as large – carrying out replicates with direct solid sampling is simpler and shorter than the preparation and the analysis of a slurry preparation so that the number of replicates will be larger – this characteristic must be judged as an advantage of between-slurry homogenity testing.

Fig. 6.17. Microhomogeneity study of Pb in a Beech Leaves reference material from ESB (reference content 4.9 mg/kg). **a** Time series of test sample measurements achieved with direct solid sampling (m=0.14 mg) **b** Time series of test sample measurements achieved with slurry solid sampling, each bars of mean and standard deviation represents 5 slurry preparations (m=5.17 mg) (from [317], with permission)

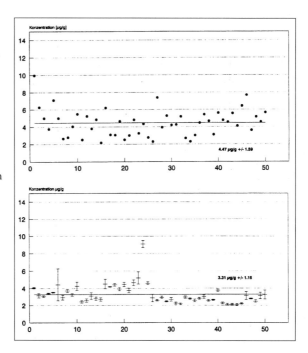

6.6.4
Assessing a Reliable Homogeneity Characterization

As shown above, solid sampling using μg- to mg-test samples allows in principle an easy characterization of sample homogeneity for trace elements. Two *conditions* must be fullfilled for the assessment of a homogeneity characterization using solid sampling data:

1. The number of replicate measurements must be large (e.g. >100), and
2. the test sample mass should only vary within a limited range (e.g. <30%).

For the first "strong" demand two reasons must be named: First, it must be realized that the estimate of a standard deviation has an uncertainty that is very large with a low number of samples. The *confidence interval for a standard deviation* at n=20 samples is approximately ±35% and even for n=100 is still around ±14% (because of the χ-distribution the range is asymmetric). Second, a possible nugget-effect can only be recognized if the *pattern of the Poisson distribution* appears that only comes out from a large(r) number of test samples (see discussion below). Obviously, the use of an automatic sampling system for dosing, transporting and introduction of solid test samples would facilitate homogeneity testing to a high extent (see Sect. 2.1.4.3). According to the experience with the use of liquid sampling systems, the influence of random errors connected with the handling procedure will be surpressed with such a system.

Furthermore, for the attainment of reliable homogeneity constants the analytical raw data must be properly treated. In this section some comments are made on *typical effects* that must be controlled or corrected.

First, the occurrence of *random effects* others than the sampling error may enlarge the calculated standard deviation from the test sample measurements. Hence, using the experimental RSD directly, the estimation of sample heterogeneity would be more or less overestimated (compare with the discussion in Sect. 2.3.1.2). If only a upper limit for the homogeneity constant or a lower limit for an acceptable test sample mass has to be fixed, the use of the experimental RSD may give a number that is suitable enough for that purpose. In that case the value gained should be stated as "lower than ..." (see Eq. 6.1).

Example

From the raw data of the measurements documented in Figure 2.14 and Table 2.8, a relative homogeneity constant of $H_{Zn}<4.2$ mg$^{1/2}$ can be calculated. Only when the noticeable contribution of the baseline fluctuation for the measurements with low test sample masses is considered, the statement of the homogeneity constant can then reasonably be given as $H_{Zn}=3.7$ mg$^{1/2}$.

The experimental design may be helpful, to keep other random effects as low as possible. For example, for the outcome of the sampling error it is useful to choose the test sample mass to be as low as possible, however, if the analyses are carried out at the limit of detection, the instrumental error may be a dominant factor – this suggests using compromise conditions. Other random effects may be significant, e.g. due to the weighing procedure or instrumentation. Consequently additionally experiments and evaluations may possibly have to be carried out in order to reduce the influence of these effects or to work out the respective effects quantitatively so that the sampling variance can deconveluted (Eq. 2.16).

Berglund and Baxter [95] pointed to the fact that the precision of the content values may be influenced also by the transformation instrument response to analyte amount, e.g. if an intercept occur or the analysis is performed in the nonlinear range. In order to avoid introducing *errors associated with the calibration function* they propose that the RSD be calculated directly from the mass-specific response data (see Eq. 2.1) according to

$$RSD = \frac{s\left(R'_i\right)}{\overline{R'}}\left(\times 100\%\right)$$
Eq. 6.3

Secondly, the calculated standard deviation for the test samples may also be enlarged due to *instability in the instrumental conditions* which leads to a (continious or discontinious) shift in the analytical sensitivity. The origin of *drift effects* are well known in analytical atomic spectrometry and are caused e.g. in GF-AAS by thermal drift of the lamp or a change in the properties of the graphite tube. If the measurements are carried out on different days using different calibration curves, *step effects* can occur. In the course of a homogeneity study which can last several hours or even several days because of the very large number of replicates required, the measurements must be monitored for such effects.

Figure 6.18 shows a time series of Cu measurements that Ihnat and Stoeppler [663] performed as a preliminary assessment of homogeneity of candidate agri-

Fig. 6.18. Time series of a microhomogeneity study of Cu in a Bovine Muscle powder 136 reference material, (from [663], with permission)

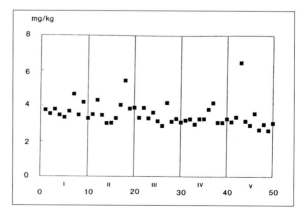

culture/food reference materials. It shows a distinct drift in the course of the 50 measurements. The authors stated: "This work is based on experiments with solid sampling graphite furnace atomic absorption spectrometry. The slight downward trend of the points is due to uncorrected instrumental drift and does not reflect on the material." A quantitative evaluation of a homogeneity constant was not performed. However, it would be possible to correct these measurements for the drift effect, because of the large number of replicates a "drift rate" can be evaluated. By a trend analysis, a straight line (in this case descending) could be estimated. Then each point can be corrected according to its chronological-position on that line, so that a horizontal line results. When this correction is performed with the data of Fig. 6.18 b (reconstructed from [663]) reduction of the standard deviation results, yielding a homogeneity constant of $H_{Cu}=7.8$ mg$^{1/2}$ instead of 9.5 mg$^{1/2}$ when the original data are used (for this estimation the largest values is not considered, see discussion below in c.).

"Step-effects" can be recognized in Figure 6.17 b showing the Pb determination in Beech Leaves using slurry sampling (about the origin of this steps see discussion about slurry below). It can be seen that the steps occur in groups of 5 points because the measurements were carried out on 10 different days [317, 567]. For the correction of this effect, two methods can be applied after calculating separately the mean values of the different groups of data: (1) The data of "shifted" groups can be corrected according to the relation of the mean values. Then the standard deviation of the entire data set can be used for the estimation of the homogeneity constant. (2) The *relative* standard deviation from every group is calculated separately. For the estimation of the homogeneity constant the average of these values can be used (if both groups comprise a sufficient number of replicates, a consistent measurement would result in close values for the RSDs of all groups). The evaluation of data visualized in Fig. 6.17 b according to the second method yields a mean standard deviation of ~10% (range 4–17%) for 10 groups of 5 test sample each. Considering the test sample weight of 5.2 mg (used for 50 slurry preparations) a homogeneity constant of $H_{Pb}<23$ mg$^{1/2}$ can be estimated (instead of 83 mg$^{1/2}$ using the "overall" standard deviation). The corrected value is still significantly larger as the value achieve with direct solid sampling (Fig 6.17 a) that is estimated to $H_{Pb}=13$ mg$^{1/2}$ for this material. A deconvolution of imprecision due to instrumental or preparation effects from the overall RSD might lead to comparable figures.

Third, special attention must be paid, if *outliers* occur in the course of the homogeneity test measurements. While, for the estimation of the mean content, it may be recommended an outlying value be deleted (or "winsorized", see Sect. 2.4.4), for obvious reasons this can lead to an underestimation of the sampling error. If one or two outlying values are observed in a relatively small series of measurements (e.g. n~20) it must be investigated thoroughly to see if this value(s) is the *result of extreme analyte heterogeneity* in the laboratory sample, i.e. is the outcome of a nugget-effect (see Sect. 2.3.2) or is caused by a contamination in the course of the sample handling during analysis. Certainty for the first can only be gained by increasing the number of replicate measurements to such an extent that the regularity of outlying values can be recognized, possibly revealing a Poisson distribution of the "outlying" values (compare Figs. 2.16–2.19). In contrast, the chronological series documented in Fig. 6.18 is too small for a decision to be made as to whether the outlying value must be considered or not, although the distribution of the data points reveals a slight skewness to larger content suggesting a nugget-effect. For the exclusion of *(secondary) contamination* a careful inspection of the analytical procedure is required.

Example

In [663] time series of 100–160 determinations for Pb are documented that show a large number of extremely outlying values. For all three bovine muscle samples, a mode below 0.3 mg/kg was found and a distinct skewness showing values up to 2 mg/kg. However, more than 10% of the values are in the range between 4 to 8 mg/kg! The occurrence of a "super nugget" which would be the only explanation of such a pattern, can be excluded by several reasons: (1) One of the samples is the well investigated sample BCR 184 that in fact shows the nugget effect, but only to values below 2 mg/kg (see Fig. 2.18). (2) If the extreme values are considered a mean content of >1 mg/kg would result for all three muscle samples. Such a content is one order of magnitude larger than what is found in various investigations of Pb content in bovine muscle (compare [643]). (3) Most of the extreme outliers lie on a "straight line", showing no statistical distribution. From this evidence, an inspection of the analytical procedure was thought necessary and the instrumentation revealed a serious source of contamination that had occurred during the analysis of these samples: The mechanical device that should guide the sample boat axially into the graphite tube was maladjusted, so that the boat frequently touched the cold end of the tube and becoming contaminated with large amounts of condensed analyte resulting in extreme peaks with the atomization!

Because in the literature, the concept of the homogeneity constant is sometimes applied without the above described critical evaluation of the data, values for the homogeneity constant can often be regarded only as the upper limit of sample heterogeneity.

6.6.5
The Potential Relevance of Heterogeneity Effects

Finally a brief remark should be made about the potential relevance of imprecise solid sampling data due to analyte heterogeneity [22, 666]: Because such a scatter of test sample results reflects the *micro-distribution* of the analyte in the sample, it can give important information which gets lost if a larger sample mass is used for the analysis.

Normally "distribution analysis" is aimed at the analysis of different parts or fractions of a sample that can be separated, e.g. morphologically defined parts of an organism. Examples for such investigations of the distributions of trace elements using solid sampling are described in Sect. 6.1.2. However, if those fractions that are different in analyte content cannot be identified or are extremely small (single particles) the analysis of the (pulverized) sample leads to an increase in the imprecision. Using small test samples, such effects came out more pronounced, in the case of a small fraction containing a very large analyte amount the "nugget effect" appears. Consequently imprecise data or the appearance of "outliers" that are the result of analyte nuggets indicate micro-heterogeneity of the original sample. (In Sect. 2.3.2, overcoming the "nugget effect" is discussed with the aim of fixing one value for the overall content of a laboratory sample.)

Examples

The existence of nuggets in the bovine muscle sample are postulated only by analytical results and the statistical evaluation (see the example in Sect. 2.3.2). It was a scientific coincidence that for bovine muscle materials, the endogenous origin of nuggets were identified. Lücker et al. [128, 643] investigated the calcificated capsules of dead *Cysticercus bovis* (larval stage of the beef tapeworm, see Fig. 6.7). They found that these capsules accumulate lead with a content up to 700 times higher than the surrounding muscle tissue. From solid sampling analysis of pulverized bovine muscle samples, it becomes possible to discriminate between the lead content of the tissue and the contribution from the calcificated material. For the BCR CRM-184 the lead content was certified to 0.24 mg/kg. By the solid sampling data given in Fig. 2.18 and the evaluation by the nugget model, the content of the muscle tissue can be estimated to be 0.14 mg/kg. Obviously, the contribution of the calcified capsules of the ox tapeworm to the lead content is notable. (It is doubtful that a consumer will prepare and eat these parts of bovine meat.)

The occurrence of skewed distributions or "outliers" may also indicate exogenous contamination of a material as was shown for the analysis of spruce needles (compare Fig. 2.16 a). Wittenbach et al. [129] showed that the inclusion of particles from anthropogenic aerosols leads to a significant increase in the overall analyte content in the material.

Baumgardt [667] investigated the heavy metal content of lung tissues from exposed and unexposed people (post mortem). The results of conventional methods (sample digestion, analyte enrichment, flame-AAS) are compared to those achieved with direct solid sampling GF-AAS. While good agreement in the mean values and the standard deviations was found for Cu, Cd, and Pb the solid sampling test sample results for Cr showed a scatter of about 100%. This effect was interpreted as being the result from inhaled and incorporated particles. Although the data set was too small to evaluate the distribution on the basis of the nugget-model, this example shows that more extended direct solid sample analysis would give the possibility of distinguishing between trace metals of the tissue (normally a low content) and trace metals amount contained in exogeneous particles (if these are relatively rare and the content is considerable larger).

6.7
Further Advantageous Fields of Application

6.7.1
Clinical/Medical Applications

In investigations of intoxication, and poisoning as well as of long-term exposure to toxic metals, direct SS-GF-AAS has proven to be a very effective and quick approach for autopsy and biopsy samples. Analysis has been reported for a broad selection of solid materials such as hair, finger and toe nails, skin, teeth, bone and various organs as well as for dried body fluids. Very often, because of the availability of only small sample amounts, solid sampling was the method of choice [668]. Several examples will be described in some detail below.

The analysis of trace element distribution along strands of *hair* or in single hairs in order to confirm e.g. occupational exposure or to identify the time of poisoning is very promising. Spruit and Bongaarts [669] compared the nickel concentration in human hair of a group of occupationally exposed employees of a nickel factory with that in a nonexposed control group by direct introduction of hair sections of 7 mm length taken at a distance of 1 cm from the scalp into the graphite furnace. The mean nickel content for the exposed group was found to be 14.5 mg/kg, whereas the control group had a mean of 0.6 mg/kg.

Pantikow-Voßgätter and Denkhaus [475] carried out a methodological study of the analysis of human hair using a ETV-ICP-AES system. Sample preparation and optimum operation parameters are described. For controlling the sampling error, the hair samples were ground in a ceramic mortar with liquid nitrogen and large test samples (8–10 mg) were introduced into the large graphite cup. The limit of analysis were found for Zn <0.8 mg g^{-1}, for Mn <0.3 mg g^{-1}, and for Mg <6.6 mg g^{-1}. The concentration values found showed good agreement with the corresponding certified values of the reference materials, relative standard deviations lay around 10%.

Stupar and Dolinsek [52] pointed out the feasibility of using direct solid sampling GF-AAS for single hair segments (1 cm), also for *in vivo plant tissue biopsy*. They studied the element incorporation into the *hair* structure. Amounts between 0.2 and 4 mg of hair were direct introduced into the graphite cup by means of the"tape sandwich" technique (see Figure 6.20 and Sect. 2.1.3.1). By means of an oxygen stream for 12 s during the ashing step, most of the organic matter was destroyed and the smoke interference during the atomization step was reduced, so that it could be corrected by the D_2 technique. The longitudinal concentrations of Cr in the scalp hair of tannery workers were found in the range from 6 to 11 µg/g at a distance from the scalp of up to 50 mm. "An interesting fea-

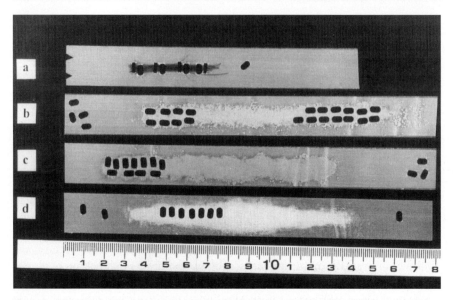

Fig. 6.19. "Tape Sandwich technique" developed for the longitudinal sampling of hair (extended for the direct introduction of pulverized test samples into the furnace, see Sect. 2.1.3.1) **a** undisturbed hair bundle, 2 mm segments, **b** hair powder, **c** plant material powder, **d** tissue sample powder. Segments were punched from the sample areas and at both sides for analyte blank measurements and the determination of the blank weight (from [52], with permission)

ture relating to the difference between washed and unwashed hair can be observed. It appears that a substantial amount of Cr can be washed from the hair by *n*-hexane or ethanol only from the proximal part of the and very little or none from the distal part." ... "The different Cr distributions of unwashed and washed hair may thus suggest slow kinetics of Cr binding to hair proteins, which is characteristic for Cr-complexes."

Bermejo-Barrera et al. [715] used the slurry sampling technique for the determination of pulverized *human scalp hair*. The effect of different inorganic constituents was studied and no significant interferences were found.

Alder and Batoreu [670] determined lead and nickel in *epithelial tissue* (hair, nail and skin) using doped gelatin powder for nickel and NIST SRM Bovine Liver reference material for lead calibration. The variation of nickel concentrations along individual hairs from root to tip and in pooled sections of finger- and toenails, as well as in ordered layers of the same fingernail was studied by direct solid sampling in cases of contact dermatitis after occupational or accidental exposure. Nickel values in pooled fingernails were 2.4 and 2.9 mg/kg. Nickel in sets of skin biopsy samples were 2.9 and 4.4 mg/kg. Lead in several nail pools ranged from 1.95 to 5.1 mg/kg and lead showed in hair some increase from root to tip from 4 to 7 mg/kg and 1 to 3 mg/kg respectively.

Barnett and Kahn [25] compared acid digestion/flame AAS and direct solid sampling for the determination of copper in *fingernails*. Nail samples from 11 people, containing between 5 and 50 mg/kg of copper showed a good correlation between the two methods. The authors mentioned that the direct method with sample amounts no larger than 0.3 mg by use of a "solid sampling spoon" required less sample preparation and had a good sensitivity. If the secondary Cu line of 249.2 nm was used, sample amounts in between 5 and 10 mg could be used in order to minimize inhomogeneity. Calibration was performed using aqueous reference solutions.

Alder et al. [316] determined nickel and aluminium at distinct depths within the *skin*. This was achieved by horizontal sectioning of skin biopsies and direct analysis of the 20–30 µg skin subsamples obtained by dry ashing and atomizing the samples in the furnace. Validation of the method was done by using wet-ashed guinea pig skin samples and the method of analyte (standard) addition with gelatine samples. Nickel values in skin were 2.9 mg/kg for non sensitive and 4.4 mg/kg for nickel sensitive persons. The alumimium value in a human skin sample was 13 mg/kg. Also skin samples from penetration studies with nickel were analyzed and resulted in values ranging from 1.39 to 12 mg/kg.

Langmyhr et al. [324] determined cadmium and lead in *dental* material by direct analysis of solid subsamples. Fourteen teeth were crushed manually in an agate mortar and pestle and subsequently pulverized either manually or automatically in equipment of the same material. To prevent losses, the mortar was partly covered during these procedures. Typically, calibration was done by using pre-analyzed hydroxyapatite. The values for cadmium in these samples ranged from 0.07 to 2.2 mg/kg, for lead from 1.1 to 6.4 mg/kg. The authors stated that the main advantages of the technique compared to other methods for trace metals in dental materials, are that the decomposition step is omitted, that no reagents are added, and that only small sample amounts are required.

Particularly useful was the application of direct solid sampling for trace element determination in small *kidney and saliva stones*. Strübel et al. [671] determined Cr, Pb, Cd, and Hg in human urinary calculi with the Grün SM 20 system using subsamples weighing between 0.1 and 50 mg. Calibration was performed by solid subsamples of certified reference materials i.e. BCR 142 Light Sandy Soil and IAEA Animal Bone H-5. Nickel in these samples was determined by the conventional approach after decomposition. The typical distribution pattern of trace metals in urinary calculi was found in the following concentration sequence; mean values in mg/kg in brackets: Pb(68) >Cd(0.49) >Cr(0.48) >Ni(0.35) >Hg(0.12). The authors stated that both analytical methods, solid sampling and wet digestion/AAS are useful for the investigation of trace metals in urinary calculi. These methods appeared very promising for the analysis of 200 urinary calculi within a research project to clarify the role of trace metals in pathogenesis and therapy. A further study on the content of Cd, Pb and Zn in human salivari calculi was performed by the same methodological approach and the same working group by Strübel et al. [672]. The study was performed as a contribution to research on diseases of salivary glands which are often caused by salivary calculi. Calculi weights ranged from 0.0479 to 0.5328 g. Grinding of the calculi was done by a vibrating mill and the subsample amounts taken for measurement ranged from 0.2 to 10 mg. Calibration was carried out using aqueous reference solutions. The typical distribution pattern (mean values in mg/kg in brackets) was found in the following sequence: Zn(45) >Pb(4.5) >Cd(0.11). The authors stated that the advantages of this approach were i) minimal sample consumption (the size of salivary calculi is often very small), ii) comparatively high sample throughput, iii) ease of operation and iv) no chemical pretreatment needed.

Another typical application of the technique was the determination of selenium content in insulin-producing *human cells* by Lindberg et al [284]. The optimum working conditions, i.e. use of modifier in nitric acid dispensed onto the solid cell material and the temperature programme for the graphite furnace (HGA-600) was evaluated by use of a ^{75}Se radiotracer. It was found that Pd modifier had the best stabilizing effect on Se up to about 1300 K. Validation of the method was performed by analysis of several reference materials (IAEA H-4, Animal Muscle; NIST SRM 1577a, Bovine Liver; NIST SRM 1566, Oyster Tissue) certified for Se. The Se content of the insulin-producing cells was found to be 1.29 mg/kg. Evaluation was done against an aqueous reference solution.

Various direct solid sampling studies have been performed on trace metal content in *human biopsy and autopsy samples* in different organ tissues. Herber et al. [673] compared the performance characteristics of a decomposition GF-AAS method and direct solid sampling for cadmium determination in the placenta using rat placenta in order to show the feasibility for human organs. The authors used a Grün SM1 spectrometer for the direct solid sampling method in freeze dried materials. The values found ranged from <0.005 to 26 mg/kg. Calibration was performed against aqueous reference solutions. Validation of the method was done by analysis of appropriate certified reference materials (NIST SRM 1577, 1577a and 1566). The authors stated that the direct method was preferable for the determination of cadmium in placenta and other tissues since analysis was faster and was less sensitive to contamination.

Nilsson and Berggren [321] determined cadmium in freeze-dried microgram samples of *pancreatic tissue* by direct insertion into a carbon rod atomizer. At 228.8 nm >10 fmol cadmium could be measured. The endogenous cadmium content in the endocrine and exocrine parts of the pancreas were 8.5±2.0 and 15.1±1.4 µmol (dry wt.) respectively. For higher concentrations the less sensitive wavelength at 326.1 nm was used. Calibration was performed using aqueous reference solutions. The method was validated by analysis of NIST SRM Bovine Liver.

Erler et al. [220] used the cup-in-tube technique with Zeeman background correction for the determination of cadmium and lead in human biological materials (*fresh liver, dried liver, and dried blood*) after addition of a mixed modifier consisting of ammonium hydrogen phosphate and magnesium nitrate with addition of diluted nitric acid and Triton X-100 to the sample in the cup. This allowed calibration by aqueous reference solutions. The method was validated by determination of Pb and Cd in NIST SRM Bovine Liver 1577 and 1577a. The authors mentioned that "solid sampling offers several advantages in comparison with the analysis of samples after wet digestion. No time consuming digestion procedures or separation/dilution or preconcentration steps are applied. This reduces the risk of contamination by chemicals, solvents or beakers. Furthermore the analysis of microsamples becomes possible. Thus the local distribution of trace elements within a sample can be determined." The authors mentioned further that "in the past failures in the analysis of solid samples often were caused by complexity of the processes occurring in the furnace, and lack of knowledge about these processes. Also the instrumentation did not offer the possibility to recognize and overcome interferences as modern spectrometers do. With the development of the Stabilized Temperature Furnace Concept (STPF) the introduction of AA-spectrometers with fast electronics, and undistorted graphic display of atomization peaks and effective background correction, the solid sampling technique became a reliable tool besides the conventional analysis of dissolved samples."

Aadland et al. [674] determined selenium in human liver *biopsy specimens* using the cup-in-tube technique and an AAS instrument with Zeeman background correction. For stabilization of selenium the addition of a mixed matrix modifier consisting of nickel nitrate/magnesium nitrate and Triton X-100 was necessary, based on previous experiments with the radionuclide ^{75}Se. Typical sample intakes ranged from 0.4 to 2.35 mg. Calibration was done against the two certified reference materials NIST SRM Bovine Liver 1577 and 1577a which were expected to have a matrix composition similar to that of the samples. The certified values for selenium of these materials were 1.1 and 0.7 mg/kg respectively. Selenium concentrations in 14 human liver biopsy specimens from patients with sclerosing cholangitis (a chronic progressive disease) ranged from 0.86 to 3.4 mg/kg with an average value of 2.2 mg/kg (dry weight), whereas the average reference value for 31 healthy Norwegian adults was 1.3 mg/kg.

Aaseth et al [675] used the same methodological approach for the determination of copper and selenium in *liver biopsy specimens* from 10 patients with primary sclerosing cholangitis (PSC). The study resulted in the following findings for n=10 patients with PSC: the average copper concentration in liver specimens of patients was 74±40 mg/kg, for n=31 healthy persons only 20±8 mg/kg. Also for selenium the average selenium concentration was 2.2±0.7 mg/kg, for the healthy controls only 1.3±0.3 mg/kg. The direct solid sampling method was verified by analysis of appropriate certified reference materials.

Oilunkaniemi et al. [347] determined selenium in *human heart muscle* samples using also the cup-in-tube technique and Zeeman background correction. However, in contrast to others, they reported optimal results only if a Pd-Cu modifier in Triton X-100 was applied. The modifier solution was introduced into the cup containing the solid sample by using an autosampler. This procedure allowed calibration by aqueous reference solutions. The method was validated by analyzing two certified biological reference materials (NIST SRM 1577a

Bovine Liver and NRC DORM-1 Dogfish Muscle). The Se content in freeze-dried human heart samples was found to vary between 0.62 and 1.14 mg/kg which was in good agreement with data from other authors using wet decomposition/ hydride AAS. The authors mentioned that the method was relatively fast and allowed Se determinations with acceptable accuracy and precision.

Frech and Baxter [156] evaluated the solid sampling technique for the determination of aluminium in biological materials by atomic absorption and emission spectrometry using a constant temperature atomizer equipped with a sample cup. The importance of this work was, besides the results with the new technical equipment, a scrupulous study of contamination sources for aluminium resulting in some improvements for sample preparation and sample introduction into the cup. The latter was performed with acid washed Teflon tubing instead of the standard glass capillary to avoid contamination, but also contamination from laboratory air had to be considered. For solid samples and the AAS approach using the cup, an absolute detection limit of 30 pg was reported. Compared to the better value for GF-AES (8 pg) this has, however, the advantage that higher aluminium content can be analyzed by the furnace technique without saturating the detector. The materials used for the study were lyophilized human serum, and three reference materials (IEAE H-4 Animal Muscle; NIST SRM 1577 and 1577a Bovine Liver).

Nordahl et al [155] determined aluminium in human *biopsy and necropsy specimens* using the same methodological approach as already mentioned above [220, 641, 675] using the cup-in-tube technique. For aluminium determination the particular advantage of the solid sampling technique is the fact that this approach allows better contamination control compared to digestion-GF-AAS, a method which is not only more prone to aluminium contamination but also significantly more time consuming. By using clean room facilities for sampling and sample preparation and the improvements recommended by Frech and Baxter [156] the working range of the procedure was from 0.2 to 100 mg/kg. The method includes the use of alternative, less sensitive lines and includes an oxygen-ashing step in the furnace programme. It was also necessary to add magnesium nitrate in nitric acid containing Triton X-100. Validation of the method and results was possible by analyzing several reference materials (IAEA H-4 Animal Muscle; NIST SRM 1577a Bovine Liver; IAEA A-11 Milk Powder; STE-105 Human Serum) with information values for aluminium. The authors stated that solid sampling in general finds its most important application for samples present only in minute quantities for which by definition homogeneity is not a factor.

A series of *post-mortem studies* for the evaluation of trace element levels in differently exposed groups of persons from some areas in Germany was performed by Pesch and coworkers [676, 677, 678, 679, 680] after scrupulously performed sampling procedures using the Grün SM1 instrument and calibration either with aqueous reference solutions or appropriate certified reference materials. The latter served also for method validation. Two of these studies presented data on lead and cadmium content of different human organs (vocal cords, renal cortex, lung tissue, skeleton muscle, bladder and liver). Lowest values for lead and cadmium were found in the vocal cords and highest for lead in the liver, for cadmium in the renal cortex. The authors found an increase in exposure during the last decades and also the expected significant differences in cadmium values of all critical organs between smokers and nonsmokers [678, 676]. In a more recent pilot study with a limited number of samples, the renal post-mortem cadmium content between Erlangen (West-Germany) and Leipzig (East-Germany) was compared [679]. The authors found in this study that for smokers (factor 1.5) as well as for nonsmokers (factor 1.8) the average cadmium content in samples from Leipzig were somewhat higher compared with that from Erlangen. However, these preliminary findings have to be confirmed by analysis of a larger number of samples. In another pilot study [680] the concentration of copper in the placenta and liver of foetuses, infants and young children was studied in order to find out if there are influences of drinking water with high Cu content.

6.7.2
Water Analysis by Use of Solid Enrichment Materials

For the determination of trace elements in water at natural background levels even the high sensitivity of modern graphite furnace AAS is often not sufficient. Numerious workers proposed the preconcentration of trace analytes on solid materials using ion exchange materials, precipitation or coprecipitation. With the use of direct solid sampling the preconcentration factor gained with the enrichment material can be fully utilized because no dilution with a subsequent dissolving process arises and the danger of contamination in the "ultra-trace" determination in water is reduced.

De Kersabiec et al. [682] compared different preconcentration techniques and the subsequent direct analysis with GF-ZAAS. As the simplest enrichment process they evaporated or freeze-dried water samples and analyzed the *dry residue*. The authors pointed that the limits of analysis, quoted in Table 6.x, are valid for waters with a mean mineralization of 0.5–1 g/l (e.g. rain water). The detection limits obtained in Zeeman spectrometry are then 10 to 20 times lower than in classical spectrometry. However, the enrichment of the minerals in the dry residue that are responsible for high background prevented an application to water heavily burdened with mineral salts (marinewaters - brines).

This problem was partly overcome with the extraction of the analyte traces by *ion-exchange resins*. Sansoni and Brunner [681] have shown that the direct determination of Pb adsorbed on resin spheres (Dowex, Levatit) is possible. De Kersabiec et al. [682] used extraction by stirring the water (50 ml) with the resin (0.5 g) for 5 min. the resin is then isolated, washed, dried at 100 °C, ground and analyzed directly. The extraction coefficient was found to be 85–95%.

Takada and Koide [683] used this technique for the analysis of tap and distilled waters, however they analyzed the resin particles as a whole. These were sieved for a mean particle size of 0.85 mm (mean mass 0.6 mg) in order avoid the weighing step ("numerometric" dosing, see also [681]) and to get standardized thermal conditions in the furnace. They observed no distinct heterogeneity effects (RSD typ. 4%). They studied the effect of pH, stirring temperature, and time on the adsorption behavior. A detailed description of the preparation, the handling of the resin, and the procedure is given. They obtained 2 ng/l for 1% absorption, thus the limit of analysis will be even lower.

Van Berkel and Maessen [392] evaluated the use of resins for the multielement determination with ETV-ICP-AES. They evaluated the analytical method [see also 395] and with the established experimental conditions they achieved good agreement with the reference value of a standard reference urine sample. They gave only the minimum (detectable) quantity of analyte mass (in ng), from which the limit of analysis for the original sample in Table 6.x is calculated by the division by the maximum amount of resin sample that can be introduced into the furnace (20 mg) and the enrichment factor water/resin (50 for the urine sample).

In a series of papers, Slovák et al. studied the sorption of Hg [684 ,685] and As, Sb, and Bi [686] on Spheron thiol for the slurry determination in GF-AAS. They investigated the optimum conditions for the sorption and the atomization process. They give the following recommendation for the procedure [686]: "Acidify 50 ml of sample solution with concentrated sulphuric acid to give a concentration of 1 M of the acid. Add 50 mg of Spheron Thiol, and stir or shake the suspension for 2 h. Transfer the remaining slurry to a centrifuge tube, centrifuge, and pour off the clear supernatant liquid. After washing twice with 5-ml portions of distilled water, transfer the resin to a small beaker and dilute to a final volume of 5 ml by adding 0.5 M sulphuric acid." This procedure was used for 8 experiments of As added to waters and to 5% salt solutions. The recovery was found to be between 97% to 122% (5% KCL).

Sedykh et al. [237] analyzed polymer sorbents that are introduced into the furnace as a slurry. The influence of the suspension particle size and density and furnace conditions on the analytical signal has been studied. They showed that the analyte-carrying resins that are introduced as a slurry can also be analyzed.

Akatsuka et al. [687, 688, 689, 690] applied *coprecipitation* with 8-hydroxy-quinoline [691] and 8-quinolinol and direct analysis of the precipitate for the analysis of waters. With the addition of magnesium as carrier ions, they achieved a recovery of nearly 100%. They described the procedure as follows: "In a standard procedure, an aliquot of sample solution (100–400 cm^3) was placed in a Teflon beaker, into which 20 mg of magnesium ions and 5 cm^3 of the 8-Q solution (100 mg as 8-Q) were added. This solution was adjusted to pH 9 by adding an aqueous ammonia solution, and allowed to age for 1 h at 70 °C on a hot plate. It was then filtered by a glass filter (no. 4G). The filter was dried at 110 °C for 1 h in a drying oven, and then weighed accurately. With a portion of the precipitate, the quantitative analysis was performed directly with AAS." With the extreme large concentration factors of around 5000 (mass relation of water sample and prepicitate) the attained limits of analysis are in the ppt-range (see Table 6.x, 8-quinolinol). Obviously with the use of reagents in the process the appearance of blanks must be taken into account. "Examinations on sample volumes larger than 500 cm3 were not made because high signals from reagent blanks were found …" With the standard procedure, the determination of Zn, they found a blank of 11 ng that was approx. 25% of the total amount of Zn when 400 ml water was used.

Atsuya et al. [692] applied coprecipitation for preconcentration using dimethyl-glyoxime+1-(2-)pyridylazo-2-naphthol with nickel as a carrier element. They achieve very high concentration factors (at least 18,000) and recoveries at 100%, but because of high blanks (Fe: 120 ng at 100 g sample) it could not be took advantage fully for the limit of analysis (see Table 6.x). For the determination of selenium and lead with this method [693] they used an internal standard for calibration with the simultaneous detection (H Z9000) of both absorption signal. As the consequence the solid test samples have not to be weighed, thus no expensive microbalance is required. As the internal standard element In was used because of the low background level of this element in water and it is not apt to contamination.

Nakamura et al. [694] used preconcentration by coprecipitation with $Zr(OH)_4$ for the determination of trace elements in natural waters. The developed method is similar to that described above. However, in order to dehydrate the precipitate it was dried at 600 °C and then analyzed directly in the graphite furnace. Recovery experiments showed that the analyte was found nearly totally in the precipitate. No problems caused by blanks were reported. An extremely large concentration factor (14,500 using 1 l water) resulted in excellent detection limits that are given in Table 6.x for several elements.

Biological organisms have the potential ability to adsorb selectively specific elements without concentrating the matrix. Shengjun and Holcombe [695] have utilized unicellular *green algae* to concentrate Ni and Co ions from sea-water and riverine water samples. Preconcentration was achieved by mixing 6 mg of the algae with 50–100 ml of water samples, and subsequently isolation of the algae by centrifugation. The pellet of algae is then resuspended in 1 ml of 0.08 M nitric acid, and analyzed as a slurry by graphite-furnace AAS. Their study show that rinsing the algae with 0.12 M hydrochloric acid improves the adsorption, and appear relatively insensitive to solution pH in the range 6–9. The maximum extraction efficiencies were 87 and 73% for Ni and Co, respectively, at ng/mg levels. However, the extraction efficiency appears sensitive to the presence of high concentrations of alkali and alkaline-earth salts, and to low concentrations of many accompanying impurities. Accurate analysis of riverine and sea-water reference samples were attained, considering the reduced uptake efficiency for the sea-water that are determined for the matrix composition (Ni 75%, Co 62%).

Dobrowolski and Mierzwa [696, 149] studied the suitability of *activated carbon* for the enrichment of some heavy metals. They concluded that in comparison to synthetic ion exchangers carbon adsorbents allow determination of concentrated impurities without the necessity of their isolation from the sorbent. These impurities can be determined directly after their introduction to the electrothermal atomizer, e.g. in form of a slurry. They recommend that this technique can be especially recommended for the concentration of mercury (and methyl mercury) before their determination.

6.7.3
Forensic Investigations

In criminal cases in which firearms were used and frequently where one or more persons have been wounded or killed, forensic analytical institutes often have to contribute significantly to solving the case by the analysis of gunshot residues [697].

Gunshot residues (GSR) are a number of substances which are released from the weapon. After firing a handgun, for example, a cloud of GSR appears at the muzzle. The GSR cloud leaves the muzzle first and the projectile then overtakes it. After the projectile has left the barrel, a further cloud expands behind it. It is the GSR that produce the traces, which are then analyzed by different analytical methods to answer the questions mentioned below. The GSR consist mainly of the residues of the burnt nitrocellulose propellant in the cartridge. More important, however, for the analytical proof are the components of the priming charge that the firing pin causes to react in the percussion cap of the cartridge. Depending on the individual manufacturer, some of the following elements may primarily be contained therein: Pb, Ba and Sb being the so-called "conventional elements of GSR" but also Sn, Ti, Mn, Hg, Al, Cu, Fe, Si, K and S.

The reaction of the priming charge is transmitted to the propellant by means of glowing or burning particles, since priming efficacy is not ensured solely by hot gases. Due to the high temperatures caused by the reactions and the mechanical abrasion the GSR also contain components of the projectile, the cartridge case and the priming case as well as particles of the material of the weapon. These residues can be found on the object (person) fired at and on the person firing the weapon, mainly on his hand.

The analysis of GSR thus is a means of clarify the three usually most important questions: (1) Shot range determination for the reconstruction of the sequence of events, (2) to find out any indication as to who fired the shot(s) in order to rule out the possibility of a certain person being the possible offender and (3) to provide information with regard to the weapon and/or the ammunition used.

Gunshot residue particles can be made visible or quantitatively determined by various methods such as scanning electron microscopy (SEM), energy dispersive X-ray spectroscopy (EDXS), auger electron spectroscopy (AES) and secondary ion mass spectroscopy (SIMS) etc.. In addition, the local distribution of GSR elements with regard to their quantity can be very sensitively determined as described first by Lichtenberg for Sb in Pb free ammunition [698, 711] by direct solid sampling using graphite furnace atomic absorption spectrometry equipped with Direct Zeeman background correction (SM 1/20/30). This methodological approach could until now be very successfully applied by the same author and his coworkers of the Bundeskriminalamt (BKA, German Federal Office of Criminal Investigation).

A method, already used for some time with good results, to obtain a sample for the subsequent application of several detection methods was developed earlier by Leszczynski [699]. He used an acetic acid-soaked cellophane foil by means of compression and diffusion to transmit the superficial adherents around the entrance hole of the projectile. In the case of determinations of the shot range, the entrance-hole can be distinguished from the exit-hole by means of the so-called bullet-wipe ring, which is caused by the GSR particles adherent to the projectile that are "wiped off" at the entrance hole [697]. If this foil-printing replica is used and small sample fragments are taken from the material hit by the projectile these can be analyzed for e.g. Pb, Zn, Sb, Cu etc. with GF-AAS. Because of the low detection limit of this method the former detection limit reached by other methods was able to be extended to 20 cm or even greater, see below [700,701]. A similar simple but fully efficient technique for this purpose can be used to lift the GSR deposit from the hand of person suspected of firing the weapon and also from the skin of the victim by using an adhesive tape. If the GF-AAS method is applied to the forefinger of the suspect's hand, small sample fragments of the adhesive tape can be analyzed for the element under question (e.g. Pb [697, 700, 698, 702], Zn [698] or Sb [711] so that a complete picture of the element distribution can be obtained and the weapon firer positively identified (see Fig. 6.20).

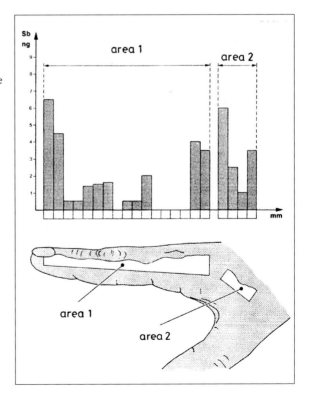

Fig. 6.20. Distribution pattern of the GSR element antimony on the forefinger of the hand of a person who had fired a gun. Analysis was performed by direct SS-GF-AAS of the the tape which was used to collect the Sb traces from the forefinger (from [711], with permission)

Direct solid sampling AAS was also successfully used for determination of the shooting distance in a murder case by analyzing the lead concentration on the skin of the victim. This could be evaluated in comparison with shooting experiments with a gun and lead containing ammunition similar to that used in the criminal case by using pig skin samples. Based on these experiments the shot range could be determined as being below 60 cm [701].

6.7.4
Analysis of Rare or Precious Sample Materials

A distinctive feature of all direct solid sampling techniques with atomic spectrometry is the requirement of comparatively small sample quantities for an entire analysis. Thus, solid sampling may be the method of choice if the material to analyze is difficult to sample, is scarce, or precious.

Freedman et al. [334] gave an example for the advantage of direct solid sampling, if the *sampling is laborious or time consuming*. They used direct solid sampling GF-AAS for the analysis of groundwater colloids. A multi-layer sampler was used for the sampling of the colloids of the aquifer. The passive sampling is difficult and time consuming. So the significant lower sample amount that is required for a reliable analysis by direct solid sampling allows a substantial reduction of the duration of the sampling period. The authors make the situation clear as follows: "For example, if we assume the concentration of colloids in groundwater to be $20 \, mg \, l^{-1}$, ... , groundwater flow velocity to be 10 m per year, the sampling vertival interval to be 30 cm, and the well diameter to be 5 cm, then the time for collection of 200 mg of colloids under natural gradient flow conditions should be 31 days ... Under the same conditions, the time needed for collecting colloids in sufficient quantities for GF-AAS analysis of Cu and Cd by solid sampling (10 mg) is only two days."

Similar condition may be given with the sampling of particulate matter from air or water (compare Sect. 6.5) or sampling of the biosphere e.g. for small insects or microorganisms (compare Sect. 6.2).

Especially in medical and clinical research for effects of trace elements often the *sample material is scarce* if the sample is taken from the living organism. The preference of direct solid sample analysis in this field lies in the small sample amount that is required in connection with low limits of analysis. Examples are given in Sect. 6.7.1 for analysis of biopsy specimens, dental materials, saliva, and human cells.

The attempts to assay *precious metals* such as gold, silver, or platinum using solid sampling atomic spectrometry [472] are based on the fact that - although most of these are destructive - the loss of sample material is small. Hinds and coworkers at the Royal Canadian Mint have done basic methodological investigations [80, 54, 63, 266] into the application of solid sampling in this field and these are described in other chapters of this book (e.g. see Sect. 3.3.1)

6.7.5
A Mobile Laboratory

Last, an application of solid sampling with GF-AAS should be discussed that is a contentious point within the "analytical community": An GF-AAS in a mobile laboratory for use in remote areas. The most serious limitation for "on site" analysis without a developed "laboratory environment" is surmounted if no sample decomposition is required for analysis, besides the demand that the analytical results must be available in a short time. Based on the methodological properties demonstrated in this book, it is obvious that solid sampling GF-AAS has a high potential for use in a mobile laboratory. The reason is that only relatively simple equipment is sufficient. It may consist, in addition to the spectrometer and the microbalance, of a diesel-powered current generator and a few simple tools for sampling and homogenization. This would allow the rapid investigation of highly polluted sites in the vicinity of metal processing plants, aquatic and marine sediments and soils, the latter also fertilized with sewage sludge, or eco-toxic trace metals such as Hg, Cd, Pb, Zn, Cr, Ni, and TL. Such quick and relatively inexpensive procedures are of particular benefit if it is suspected that legal limits are being exceeded or spots of extremely high concentrations have to be identified. The relatively short time needed for a single determination, of course, guarantees a minimum of statistically relevant data for site evaluation by using common arrangements of sampling points over the area to be checked.

Thus, the use of a developed solid sampling system in a mobile "on-site" laboratory seems to be attractive and conceivable.

Some attempts for special applications have been made over the last few decades for designing GF-AAS instruments that are suitable for such purpose. The developement of the Zeeman background correction technique by Hadeishi was originally aimed at the direct analysis of Hg in tuna unloaded from the fishing boats. The first experiences with the new technique (using a special Hg furnace (compare Sect. 3.3.3.2) were so promising that the authors emphasized in Science 1971 [289]: "Our objective was to develop an instrument that can be operated by completely inexperienced personnel (such as fishermen), has high accuracy in mercury detection, ... is very rapid in analysis without chemical separation from the host material, and is inexpensive." Twenty five years later this has become a realistic perspective.

Castledine and Robbins [703] described a portable atomic absorption analyzer (AAZ-2, Scintrex Ltd., Canada) that also took advantage of the Zeeman background correction technique. As an atomizer, a tungsten ribbon is used to limit the mass of the magnet (inverse ZAAS) and to avoid the need for high power electricity. Thus, no direct analysis could be performed, however, the sample preparation was reduced to a simple extraction procedure. The authors gave practical examples for geochemical analysis, e.g. Au and Ag in ore pulps. They conclude: "When used for analysis of samples in a base camp, close to the scene of a geochemical drilling program, the AAZ-2 can provide quick turnaround of the results and therefore on-the-spot guidance of these programs without loss of sensitivity or data quality." To the authors' knowledge, also Perkin Elmer (US) has developed a compact portable GF-AAS instrument for the aircraft industry. The (simultaneous) determination of wear metals from jet engine lubrication oils was used for indirect detection of worn machine parts.

As suggested in the statements above, further problems must be considered – (i) mechanical and electronical stability under rough transportation conditions, (ii) power requirements and its technical supply, and last but not least the (iii) analytical quality control for accuracy that must be ensured also in such operation. Furthermore, the ability of multielement determination would support the idea of "in-site" analysis. The recent experiences with the Advanced Mobile Analytical Laboratory (AMAL), which has been built up within the EUREKA umbrella project EUROENVIRON and which is deployed at the EC-Joint Research Centre/Environment Institute in Ispra have shown that complex ICP-MS systems at the moment do not attain the required stability. Alternatively, the use of a Zeeman-AAS that is mechanically and optically simpler to construct, having an advanced "oligo-element" (3–5 elements) capability and equipped with a modern side heated furnace that allows the introduction of solid test samples is considered (H. Muntau, personal communication).

Undoubtly, the instrumental and methodological preconditions for "on-site" solid sampling trace element analysis are already available. This should lead to new impelling forces for this application in the future because a growing need of "field screening" methods is obvious (e.g. [704]).

6.8
State and Role of SS-GF-AAS in Control and Research Laboratories

The discussions and the practical examples presented in this book show the often, regrettably, up to the present day it is still not recognized that the advantages of this method can significantly extend the available analytical potential when the often difficult and new analytical tasks are considered. In this section the different reasons are summarized which recommend the application of solid sampling. The points of view are different for specialized analytical laboratories for which solid sampling can be a valuable supplement than for institutions working on other scientific areas for which solid sampling can be the method of choice for the solution of specific analytical tasks.

6.8.1
Utilization in an Analytical Laboratory

All established analytical methods for trace elements show limitations for the analysis of certain materials or the determination of certain elements. The introduction of the solid sampling method with the graphite furnace can supplement the capability of an analytical laboratory by improving the analytical performances for "difficult" elements or matrices and extending the analytical tasks which can be solved more satisfactorily or more easily with this method.

Here, since these limitations cannot be discussed in detail, only some examples will be given:

X-ray fluorescence is also a solid sampling method for multielement determination, however, its application is restricted to materials with relatively high contents, e.g. to the analysis of soils or sediments. Using this method, one or two elements that must be determined may be below the limit of detection. Or, if the laboratory is in the position to analyse solid samples directly with neutron activation analysis (NAA) the determinantion of Pb is difficult because of physical problems. In these cases the sample can be additionally analysed with the solid sampling GF-AAS, avoiding sample decomposition only for the determination of a single element. This viewpoint can be applied analogous if problems occur with other methods. For example. with the analysis using ICP-AES after sample decomposition, the limit of detection is often not sufficient if low contaminated materials are to be analysed (Cd in foodstuff).

Furthermore, the influence of matrix concomitants or decomposition methods often can can not be quantified or cannot even be recognized. Analytical results of GF-AAS are independent of the matrix to a greater extent than other methods. For example, using XRF the analyst must be aware that the demands for the calibration materials are much more stringent with respect to the sample properties. The application of various electrochemical approaches for the analysis of organic materials require a carefully performed digestion. The application of ICP-MS additionally is extremely sensitive for contamination due to in-door pollution and its elimination often requires large scale investigations. In all of these cases, *solid sampling GF-AAS offers the chance of a convenient and reliable in-laboratory analysis for method development and analytical quality control.*

As demonstrated above in this chapter, solid sampling using the graphite furnace will generally be the method of choice whenever one can profit from its manifold advantages. Examples should be given here, again those that are of general importance also for instrumentally well-equipped laboratories:

If only few elements in many samples have to be analyzed, the costs and the effectiveness of the method applied have to be carefully evaluated. Particularly in cases in which only one element has to be determined in a great number of similar or different samples direct or slurry sampling may be the method of choice by omitting digestion procedures, thus also saving chemicals and lowering or even avoiding waste solutions.

Often only small amounts of sample materials are available, which would need the use of very expensive and lengthy procedures, e.g. by radiochemical techniques. In this case - if not too many elements have to be determined - an analysis by solid sampling is to be recommended (possibly prior to the application of quite sophisticated approaches or even complete analysis).

The advantage of the increased "detection power" (limit of analysis) of direct solid sampling compared to that of the method requiring sample decomposition will be used for the analysis of materials with low analyte contents as it is given often for foodstuff or high purity materials.

In-house or internal reference materials completely matching the matrix composition of samples to be routinely analysed are of increasing importance for current analytical quality control. The fastest and most reliable approach by far for checking of homogeneity and thus control of the success in preparation of these materials cannot be performed by any other method better than solid sampling GF-AAS.

It should be pointed out here again: If the automation of direct solid sampling has been achieved, a commercial and technical state comparable to that of liquid sampling, then direct solid sampling in connection with the different element detection methods will no longer have the role as a supplementary method, but will have the potential to become a general method that will be preferably applied also in routine analysis.

6.8.2
Method of Choice in a "Non-Analytical" Laboratory

The popularity of direct and slurry solid sample analysis in "non-analytical" institutions e.g. institutes of ecological, biological or geological research is due to further advantages of this method compared to the analytical methods that require sample decomposition:

First, the expenditure in laboratory infrastructure, equipment and consumables is much less. That gives more flexibility for the use of laboratory space and in the personnel and budget planning. Second, the methodological know-how that is required for obtaining reliable results is significantly reduced in its extent (not in its quality!) and the danger of analyte contamination is drastically reduced. Consequently, the deployment of "non-analysts" for carrying out trace analyses is more practicable.

Furthermore, using solid sampling, the analytical process is definitely closer to the initial goal of the research, i.e. the direct performance of the analytical procedures, the reduced duration of the entire analysis offers the operator or the scientists an intimate relation to the analytical result and thus actually also to the solution of the problem. From the author's experience, single-handed continuous analysis facilitates interactive and interdisciplinary work. Especially students who are engaged in "non analytical" scientific problems readily accepted solid sampling GF-AAS for their examination theses since they were able to have a complete overview of their analytical tasks and their handling.

Meanwhile, solid sampling GF-AAS has found acceptance and acknowledgement in the analytical community. It has been proven, that with this method reliable results are achievable if adequate instrumental, methodological tools are appplied. On this basis "non-analytical" laboratories especially can be encouraged to take advantage of this method if the solution of scientific or practical problems require trace element determination in solid materials.

References

1. L'vov BV (1976) Trace characterization of powders by atomic-absorption spectrometry. Talanta 23: 109–118
2. L'vov BV (1959) Inzh. Fiz. Zh. 2: 44
3. L'vov BV (1970) Atomic absorption spectrochemical analysis. Adam Hilger, London
4. Langmyhr FJ (1977) Direct atomic-absorption spectrometric analysis of geological materials – a review. Talanta 24: 277–282
5. Headridge JB (1980) Determination of trace elements in metals by atomic absorption spectrometry with introduction of solid samples into furnace; an appraisal. Spectrochim. Acta 35B: 785–793
6. Hadeishi T, McLaughlin R (1985) Direct ZAAS analysis of solid samples: Early development. Fresenius Z. Anal. Chem. 322: 657–659
7. Kurfürst U, Grobecker KH (1981) Feststoffanalytik mit der Zeeman-AAS. Laborpraxis 5: 28–31
8. Rosopulo A, Grobecker KH, Kurfürst U (1984) Untersuchungen über die Schwermetallanalyse in Feststoffen mit der direkten Zeeman-Atom-Absorptionsspektroskopie. Teil IV. Methodik der direkten Feststoffanalyse von biologischen Materialien. Fresenius Z. Anal. Chem. 219: 540–546
9. Atsuya I, Itoh K (1983) The use of an inner miniature cup for direct determination of powdered biological samples by atomic absorption spectrometry. Spectrochim. Acta 38B: 1259–1264
10. Chakrabarti CL, Wan CC, Li WC (1980) Direct determination of traces of copper, zinc, lead, cobalt, iron and cadmium in bovine liver by graphite furnace atomic absorption spectrometry using the solid sampling and the platform technique. Spectrochim. Acta 35B: 93–105
11. Chakrabarti CL, Wan CC, Li WC (1980) Atomic absorption spectrometric determination of Cd, Pb, Zn, Cu, Co and Fe in oyster tissue by direct atomization from the solid state using the graphite furnace platform technique. Spectrochim. Acta 35B: 547–560
12. Kurfürst U, Grobecker KH, Stoeppler M (1984) Homogeneity studies in biological reference and control materials with solid sampling and direkt Zeeman-AAS. In: Schramel P, Brätter P (eds) Trace Element Analytical Chemistry in Medicine and Biology. Vol. 3, de Gruyter, Berlin, pp 591–601
13. Kurfürst U, Pauwels J, Grobecker KH, Stoeppler M, Muntau H (1993) Micro-heterogeneity of trace elements in reference materials – determination and statistical evaluation. Fresenius J. Anal. Chem. 345: 112–120
14. Baxter DC, Frech W (1990) On the direct analysis of solid samples by graphite furnace atomic absorption spectrometry. Fresenius J. Anal. Chem. 337: 253–263
15. Brady DV, Montalvo JG, Glowacki G, Piscotta A (1974) Direct determination of zinc in sea-bottom sediment by carbon tube atomic absorption spectrometry. Anal. Chim. Acta 70: 448–452
16. Miller-Ihli NJ (1992) A systematic approach to ultrasonic slurry GFAAS. At. Spectrosc. 13: 1–6
17. Langmyhr FJ (1979) Direct analysis of solids by atomic-absorption spectrometry. Analyst 104: 993–1016
18. Kurfürst U, Grobecker KH (1982) Automatisierte Schwermetallanalyse mit der AAS-naßchemisch oder direkt aus dem Feststoff? GIT Fachz. Lab. 26: 825–830
19. Atsuya I (1991) Direct analysis of biological samples. Simultaneous multielement analysis-atomic absorption spectrometry with miniature cup solid sampling. In: Subramanian KS, Iyengar GV, Okamoto K (eds) Biological trace element research. Multidisciplinary perspectives. 445. American Chemical Society, Washington DC, pp 196–205

20. Belarra MA, Lavilla I, Anzano JM, Castillo JR (1992) Rapid determination of lead by analysis of solid samples using graphite furnace atomic absorption spectrometry. J. Anal. At. Spectrom. 7: 1075–1078
21. IUPAC (1978) Nomenclature, symbols, units and their usage in spectrochemical analysis. Spectrochim. Acta 33B, No. 6: 218–282
22. Bäckman F, Karlsson RW (1979) Determination of lead, bismuth, zinc, silver and antimony in steel and nickel-base alloys by atomic-absorption spectrometry using direct atomisation of solid samples in a graphite furnace. Analyst 104: 1017–1029
23. Welz B, Sperling M (1997) Atomic absorption spectrometry. Willy-VCH, Weinheim
24. Rettberg TM, Holcombe JA (1986) Direct analysis of solids by graphite furnace atomic absorption spectrometry using a second surface atomizer. Anal. Chem. 58: 1462–1467
25. Barnett WB, Kahn HL (1972) Determination of copper in fingernails by atomic absorption with the graphite furnace. Clin. Chem. 18: 923–927
26. Kerber JD, Koch A, Peterson GE (1973) The direct analysis of solid samples by atomic absorption using a graphite furnace. At. Absorp. Newsletter 12: 104–105
27. Kimbrough DE, Wakakuwa J (1994) Interlaboratory comparison of instruments used for the determination of elements in acid digestates of solids. Analyst 119: 383–388
28. Frech W, Welz B (1992) XXVII-CSI Pre-Symposium: Graphite atomizer techniques in analytical spectroscopy. Discussion – Modelling of gaphite furnace processes: What do we know? J. Anal. At. Spectrom. 7: 471–477
29. Schauenburg H, Weigert P (1992) Determination of element concentrations in biological reference materials by solid sampling and other analytical methods. Fresenius J. Anal. Chem. 342: 950–956
30. Miller-Ihli NJ (1995) Slurry sampling graphite furnace atomic absorption spectrometry: a preliminary examination of results from an international collaborative study. Spectrochim. Acta 50B: 477–488
31. Herber RFM, Grobecker KH (1995) A collaborative study using solid sampling graphite furnace atomic absorption spectrometry. Fresenius J. Anal. Chem. 351: 577–582
32. Miller-Ihli NJ (1990) Simultaneous multielement atomic-absorption analysis of biological materials. Talanta 37: 119–125
33. Van Loon JC (1983) Bridging the gap in analytical atomic absorption spectrometry. Spectrochim. Acta 38B: 1509–1524
34. Langmyhr FJ, Wibetoe G (1985) Direct Analysis of solids by atomic absorption spectrometry. Prog. Analyt. Atom. Spectrosc. 8: 193–256
35. E. Lücker (1996) Direkte Feststoffanalyse mit Elektrothermaler-Atomisierungs-Atomabsorptionsspektrometrie (FA-ETA-AAS) in der Lebensmittelüberwachung. Habilitations thesis, Justus-Liebig-Universität, Gießen
36. Anderson DA, Skelly Frame EM (1992) Direct analysis of polymers by graphite furnace AAS. FACSS abstract 105
37. Schothorst RC, Geron HMA, Spitsbergen D, Herber RFM (1987) Determination of heavy metals on filter material by solid sampling direct Zeeman AAS. Fresenius Z. Anal. Chem. 328: 393–395
38. Carrondo MJT, Perry R, Lester JN (1979) Comparison of electrothermal atomic absorption spectrometry of the metal content of sewage sludge with flame atomic absorption spectrometry in conjuction with different pretreatment methods. Anal. Chim. Acta 106: 309–317
39. Gries WH, Noval E (1975) New solid standard for the determination of trace impurities in metals by flameless atomic absorption spectrometry. Anal. Chim. Acta 75: 289–296
40. Siemer DD, Woodriff R (1974) Direct AA determination of metallic pollutants in air with a carbon rod atomizer. Spectrochim. Acta 29B: 269–276
41. Grobecker KH, Muntau H (1985) Spurenelementbestimmung in Rückständen in und auf organischen Filtermaterialien mit der Direkten Zeeman-Atom-Absorptions-Spektrometrie. Fresenius Z. Anal. Chem. 322: 728–730
42. Van Son M, Muntau H (1987) Determination by direct Zeeman-AAS of cadmium, lead and copper in water-borne suspended matter collected on filters. Fresenius Z. Anal. Chem. 328: 390–392
43. Chakrabarti CL, Xiuren H, Shaole W, Schroeder WH (1987) Direct determination of metals associated with airborne particulates using a graphite probe collection technique and graphite probe atomic absorption spectrometric analysis. Spectrochim. Acta 42B: 1227–1233

44. Low PS, Hsu GJ (1990) Direct determination of atmospheric lead by Zeeman solid sampling graphite furnace atomic absorption spectrometry (GFAAS). Fresenius J. Anal. Chem. 337: 299–305
45. Kurfürst U, Kempeneer M, Stoeppler M, Schuierer O (1990) An automated solid sample analysis system. Fresenius J. Anal. Chem. 337: 248–252
46. Lord DA, McLaren JW, Wheeler RC (1977) Determination of trace metals in fresh water mussels by atomic absorption spectrometry with direct solid sample injection. Anal. Chem. 49: 257–261
47. Price WJ, Dymott TC, Whiteside PJ (1980) The use of graphite cups for introduction of solid samples with an electrothermal atomizer. Spectrochim. Acta 35B: 3–10
48. Marks JY, Welcher GG, Spellman RJ (1977) Atomic absorption determination of lead, bismuth, selenium, tellurium, thallium and tin in complex alloys using direct atomization from metal chips in the graphite furnace. Appl. Spectroscopy 31: 9–11
49. Holcombe JA, Wang P (1993) Direct solid sample analysis using pressure regulated electrothermal atomization with atomic absorption spectrometry. Fresenius Z. Anal. Chem. 346: 1047–1053
50. Grobenski Z, R.Lehmann, Tamm R, Welz B (1982) Improvements in graphite-furnace atomic-absorption microanalysis with solid sampling. Microchim. Acta 1: 115–125
51. Ali AH, Smith BW, Winefordner JD (1989) Direct analysis of coal by electrothermal atomization atomic-absorption spectrometry. Talanta 36: 893–896
52. Štupar J, Dolinšek F (1996) Determination of chromium, manganese, lead and cadmium in botanical samples including hair using direct electrothermal atomic absorption spectrometry. Spectrochim. Acta B 51: 665–683
53. Lundberg E, Frech W (1979) Direct determination of trace metals in solid samples by atomic absorption spectrometry with electrothermal atomizers. Part 3. Application of an autosampler to the determination of silver, bismuth, cadmium and zinc in steel. Anal. Chim. Acta 108: 75–85
54. Hinds MW, Brown GN, Styris DL (1994) Mechanisms controlling the direct solid sampling of silicon from gold samples by atomic absorption spectrometry with electrothermal atomization – Part 1. Analyte migration to the solid surface. J. Anal. At. Spectrom. 9: 1411–1416
55. Hadeishi T, Kimura H (1979) Direct measurements of concentration of trace elements in GaAs crystals by Zeeman atomic absorption spectroscopy. J. Electrochem. Soc. 126: 1988–1992
56. Anzano JM, Martinez-Garbayo MP, Belarra MA, Castillo JR (1994) Direct determination of copper at trace levels in solid samples of animal feed using electrothermal atomic absorption spectrometry. J. Anal. At. Spectrom. 9: 125–128
57. Kurfürst U, Rues B, Wachter K-H (1983) Untersuchungen über die Schwermetallanalyse in Feststoffen mit der Direkten Zeeman-Atom-Absorptionsspektroskopie; Teil I. Ein automatischer Probengeber für die Feststoffanalyse. Fresenius Z. Anal. Chem. 314: 1–5
58. Karanassios V, Horlick G (1989) A computer-controlled direct sample insertion device for inductively coupled plasma-mass spectromtry. Spectrochim. Acta 44B: 1345–1360
59. Karanassios V, Horlick G (1990) A computer-controlled direct sample insertion device for inductively coupled plasma-atomic emission spectrometry. Spectrochim. Acta 45B: 85–104
60. ISO (1993) Guide to the expression of uncertainty in measurement. Genève, Swizerland ISBN 92-67-10188-9
61. Kurfürst U (1991) Statistical treatment of ETA-AAS (electrothermal atomisation-atomic absorption spectrometry) solid sampling data of heterogeneous samples. Pure & Appl. Chem. 63: 1205–1211
62. Lücker E, Schuierer O (1996) Sources of error in direct solid sampling Zeeman atomic absorption spectrometry analyses of biological samples with high water content. Spectrochim. Acta B51: 201–210
63. Brown GN, Styris DL, Hinds MW (1995) Mechanisms controlling direct solid sampling of silicon from gold samples by electrothermal atomic absorption spectrometry. – Part 2. Atomization from aqueous and solid samples. J. Anal. At. Spectrom. 10: 527–530
64. PROMOCHEM (1995) Reference Materials for Macro-, Micro- and Trace Element Analysis. D-46469 Wesel
65. Zeisler R, Stone SF, Parr RM, Bel-Amakeletch T (1995) Reference materials for trace element, nuclides and organic microcontaminants. IAEA, A-1400 Vienna
66. Roelandts I (1989) Biological reference materials. Spectrochim. Acta 44B: 281–290
67. Roelandts I (1989) Environmental reference materials. Spectrochim. Acta 44B: 925–934

68. Roelandts I (1991) Fourth international symposium on biological and environmental reference materials (BERM-4). Spectrochim. Acta 46B: 1299–1303
69. Roelandts I (1993) Fifth international symposium on biological and environmental reference materials (BERM-5. Spectrochim. Acta 48B: 1291–1295
70. Roelandts I (1989) Geological reference materials. Spectrochim. Acta 44B: 5–29
71. Roelandts I (1992) Additional geochemical reference materials. Spectrochim. Acta 47B: 935–945
72. Roelandts I (1991) Aluminium and aluminium alloy reference materials. Spectrochim. Acta 46B: 1101–1119
73. Roelandts I (1992) Copper and copper alloy reference materials. Spectrochim. Acta 47B: 749–758
74. Roelandts I (1993) Zinc and zinc alloy reference materials. Spectrochim. Acta 48B: 461–464
75. Roelandts I (1993) Magnesium and titanium alloy reference materials. Spectrochim. Acta 48B: 465–471
76. Roelandts I (1994) Nickel and nickel alloy reference materials. Spectrochim. Acta 49B: 1039–1048
77. Roelandts I (1994) Cobalt and zirconium alloy reference materials. Spectrochim. Acta 49B: 1097–1101
78. Roelandts I (1994) Lead and tin alloy reference materials. Spectrochim. Acta 49B: 1103–1109
79. Hofmann C, Vandecasteele C, Pauwels J (1992) New calibration method for solid sampling Zeeman atomic absorption spectrometry (SS-ZAAS) for cadmium. Fresenius J. Anal. Chem. 342: 936–940
80. Hinds MW (1993) Determination of gold, palladium and platinum in high purity silver by different solid sampling graphite furnace atomic absorption spectrometry methods. Spectrochim. Acta 48B: 435–445
81. Langmyhr FJ, Solberg R, Wold LT (1974) Atomic-absorption spectrometric determination of silver, bismuth and cadmium in sulfide ores by direct atomization from the solid state. Anal. Chim. Acta 69: 267–273
82. Noval E, Gries WH (1976) Direct determination of thallium in metallic cadmium by flameless atomic absorption spectrometry. Anal. Chim. Acta 83: 393–395
83. Hiltenkamp E, Jackwerth E (1988) Untersuchungen zur Bestimmung von Bi, Cd, Hg, Pb und Tl in hochreinem Gallium durch Graphitrohr-AAS bei Verdampfung fester Proben. Fresenius Z. Anal. Chem. 332: 134–139
84. Pauwels J, De Angelis L, Peetermans F, Ingelbrecht C (1990) Determination of traces of silver in copper by direct Zeeman graphite furnace atomic absorption spectrometry. Fresenius J. Anal. Chem. 337: 290–293
85. Janßen A, Brückner B, Grobecker KH, Kurfürst U (1985) Bestimmung von Blei, Chrom, Kupfer und Nickel in Polyethylen mit der direkten Zeeman–Atomabsorptionsspektralphotometrie. Fresenius Z. Anal. Chem. 322: 713–716
86. Akatsuka K, Atsuya I (1989) Synthetic reference material for direct analysis of solid biological samples by electrothermal atomic absorption spectrometry. Anal. Chem. 61: 216–220
87. Atsuya I, Aryu K, Zhang Q (1992) Direct determination of cobalt and nickel in powdered biological samples by graphite atomic absorption spectrometry. Anal. Sci. 8: 433–436
88. Atsuya I, Akatsuka K, Itoh K (1990) Comparison of standard materials for the establishment of calibration curves in solid sampling graphite furnace atomic absorption spectrometry. Direct determination of copper in powdered biological samples. Fresenius J. Anal. Chem. 337: 294–298
89. LABS VHG (1989) New method for measuring trace elements in II-VI semiconducting materials. Report No. B001. VHG Labs Inc., Manchester USA
90. Zàray G, Kantor T, Hassler J, Leis F, Tölg G (1994) Feststoffanalytik mittels ETV-ICP-AES. Dittrich K, Welz B: CANAS'93. Universität Leipzig, Leipzig, pp 145–152
91. Záray G, Leis F, Kántor T, Hassler J, Tölg G (1993) Analysis of silicon carbide powder by ETV-ICP-AES. Fresenius Z. Anal. Chem. 346: 1042–1046
92. Nickel H, Zadgorska Z (1995) A new electrothermal-vaporization device for direct sampling of ceramic powders for inductively coupled plasma optical emission spectrometry. Spectrochim. Acta 50 B: 527–535
93. Nickel H, Zadgorska Z (1995) A strategy for calibrating direct ETV ICP OES analysis of industrial ceramics in powder form. Fresenius J. Anal. Chem. 35: 158
94. Langmyhr FJ, Thomassen Y (1973) Atomic absorption spectrophotometric analysis by direct atomization from the solid phase. Fresenius Z. Anal. Chem. 264: 122–127

95. Berglund M, Baxter DC (1992) Methods of calibration in the direct analysis of solid samples by electrothermal atomic absorption spectrometry. Spectrochim. Acta 47B: E1567–1586
96. Eames JC, Matousek JP (1980) Determination of silver in silicate rocks by furnace atomic absorption spectrometry. Anal. Chem. 52: 1248–1251
97. Headridge JB, Johnson D, Jackson KW, Roberts JA (1987) Determination of antimony dopant and some ultratrace elements in semiconductor silicon by atomic absorption spectrometry with introduction of solid samples into the furnace. Anal. Chim. Acta 201: 311–315
98. Busheina IS, Headridge JB, Johnson D, Jackson KW, McLeod CW, Roberts JA (1987) Determination of trace elements in gallium arsenide by graphite-furnace atomic absorption spectrometry after pretreatment in gas streams. Anal. Chim. Acta 197: 87–95
99. Welz B (1986) Abuse of the analyte addition technique in atomic absorption spectrometry. Fresenius Z. Anal. Chem. 325: 95–101
100. Baxter DC (1989) Evolution of the simplified generalised standard additions method for calibration in the direct analysis of solid samples by graphite furnace atomic spectrometric technique. J. Anal. At. Spectrom. 4: 415–421
101. Atsuya I, Minami H, Zhang Q (1993) A new preparation method of calibration graphs for direct analysis of powdered biological samples by solid sampling atomic absorption spectrometry; standard addition method for solid sampling technique. Bunseki Kagaku 42: 167–172
102. Minami H, Zhang Q, Itoh H, Atsuya I (1994) Direct determination of chromium in biological materials by solid-sampling atomic absorption spectrometry: Application to three-point estimation-standard addition method. Microchem. J. 49: 126–135
103. Boonen S, Verrept P, Moens LJ, Dams RFJ (1993) Use of the simplified generalized standard addition method for calibration in solid sampling electrothermal vaporization inductively coupled plasma atomic emisssion spectrometry. J. Anal. At. Spectrom. 8: 711–714
104. Pauwels J, Hofmann C, Vandecasteele C (1994) Calibration of solid sampling Zeeman atomic absorption spectrophotometry by extrapolation to zero matrix. Fresenius J. Anal. Chem. 348: 411–417
105. Thompson M, Ramsey MH (1990) Extrapolation to infinite dilution: a method for overcoming matrix effects. J. Anal. At. Spectrom. 5: 701–704
106. Rettberg TM, Holcombe JA (1988) Vaporisation kinetics for solids analysis with electrothermal atomic absorption spectrometry: Determination of lead in metal samples. Anal. Chem. 60: 600–605
107. Sommer D, Ohls KD (1979) Spurenanalyse mit der flammenlosen AAS unter Verwendung fester Proben. Fresenius Z. Anal. Chem. 298: 123–127
108. Takada K, Hirokawa K (1982) Origin of douple-peak signals for trace lead, bismuth, silver and zinc in a microamount of steel in atomic-absorption spectrometry with direct electrothermal atomization of a solid sample in a graphite-cup cuvette. Talanta 29: 849–855
109. Hinds MW, Jackson KW, Newman AP (1985) Electrothermal atomisation atomic–absorption spectrometry with the direct introduction of slurries. Determination of trace metals in soil. Analyst 110: 947–950
110. Sonntag T-M (1996) Homogenitätsstudien für ausgewählte Elemente in verschiedenen Materialien der Bank für Umweltproben und der Internationalen Atomenergiebehörde mit der Solid-sampling-Zeeman-Atomabsorptionsspektrometrie. Diplomarbeit, Fachhochschule Aachen, Abt. Jülich
111. Benedetti-Pichler AA (1956) Essentials of quantitative analysis. Ronald Press, New York, chap 19
112. Wilson AD (1964) The sampling of silicate rock powders for chemical analysis. Analyst 89: 18–30
113. Pieczonka K, Rosopulo A (1985) Distribution of cadmium, copper and zinc in the caryopsis of wheat (*Triticum aestivum* L.). Fresenius Z. Anal. Chem. 322: 697–699
114. Branstetter B (1979) Application of Zeeman atomic absorption spectroscopy for measurements of cadmium in wheat. Laurence Berkeley Laboratory, Report LBL-9831
115. Ingamells CO (1978) A further note on the sampling constant equation. Talanta 25: 731–732
116. Ingamells CO, Switzer P (1973) A proposed sampling constant for use in geochemical analysis. Talanta 20: 547–568
117. Stoeppler M, Kurfürst U, Grobecker KH (1985) Untersuchungen über die Schwermetallanalyse in Feststoffen mit der direkten Zeeman-Atomabsorptions-Spektroskopie. Teil V: Der Homogenitätsfaktor als Kenngröße für pulverisierte Feststoffproben. Fresenius Z. Anal. Chem. 322: 687–691

118. Pauwels J, Vandecasteele C (1993) Determination of the minimum sample mass of a solid CRM to be used in chemical analysis. Fresenius J. Anal. Chem. 345: pp 121–123

119. Muntau H (1975) The role of interlaboratory studies in analytical quality control. In: Facchetti S, Pitea D (eds) Chemistry and environment: Legislation, methodologies and applications. ECSC/EEC/EAEC, Brussels and Luxembourg, 285–293

120. Lücker E, Rosopulo A, Kreuzer W (1991) Analytical quality control by solid sampling GFAAS in the production of animal tissue reference materials: Lead and cadmium in bovine liver reference materials. Fresenius J. Anal. Chem. 340: 234–241

121. Lundberg E, Frech W (1979) Direct determination of trace metals in solid samples by atomic absorption spectrometry with electrothermal atomizers. Part 1. Investigation of homogeneity for lead and antimony in metallurgical materials. Anal. Chim. Acta 104: 67–74

122. De Kersabiec AM, Benedetti MF (1987) Problems encountered in solid sampling-trace analysis of various geological samples by ETA-ZAAS. Fresenius Z. Anal. Chem. 328: 342–345

123. Pauwels J, Hofmann C, Grobecker KH (1993) Homogeneity determination of Cd in plastic CRMs using solid sampling atomic absorption spectrometry. Fresenius Z. Anal. Chem. 345: 475–477

124. Mohl C, Grobecker KH, Stoeppler M (1987) Homogeneity studies in reference materials with Zeeman solid sampling GFAAS. Fresenius Z. Anal. Chem. 328: 413–418

125. Epstein MS, Carnrick GR, Slavin W, Miller-Ihli NJ (1989) Automated slurry introduction for analysis of a river sediment by graphite furnace atomic absorption spectrometry. Anal. Chem. 61: 1414–1419

126. Sachs L (1984) Angewandte Statistik (Applied Statistics). Springer Berlin Heidelberg New York

127. Pauwels J, Kurfürst U, Grobecker KH, Quevauviller P (1993) Microhomogeneity study of BCR candidate reference material CRM-422 – cod muscle. Fresenius J. Anal. Chem. 345: 478–481

128. Lücker E, König H, Gabriel W, Rosopulo A (1992) Analytical quality control by solid sampling GFAAS in the production of animal tissue reference materials: bovine liver, teeth, bone, muscle, blood and equine renal cortex. Fresenius J. Anal. Chem. 342: 941–949

129. Wyttenbach A, Bajo S, Tobler L (1993) Spruce needles: standards and real samples. Fresenius J. Anal. Chem. 345: 294–297

130. Nakamura T, Okubo K, Sato JT (1988) Atomic absorption spectrometric determination of copper in calcium carbonate scale and carbonate rocks by direct atomization of solid samples. Anal. Chim. Acta 209: 287–292

131. Nakamura T, Oka H, Morikawa H, Sato J (1992) Determination of lithium, beryllium, cobalt, nickel, copper, rubidium, caesium, lead and bismuth in silicate rocks by direct atomization atomic absorption spectrometry. Analyst 117: 131–135

132. Fechner P, Malcherek C (1989) Anwendung der Feststoff-AAS bei der Untersuchung von Lebensmitteln. In: Welz B (ed) 5. Colloquium Atomspektrometrische Spurenanalytik. Bodenseewerk Perkin-Elmer, Überlingen, pp 501–511

133. Pauwels J, De Angelis L, Grobecker KH (1991) Solid sampling Zeeman atomic absorption spectrometry in production and use of certified reference materials. Pure & Appl. Chem. 63: 1199–1204

134. Pauwels J, Kramer GN, De Angelis L, Grobecker KH (1990) The preparation of codfisch candidate reference material to be certified for Cd, Hg, Fe and Zn. Fresenius J. Anal. Chem. 338: 515–519

135. Kramer GN, Grobecker KH, Pauwels J (1995) Comparison of methods used for the preparation of biological CRMs. Fresenius J. Anal. Chem. 352: 125–130

136. Sonntag T-M, Rossbach M (1997) Micro-homogeneity of candidate reference materials characterised by particle size and homogeneity factor determination. Analyst 122: 27–31

137. Dolinšek F, Štupar J, Vrščaj V (1991) Direct determination of cadmium and lead in geological and plant materials by electrothermal atomic absorption spectrometry. J. Anal. At. Spectrom. 6: 653–660

138. Nichols JA, Jones RD, Woodriff R (1978) Background reduction during direct atomization of solid biological samples in atomic absorption spectrometry. Anal. Chem. 50: 2071–2076

139. Grobenski Z, Lehmann R, Welz B (1981) Inprovements in graphite furnace AA using solid sampling. At. Spectrosc. Appl. Nr. 667

140. Ohls K (1984) Die Bestimmung kleiner Cadmium-Anteile in verschiedenen Materialien durch Festprobeneinsatz bei ICP- und flammenloser Atomabsorptionsspektrometrie. Spectrochim. Acta 39B: 1105–1111

141. Grobecker KH, Kurfürst U, Stoeppler M (1986) Die Bedeutung der Homogenität von Feststoffproben für die direkte Analyse und für die Analyse nach Aufschluß. In: Welz B (ed) Fortschritte in der atomspektroskopischen Spurenanalytik. vol 2, Verlag Chemie, Weinheim Germany, pp 81–89

142. Jackson KW, Newman AP (1983) Determination of lead in soil by graphite furnace atomic-absorption spectrometry with the direct introduction of slurries. Analyst 108: 261–264

143. Fuller CW, Hutton RC, Preston B (1981) Comparison of flame, electrothermal and inductive-ly coupled plasma atomisation techniques for the direct analysis of slurries. Analyst 106: 913–920

144. Stephen SC, Ottaway JM, Littlejohn D (1987) Slurry atomisation of foodstuffs in electrothermal atomic absorption spectrometry. Fresenius Z. Anal. Chem. 328: 346–353

145. ISO/DIN (1991) Genauigkeit (Richtigkeit und Präzision) von Meßverfahren und Meßergebnis-sen. Accuracy (trueness and precision) of measurement methods and results. 5725–1 (proposal)

146. ISO (1989) Uses of certified reference materials. ISO-guide 33

147. Doerffel K (1994) Assuring trueness of analytical results. Fresenius J. Anal. Chem. 348: 183–187

148. Hinds MW, Chapman G (1996) Mass traceability for analytical measurements. Anal. Chem. 68: 35A–39A

149. (1996) 7th International Colloquium "Solid Sample Analysis with Atomic Spectrometry" (ICSSAS) at FACSS XXIII, Kansas City

150. Carnrick GR, Lumas BK, Barnett WB (1986) Analysis of solid samples by graphite furnace atom-ic absorption spectrometry using Zeeman background correction. J. Anal. At. Spectrom. 1: 443–447

151. Voellkopf U, Grobenski Z, Tamm R, Welz B (1985) Solid sampling in graphite furnace AAS using the central probe technique. Analyst 110: 573–577

152. Kurfürst U, Rues B (1981) Schwermetallbestimmung (Pb, Cd, Hg) in Klärschlämmen ohne chemischen Aufschluß mit der Zeeman-Atomabsorptionsspektroskopie. Fresenius Z. Anal. Chem. 308: 1–6

153. Langmyhr FJ, Aadalen U (1980) Direct atomic absorption spectrometric determination of cop-per, nickel and vanadium in coal and petroleum coke. Anal. Chim. Acta 115: 365–368

154. Langmyhr FJ, Stubergh JR, Thomassen Y, Hanssen JE, Dolezal J (1974) Atomic absorption spec-trometric determination of cadmium, lead, thallium and zinc in silicate rocks by direct atom-ization from the solid state. Anal. Chim. Acta 71: 35–42

155. Nordahl K, Radziuk B, Thomassen Y, Weberg R (1990) Determination of aluminium in human biopsy and necropsy specimens by direct solid sampling cup-in-tube electrothermal atomic absorption spectrometry. Fresenius J. Anal. Chem. 337: 310–315

156. Frech W, Baxter DC (1987) Evaluation of the solid sampling technique for the determination of aluminium in biological materials by atomic absorption and emission spectrometry. Fresenius Z. Anal. Chem. 328: 400–404

157. Miller-Ihli NJ (1990) Slurry sampling for graphite furnace atomic absorption spectrometry. Fresenius J. Anal. Chem. 337: 271–274

158. Heydorn K (1991) Quality assurance and statistical control. Microchim. Acta 3: 1–10

159. Griepink B, Muntau H (1988) The certification of the contents (mass fraction) of As, B, Cd, Cu, Hg, Mn, Mo, Ni, Pb, Sb, Se and Zn in rye grass. Community Bureau of Reference: BCR Information. EUR11839EN, Luxembourg

160. Watters RL, JR (1991) The use of standard reference materials for quality assurance in induc-tively coupled plasma optical emission and atomic absorption spectrometry. Spectrochim. Acta 46B: 1593–1603

161. Taylor JK (1988) The role of statistics in quality assurance. Fresenius Z. Anal. Chem. 332: 722–725

162. Taylor JK (1993) Standard Reference Materials. National Institute of Standards and Technologie (NIST): Handbook for SRM Users. Publication 260–100. U.S. Government Printing Office, Washington

163. Miller JC, Miller JN (1993) Statistics for analytical chemistry. Ellis Horwood PTR Prentice Hall, London

164. ISO (1991) Shewhart control charts. ISO 8258

165. Eurachem (1995) Quantifying uncertainty in analytical measurement

166. Taylor BN, Kuyatt CE (1994) Guidelines for evaluation and expressing the uncertainty of NIST measurement results. NIST Technical Note 1297

167. Williams A (1996) Measurement uncertainty in analytical chemistry. Accred. Qual. Assur. 1: 14–17
168. Ebel S (1992) Fehler und Vertrauensbereiche analytischer Ergebnisse. In: Günzler H et al. (eds) Analytiker Taschenbuch. Vol. 11. Springer, Berlin Heidelberg New York, pp 3–59
169. Griepink B, Maier EA, Quevauviller P, Muntau H (1991) Certified reference materials for the quality control of analysis in the environment. Fresenius J. Anal. Chem. 339: 599–603
170. Welz B (1972) Direkte Bestimmung von Spurenelementen in Gesteinen mit flammenloser Atom-Absorption ohne vorheriges Lösen. Fortschr. Miner. 50 (Bhl): 106
171. Verrept P, Dams R, Kurfürst U (1993) Electrothermal vaporisation inductively coupled plasma atomic emission spectrometry for the analysis of solid samples: contribution to instrumentation and methodology. Fresenius Z. Anal. Chem. 345: 1035–1041
172. Davis OL, Goldsmith PL (1988) Statistical methods in research and production. Longman Scientific & Technical, Harlow, UK
173. Massart DL, Vandeginstre BGM, Deming SN, Michotte Y, Kaufman L (1988) Chemometrics: a textbook. Elsevier, Amsterdam
174. Danzer K, Wagner M, Fischbacher C (1995) Calibration by orthogonal and common least squares – Theoretical and practical aspects. Fresenius J. Anal. Chem. 352: 407–412
175. Bendicho C, Loos-Vollebregt MTC (1991) Solid sampling in elektrothermal atomic absorption spectrometry using commercial atomizers. J. Anal. At. Spectrom. 6: 353–374
176. Griepink B, Marchandise H (1985) Referenzmaterialien. In: Günzler H et al. (eds) Analytiker Taschenbuch. Vol. 6. Springer, Berlin Heidelberg New York, pp 3–16
177. Wagstaffe PG, Griepink B, Hecht H, Muntau H, Schramel p (1986) The certification of the contents (mass fractions) of Cd, Pb, As, Hg, Se, Cu, Zn, Fe and Mn in three lyophilised animal tissue reference materials. Community Bureau of Reference, BCR Information EUR 10618 EN, Luxembourg
178. Gerwinski W (1985) Das Verhältnis Peakhöhe/Peakfläche als Auswahlkriterium für Feststoff-Standards bei der Analyse von Böden. Fresenius Z. Anal. Chem. 322: 685–686
179. Esser P (1985) Direct analysis of inorganic solid: Experiences of a laboratory in cement industries. Fresenius Z. Anal. Chem. 322: 677–680
180. Docekal B, Krivan V (1995) Determination of trace impurities in powdered molybdenum metal and molybdenum silicide by solid sampling GFAAS. Spectrochim. Acta 50B: 517–526
181. Friese KC, Krivan V, Schuierer O (1996) Solid sampling ETAAS for analysis of high purity tantalum powders using an improved solid sampling ETAAS system. Spectrochim. Acta B 51: 1223–1233
182. Henn EL (1974) Determination of trace metals in polymers by flameless atomic absorption with a solid sampling technique. Anal. Chim. Acta 73: 273–281
183. Chakrabarti CL, Karowska R, Hollebone BR, Johnson PM (1987) Direct determination of lead in bovine liver by solid sampling with graphite furnace atomic absorption spectrometry – a comparison of tube wall atomization, platform atomization and probe atomization. Spectrochim. Acta 42B: 1217–1225
184. Koshino Y, Narakawa A (1992) Direct determination of trace metals in graphite powders by electrothermal atomic absorption spectrometry. Analyst 117: 967–969
185. Zhang Q, Minami H, Atsuya I (1994) Direct determination of trace amounts of lead in biological samples by pre-ashing concentration/furnace AAS with solid sampling. Bunseki Kagaku 43: 39–43
186. Atsuya I, Minami H, Zhang Q (1993) Pre-ashing of biological samples – a sample preparation technique for solid sampling ZAAS. Fresenius Z. Anal. Chem. 346: 1054–1057
187. Baker AA, Headridge JB, Nicholson RA (1980) Determination of silver and thallium in nickel-base alloys by atomic-absorption spectrometry with introduction of solid samples into an induction furnace. Anal. Chim. Acta 113: 47–53
188. Belyaev et al. (1971) Zh. Anal. Khim. 26
189. Siemer DD, Wei H-Y (1978) Determination of lead in rocks and glasses by temperature controlled graphite cup atomic absorption spectrometry. Anal. Chem. 50: 147–151
190. Pinta M, De Kersabiec AM, Richard ML (1982) Possibilités d'exploitation de l'effect Zeeman pour la correction d'absorption non spécifiques en absorption atomique. Analusis 10: 207–215
191. Kurfürst U (1979) Arsen in Gestein (Granit). Grün & Erdmann: Application Zeeman Atomic Absorption. Nr.13–14, Wetzlar

192. Schrön W, Bombach G, Beuge P (1983) Schnellverfahren zur flammenlosen AAS-Bestimmung von Spurenelementen in geologischen Proben. Spectrochim. Acta 38B: 1269–1276
193. Bley R (1988) Direktbestimmung von Elementspuren in Feststoffen mittels Zeeman-AAS. Diplomarbeit, Universität GHS Essen
194. Nakamura T, Noike Y, Koizumi Y, Sato J (1995) Atomic absorption spectrometric determination of copper and lead in silicon nitride and silicon carbide by direct atomization. Analyst 120: 89–94
195. Takada K (1987) Atomic absorption spectrometric determination of trace copper in high-purity aluminium by direct atomization. Anal. Sci. 3: 221–224
196. Munoz FE, Calvo A, León LE (1983) Direct analysis of solid samples by flameless atomic absorption spectrometry – application to some siliceous materials. Anal. Lett. 16: 835–846
197. Alverado J, Petrola A (1989) Determination of cadmium, chromium, lead, silver and gold in venezuelan red mud by atomic absorption spectrometry. J. Anal. At. Spectrom. 4: 411–414
198. Sonntag TM, Rossbach M (1997) Improved power of determination in SS-ZAAS using pellets of 10 to 20 mg for analysis. Fresenius J. Anal. Chem., 358: 767–770
199. Hwang JY, Thomas GP (1974) New generation flameless AA atomizer. Am. Lab. 6: 42–45
200. Conley MK, Sotera JJ, Kahn HL (1981) Reduction of interferences in flameless AAS. Instrumentation Laboratory Report No. 149
201. Kurfürst U (1991) Die direkte Analyse von Feststoffen mit der Graphitrohr-AAS. In: Günzler H (ed) Analytiker Taschenbuch. Vol. 10. Springer, Berlin Heidelberg New York, pp 190–248
202. Nowka R, Müller H (1997) Direct analysis of solid samples by graphite furnace atomic absorption spectrometry with a transversaly heated graphite atomizer and D_2-background correction system (SS-GF-AAS). Fresenius J. Anal. Chem. 359: 132–137
203. Lücker E, Kreuzer W, Busche C (1989) Feststoffanalytik mit Autoprobe-GFAAS. Teil I: Zertifizierte und laborintern standardisierte Referenzmaterialien. Fresenius Z. Anal. Chem. 335: 176–188
204. Brown AA, Lee M, Küllemer G, Rosopulo A (1987) Solid sampling in graphite furnace atomic absorption spectrometry. Fresenius Z. Anal. Chem. 328: 354–358
205. Schmidt KP, Falk H (1987) Direct determination of Ag, Cu and Ni in solid materials by graphite furnace atomic absorption spectrometry using a specially designed graphite tube. Spectrochim. Acta 42B: 431–443
206. Kurfürst U (1983) Untersuchungen über die Schwermetallanalyse in Feststoffen mit der Direkten Zeeman-Atom-Absorptions-Spektroskopie, Teil III. Bedeutung der Graphit-Schiffchentechnik für die Feststoffanalyse. Fresenius Z. Anal. Chem. 316: 1–7
207. Atsuya I, Itoh K (1982) Direct determination of cadmium in the NBS bovine liver by Zeeman atiomic absorption spectrometry using the graphite miniature-cup. Bunseki Kagaku 31: 713–717
208. Atsuya I, Itoh K, Akatsuka K, Jin K (1987) Direct determination of trace amounts of arsenic in powdered biological samples by atomic absorption spectrometry using an inner miniature cup for solid sampling technique. Fresenius Z. Anal. Chem. 326: 53–56
209. Berndt H (1984) Probeneintragssystem mit Probenverbrennung oder Probenverdampfung für die direkte Feststoffanalyse und für die Lösungsspektralanalyse. Spectrochim. Acta 39B: 1121–1128
210. Takada K (1985) Enhancement of sensitivity in atomic-absorption spectrometry by addition of a graphite lid to a cup furnace. Talanta 32: 921–925
211. Brown AA (1988) Probe design considerations in graphite probe electrothermal atomisation atomic absorption spectrometry. J. Anal. At. Spectrom. 3: 67–71
212. Massmann H, Gücer S (1974) Physikalische und chemische Prozesse bei der Atomabsorptionsanalyse mit Graphitrohröfen – I. Messbedingungen, Versuchsanordung und Temperaturmessung. Spectrochim. Acta 29B: 283–300
213. De Loos-Vollebregt MTC, De Galan L (1985) Zeeman atomic absorption spectrometry. Prog. analyt. atom. Spectrosc. 8: 47–81
214. Kurfürst U (1989) 20 Jahre Zeeman-AAS. Laborpraxis Labor 2000: 80–88
215. Massmann H (1982) The origin of systematic errors in background measurements in Zeeman atomic-absorption spectrometry. Talanta 29: 1051–1055
216. Slavin W, Carnrick GR (1988) Background correction in atomic absorption spectroscopy (AAS). CCR (Critical Reviews in Anal. Chem.) 19: 95–134
217. Carnrick GR, Barnett W, Slavin W (1986) Spectral interferences using the Zeeman effect for furnace atomic absorption spectroscopy. Spectrochim. Acta 41B

218. Wibetoe G, Langmyhr FJ (1986) Spectral interferences and background overcompensation in Zeeman-corrected atomic absorption spectrometry. Part 1. The effect of iron on 30 elements and 49 element lines. Anal. Chim. Acta 186: 155–162
219. Besse A, Rosopulo A, Busche C, Küllmer G (1987) Feststoff-AAS mit Deuterium-Untergrundkompensation. Laborpraxis Special: 102–107
220. Erler W, Lehmann R, Voellkopf U (1987) Determination of cadmium and lead in animal liver and freeze dried blood by graphite furnace Zeeman atomic absorption spectrometry and solid sampling. In: Schramel P, Brätter P (eds) Trace Element – Analytical Chemistry in Medicine and Biologie. Vol. 4, W. de Gruyter, Berlin, pp 385–392
221. Frech W, Baxter DC, Lundberg E (1988) Spatial and temporal non-isothermality as limiting factors for absolute analysis by graphite furnace atomic absorption spectrometry. J. Anal. At. Spectrom 3: 21–25
222. L'vov BV (1978) Electrothermal atomization-the way toward absolute methods of atomic absorption analysis. Spectrochim. Acta 33B: 153–193
223. Chakrabarti CL, Wu Shaole, Karwowska R, Rogers JT, Haley L, Bertels PC, Dick R (1984) Temperature of platform, furnace wall and vapour in a pulse-heated electrothermal graphite furnace in atomic absorption spectrometry. Spectrochim. Acta 39B: 415–448
224. Chakrabarti CL, Wu Shaole, Karwowska R, Rogers JT, Dick R (1985) The gas temperature in and the gas expulsion from a graphite furnace used for atomic absorption spectrometry. Spectrochim. Acta 40B: 1663–1676
225. Koirtyohann SR, Giddings RC, Taylor HE (1984) Heating rates in furnace atomic absorption using the L'vov platform. Spectrochim. Acta 39B: 407–413
226. Frech W, Arshadi M, Baxter DC, Hütsch B (1989) Vapour-phase temperature measurements in the evaluation of platform designs for graphite furnace atomic absorption spectrometry. J. Anal. At. Spectrom 4: 625–634
227. Slavin W, Manning DC (1979) Reduction of matrix interferences for lead determiantion with the l'vov platform and the graphite furnace. Anal. Chem. 51: 261–278
228. Slavin W, Manning DC (1980) The L'vov platform for furnace atomic absorption analysis. Spectrochim. Acta 35B55: 701–7140
229. Siemer DD (1983) Furnace atomic absoption spectrometry atomizer with independent control of volatilization and atomization conditions. Anal. Chem. 55: 692–697
230. Baxter DC, Frech W, Lundberg E (1985) Determination of aluminium in biological materials by constant–temperature graphite furnace atomic–emission spectrometry. Analyst 110: 475–482
231. Lundberg E, Frech W, Baxter DC, Cedergren A (1988) Spatially and temporally constant-temperature graphite furnace for atomic absorption/emission spectrometry. Spectrochim. Acta 43B: 451–457
232. L'vov BV, Frech W (1993) Matrix vapours and physical interference effects in graphite furnace atomic absorption spectrometry-I. end-heated tubes. Spectrochim. Acta 48B: 425–433
233. Berglund M, Frech W, Baxter DC (1991) Achieving efficient, multi-element atomisation conditions for atomic absorption spectrometry using a platform-equipped, integrated-contact furnace and a palladium modifier. Spectrochim. Acta 46B: 1767–1777
234. Chakrabarti CL, Hamed HA, Wan CC, Li WC, Bertels PC, Gregoire DC, Lee S (1980) Capacitive discarge heating in graphite furnace atomic absorption spectrometry. Anal. Chem. 52: 167–176
235. Frech W, Lundberg E, Barbooti MM (1981) Direct determination of trace metals in solid samples by atomic absorption spectrometry with electrothermal atomizers – Part 4. Interference effects in the determination of lead and bismuth in steels. Anal. Chim. Acta 131: 45–52
236. Kurfürst U (1985) Solid sample insertion systems and L'vov platform effect. Fresenius Z. Anal. Chem. 322: 660–665
237. Sedykh EM, Myasoedova GV, Ishmiyarova GR, Kasimova OG (1990) Direct analysis of sorbent-concentrate in a graphite furnace. Zh. Anal. Khim. 45: 1895–1903
238. Friese KC, Krivan V (1995) Analysis of silicon nitride powders for Al, Cr, Cu, Fe, K, Mg, Na, and Zn by slurry-sampling electrothermal atomic absorption spectrometry. Anal. Chem. 34: 354–359
239. Sandoval L, Heraez J-C, Steadman G, Mahan KI (1992) Determination of lead and cadmium in sediment samples by ETA-AAS: a comparison of methods for the preparation and analysis of sediment slurries. Mikrochim. Acta 108: 19–27

240. Dobrowolski R, Mierzwa J (1992) Direct solid vs. slurry analysis of tobacco leaves for some trace metals by graphite furnace AAS – a comparative study. Fresenius Z. Anal. Chem. 344: 340–344
241. Schlemmer G, Welz B (1987) Determination of heavy metals in environmental reference materials using solid sampling graphite furnace AAS. Fresenius Z. Anal. Chem. 328: 405–409
242. Aadland E, Aaseth J, Radziuk B, Saeed K, Thomassen Y (1987) Direct electrothermal atomic absorption spectrometric analysis of biological samples and its application to the determination of selenium in human liver biopsy specimens. Fresenius Z. Anal. Chem. 328: 362–366
243. Irwin R, Mikkelsen A, Michel RG, Dougherty JP, Preli FR (1990) Direct solid sampling of nickel based alloys by graphite furnace atomic absorption spectrometry with aqueous calibration. Spectrochim. Acta 45B: 903–915
244. Kitagawa K, Ohta M, Kancko T, Tsuge S (1991) Packed glassy carbon tube atomizer for direct determination by atomic absorption spectrometry, free from background absorption. J. Anal. At. Spectrom. 9: 1273–1277
245. Hadeishi T, McLaughlin RD (1976) Zeeman atomic absorption determination of lead with a dual chamber furnace. Anal. Chem. 48: 1009–1011
246. Almeida AA, Lima JLFC (1995) Determination of Cd, Cr, Cu, Ni, and Pb in industrial atmospheric particulate matter by ETA-AAS using solid samples directly from trapping filters. At. Spectrosc. 16: 261–265
247. Pickford CJ, Rossi G (1975) Determination of some trace elements in NBS (SRM-1577) Bovine Liver using flameless atomic absorption and solid sampling. At. Absorpt. Newsletter 14: 78–80
248. Wang P, Holcombe JA (1992) Pressure-regulated electrothermal atomizer for atomic absorption spectrometry. Spectrochim. Acta 47B: 1277–1286
249. Hassell DC, Rettberg TM, Fort FA, Holcombe JA (1988) Low-pressure vaporisation for graphite furnace atomic absorption spectrometry. Anal. Chem. 60: 2680–2683
250. Patent Great Britain (22. Feb. 1980). Hitachi Ltd., Tokio
251. L'vov BV, Polzik LK, Kocharova NV (1992) Theoretical analysis of calibration curves for graphite furnace atomic absorption spectrometry. Spectrochim. Acta 47B: 889–895
252. L'vov BV, Polzik LK, Kocharova NV, Nemets YuA, Novichikhin AV (1992) Linearisation of calibration curves in Zeeman atomic absoption spectrometry. Spectrochim. Acta 47B: 1187–1202
253. L'vov BV, Polzik LK, Fedorov PN, Slavin W (1992) Extension of the dynamic range in Zeeman graphite furnace atomic absorption spectrometry. Spectrochim. Acta 47B: 1411–1420
254. Atsuya I, Itoh K, Akatsuka K (1987) Development of direct analysis of powder samples by atomic absorption spectrometry using the inner miniature cup technique. Fresenius Z. Anal. Chem. 328: 338–341
255. Gong H, Suhr NH (1976) The determination of cadmium in geological materials by flameless atomic absorption spectrometry. Anal. Chim. Acta 81: 297–303
256. Grobenski Z, Lehmann R (1983) Determination of lead in rock samples using solid sampling with the stabilized temperature platform furnace and Zeeman background correction. At. Spectrosc. 4: 111–112
257. Rosopulo A, Hornung E, Busche C, Küllmer G (1987) Feststoff-AAS: D2-Untergrund-Kompensation in Klärschlämmen und Böden. Laborpraxis 12: 1436–1444
258. Schrön W, Dreßler B Direkte Feststoff-Graphitrohrofen-AAS mit Gesteinen. II. Kalibration durch Variation der Einwaage. In: Welz B (ed) CANAS'95. Perkin Elmer, Überlingen, pp 121–130
259. Hilbig K (1995) Ph D Thesis, Friedrich Schiller Universität Jena
260. Schrön W, Detcheva A, Deßler B, Danzer K (1997) Determination of copper, lead, cadmiun, zink, and iron in calcium fluride and other fluoride-containing samples by means of direct solid sampling GF-AAS. Fresenius J. Anal. Chem, in press
261. Luecke W, Eschermann F, Lennartz U, Papastamataki AJ (1974) Interferenzprobleme bei der flammenlosen Atomabsorptions-Spektralanalyse geochemischer Proben. N. Jb. Miner. Abh. 120: 178–204
262. Roelandts I (1991) A look at various procedures for evaluation of "best" values in geochemical reference materials. Spectrochim. Acta 46B: 1639–1652
263. Rosopulo A (1985) Schwermetallbestimmung direkt aus dem Feststoff und nach chemischem Aufschluß – ein Methodenvergleich. Fresenius Z. Anal. Chem. 322: 669–672
264. Lundberg E, Frech W (1979) Direct determination of trace metals in solid samples by atomic absorption spectrometry with electrothermal atomizers – Part 2. Determiantion of lead in steels and nickel-based alloys. Anal. Chim. Acta 104: 75–84

265. Nikolaev GI (1973) Zh. Anal. Khim. 28: 454
266. Hinds MW, Kogan VV (1994) Determination of silicon in fine gold by solution and solid sample graphite furnace atomic absorption spectrometry and inductively coupled plasma atomic emission spectrometry. J. Anal. At. Spectrom. 9: 451–455
267. Wang P, Holcombe JA (1994) Electrothermal atomization with atomic absorption at reduced pressures for studies of analyte distribution in solids. Appl. Spectroscopy 48: 713–719
268. Andrews DG, Headridge JB (1977) Determination of bismuth in steels and cast irons by atomic-absorption spectrophotometry with an induction furnace: direct analysis of solid samples. Analyst 102: 436–445
269. Baker AA, Headridge JB (1881) Determination of bismuth, lead and tellurium in copper by atomic absorption spectrometry with introduction of solid samples into an induction furnace. Anal. Chim. Acta 125: 93–99
270. Busheina IS, Headridge JB (1982) Determination of cadmium, indium and zinc in nickel-base alloys by atomic absooption spectrometry with introduction of solid samples into furnaces. Anal. Chim. Acta 142: 197–205
271. Headridge JB, Nicholson RA (1882) Determination of arsenic, antimony, selenium and tellurium in nickel-base alloys by atomic-absorption sprectrometry with introduction of solid samples into furnaces. Analyst 107: 1200–1211
272. Headridge JB, Riddington IM (1982) Microchim. Acta [Wien] II: 457
273. Takada K, Hirokawa K (1982) Atomization and determination of traces of copper, manganese, silver and lead in microamounts of steel. Atomic-absorption spectrometry using direct atomization of solid sample in a graphite-cup cuvette. Fresenius Z. Anal. Chem. 312: 109–113
274. Takada K, Hirokawa K (1983) Determination of traces of lead and cadmium in high-purity tin by polarized Zeeman atomic-absorption spectrometry with direct atomization of solid sample in a graphite-cup cuvette. Talanta 30: 329–332
275. Takada K, Shoji T (1983) Trace analysis of zinc and bismuth in high purity tin by polarized Zeeman atomic absorption spectrometry with direct atomization of solid samples. Fresenius Z. Anal. Chem. 315: 34–37
276. Tsalev DL, Slaveykova VI, Mandjukov PB (1990) Chemical modification in graphite-furnace atomic absorption spectrometry. Spectrochim. Acta Rev. 13: 225–274
277. Katskov DA (1991) Analyte release and transport processes in electrothermal atomic absorption analysis. Spectrochim. Acta Rev. 14: 409–436
278. Styris DL, Redfield DA (1993) Perspectives on mechanisms of electrothermal atomization. Spectrochim. Acta Rev. 15: 71–123
279. Liu J, Sturgeon RE, Boyko VJ, Willie SN (1996) Determination of total chromium in Marine Sediment Reference Materials BCSS-1. Fresenius J. Anal. Chem. 356: 416–419
280. Ergenoglu B, Olcay A (1990) Determination of the germanium content of lignite by atomic absorption sprectrometry using a solid sample and a graphite furnace atomizer. Fuel Sci. Technol. Int. 8: 743–752
281. Zeisler B, Becker DA, Gills TE (1995) Certifying the chemical composition of a biological material – a case study. Fresenius J. Anal. Chem. 352: 111–115
282. Dürnberger R, Esser P, Janßen A (1987) Determination of selenium with solid sampling and direct Zeeman-AAS. Part 1. Fresenius Z. Anal. Chem. 327: 343–346
283. Esser P, Dürnberger R (1987) Determination of selenium with solid sampling and direct Zeeman-AAS. Part II. Fresenius Z. Anal. Chem. 328: 359–361
284. Lindberg I, Lundberg E, Arkhammar P, Berggren PO (1988) Direct determination of selenium in solid biological materials by graphite furnace atomic absorption spectrometry. J. Anal. At. Spectrom. 3: 497–501
285. Fleckenstein J (1985) Direktmessung von Quecksilber in biologischen Feststoffproben mittels Zeeman-Atomabsorption (ZAAS) im Graphitrohrofen. Fresenius Z. Anal. Chem. 322: 704–707
286. Fleckenstein J (1987) Direct measurement of cadmium, lead, copper and zinc in samples from a river ecosystem by Zeeman atomic absorption spectrometry (ZAAS) in the graphite furnace. Fresenius Z. Anal. Chem. 328: 396–399
287. Hadeishi T (1972) Isotope-shift Zeeman effect for trace element detection: an application of atomic physics to environmental problems. Appl. Phys. Lett. 21: 438–440
288. Church DA, Hadeishi T, Leong L, McLaughlin RD, Zak BD (1974) Two-chamber furnace for flameless atomic absorption spectrometry. Anal. Chem. 46: 1352–1355
289. Hadeishi T, McLaughlin RD (1971) Hyperfine Zeeman effect atomic absorption spectrometer

for mercury. Science 174: 404–407

290. Kurfürst U, Rues B (1982) Quecksilberbestimmung direkt aus dem Feststoff. Laborpraxis 6: 1098–1099

291. Campos RC, Curtius AJ, Berndt H (1990) Combustion and volatilisation of solid samples for direct atomic absorption spectrometry using silica or nickel tube furnace atomiser. J. Anal. At. Spectrom. 5: 669–673

292. Tobies KH, Großmann W (1989) Die Bestimmung von Quecksilber in Böden, Kohlen und Aschen ohne Aufschluss. In: Welz B (ed) 5. Colloquium Atomspektrometrische Spurenanalytik. Bodenseewerk Perkin-Elmer, Überlingen, pp 513–521

293. Kurfürst U (1982) Direkte Bestimmung von Schwermetallen (Pb, Cd, Ni, Cr, Hg) in Vollblut und Harn mit der Zeeman-Atom-Absorption. Fresenius Z. Anal. Chem. 313: 97–102

294. Grobecker KH, Klüßendorf B (1985) Schwermetalle in marinen Nahrungsmitteln unterschiedlicher Herkunft. Bestimmung der Gehalte von Cadmium, Blei und Quecksilber in Frisch- und Trockensubstanz durch direkte Feststoffanalyse. Fresenius Z. Anal. Chem. 322: 673–676

295. Bachmann G, Rechenberg W (1991) Aufschluss für die Atomspektrometrische Quecksilberbestimmung von Stoffen der Zementherstellung. In: Welz B (ed) 6. Colloquium Atomspektrometrische Spurenanalytik. Bodenseewerk Perkin-Elmer GmbH, Überlingen, pp 699–706

296. Kurfürst U, Pauwels J (1994) Spectral interference on lead 283.3 nm line in Zeeman-effect atomic absorption spectrometry. J. Anal. At. Sprectrom. 9: 531–534

297. Pearse RWB, Gaydon AG (1950) The identification of molecular spectra. John Wiley, New York

298. Manning DC, Slavin W (1987) Silver as a test element for Zeeman furnace AAS. Spectrochim. Acta 42B: 755–763

299. Ohlsson KEA, Frech W (1989) Photographic observation of molecular spectra in inverse Zeeman-effect graphite furnace atomic absorption spectrometry. J. Anal. At. Spectrom. 4: 379–385

300. Epstein MS, G.C.Turk, Yu LJ (1994) A spectral interference in the determination of arsenic in high purity-lead and lead-base alloys using electrothermal atomic absorption spectrometry and Zeeman-effect background correcton. Spectrochim. Acta 49B: 1681–1688

301. Wibetoe G, Langmyhr FJ (1985) Spectral interferences and background overcompensation in Zeeman-corrected atomic absorption spectrometry. Part 2. The effect of cobalt, manganese and nickel on 30 elements and 53 element lines. Anal. Chim. Acta 176: 33–40

302. Trostle D, Beals T, Kuczenski R, Shaver M (1991) Spectral interference using Zeeman effect in the determination of Bi in superalloys. At. Spectrosc. 12: 64–67

303. Wibetoe G, Langmyhr FJ (1987) Interferences in inverse Zeeman-corrected atomic absorption spectrometry caused by Zeeman splitting of molecules. Anal. Chim. Acta 198: 81–86

304. Doidge PS (1991) Baseline shift in inverse Zeeman-effect graphite furnace atomic absorption spectrometry ascribed to the Zeeman effect of the CN molecule. Spectrochim. Acta 46B: 1776–1787

305. Wibetoe G, Langmyhr FJ (1984) Spectral interferences and background overcompensation in Zeeman-corrected atomic absorption spectrometry. Part 1. The effect of iron on 30 elements and 49 element lines. Anal. Chim. Acta 165: 87–96

306. Zong VanY, Parsons PJ, Slavin W (1994) Background correction problems for lead in the presence of phospate modifiers in Zeeman graphite furnace atomic absorption spectrometry. Spectrochim. Acta 49B: 1667–1680

307. Frigge C, Jackwerth E (1991) Spectral interferences in the determination of traces of Pd in the presence of lead by atomic absorption spectroscopy. Spectrochim. Acta 47B: 787–791

308. Radziuk B, Thomassen Y (1992) Chemical modification and spectral interferences in selenium determination using Zeeman-effect electrothermal atomic absorption spectrometry. J. Anal. At. Spectrom. 7: 397–403

309. Aller AJ, García-Olalla C (1992) Spectral interferences on the determination of selenium by electrothermal atomic absorption spectrometry. J. Anal. At. Spectrom. 7: 753–760

310. Le Bihan A, Cabon JY, Elleouet C (1992) Interférences spectales liées à la présence de nitrate et de monoxyde d'azote en spectrophotométrie d'absoption atomique à atomisation électrothermique et correction Zeeman. Analusis 20: 601–604

311. Kurfürst U (1983) Untersuchungen über die Schwermetallanalyse in Feststoffen mit der Direkten Zeeman-Atom-Absorptionsspektroskopie. Teil II: Theorie, Eigenschaften und Leistungsfähigkeit der Direkten ZAAS. Fresenius Z. Anal. Chem. 325: 304–320

312. Akatsuka K, Atsuya I (1989) Determination of vitamin B_{12} as cobalt by electrothermal atomic absorption spectrometry using solid sampling technique. Fresenius Z. Anal. Chem. 335: 200–204

313. Baxter DC, Frech W (1987) Analysis of solid samples by graphite furnace atomic emission spectrometry. Fresenius Z. Anal. Chem. 328: 324–329

314. Crabi G, Cavalli P, Achilli M, Rossi G, Omenetto N (1982) Use of the HGA-500 graphite furnace as a sampling unit for ICP emission spectrometry. At. Spectr. 3: 81–88

315. Headridge JB, Riddington IM (1984) Determination of silver, lead and bismuth in glasses by atomic absorption spectrometry with introduction of solid samples into furnaces. Analyst 109: 113–118

316. Alder JF, Batoreu MCC, Pearse AD, Marks R (1986) Depth concentration profiles obtained by carbon furnace atomic absorption spectrometry for nickel and aluminium in human skin. J. Anal. At. Spectrom. 1: 365–367

317. Mohl C, Bargouth I, Dürbeck HW (1994) Vergleichende Homogenitätsmessungen in biologischen Materialien mit der Feststoff- und Slurry Sampling GFAAS. In: Dittrich K, Welz B (eds) CANAS'93. Universität Leipzig, Leipzig, pp 423–429

318. Wagley D, Schmiedel G, Mainka E, Ache HJ (1989) Direct determination of some essential and toxic elements in milk and milk powder by graphite furnace atomic absorption spectrometry. At. Spectrosc. 10: 106–111

319. Mohl C, Narres HD, Stoeppler M (1986) Sauerstoffveraschung bei der Bestimmung von Blei und Cadmium in schwierigen Materialien mit Graphitrohrofen–AAS. In: Fortschritte in der atomspektrometrischen Spurenanalytik. vol 2, Verlag Chemie, Weinheim, Germany, pp 439–446

320. Langmyhr FJ, Aamodt J (1976) Atomic absorption spectrometric determination of some trace metals in fish meal and bovine liver by the solid sampling technique. Anal. Chim. Acta 87: 483–486

321. Nilsson T, Berggren P-O (1984) The determination of cadmium in microgram amounts of pancreatic tissue by electrothermal atomic absorption spectrometry. Anal. Chim. Acta 159: 381–385

322. Ebdon L, Evans EH (1987) Determination of copper in biological microsamples by direct solid sampling graphite furnace atomic absorption spectrometry. J. Anal. At. Spectrom. 2: 317–320

323. Lehmann C, Lorber K-E, Ruthenberg K, Schulze G (1989) Heavy metal status of the Plötzensee in Berlin – Comparison of solid sample determination and measurements after digestion with Zeeman atomic absorption spectrometry. Fresenius Z. Anal. Chem. 333: 707–708

324. Langmyhr FJ, Sundli A, Jonsen J (1974) Atomic absorption spectrometric determination of cadmium and lead in dental material by atomisation directly from the solid state. Anal. Chim. Acta 73: 81–85

325. Headridge JB, Smith DR (1973) An inductive furnace for the determination of cadmium in solutions and zinc-base metals by atiomic-absorption spectrometry. Talanta 18: 247–251

326. Pauwels J, Lamberty A, De Bièvre P, Grobecker K-H, Bauspiess C (1994) Certified reference materials for the determination of cadmium in polyethylene. Fresenius J. Anal. Chem. 349: 409–411

327. Brunner G, Korneck F, Müller-Vogt G, Wendl W, Send W (1992) The application of sputtering as new sampling technique in graphite furnace atomic absorption spectrometry. Spectrochim. Acta 47B: 1097–1105

328. Knezevic G, Kurfürst U (1985) Schwermetallbestimmung in Papier – ein Methodenvergleich (Feststoff- und Aufschlußanalyse mit Graphitrohr-AAS). Fresenius Z. Anal. Chem. 322: 717–718

329. Völlkopf U, Lehmann R, Weber D (1985) Metallspurenbestimmung in Kunststoffen mit der "Cup-in-Tube"-Technik. Reprint from: Laborpraxis 9

330. Völlkopf U, Lehmann R, Weber D (1987) Determination of cadmium, copper, manganese and rubidium in plastic materials by graphite furnace atomic absorption spectrometry using solid sampling. J. Anal. At. Spectrom. 2: 455–458

331. Itho K, Akatsuka K, Atsuya I (1984) Direct determination of chromium in environmental samples by Zeeman atomic absorption spectrometry with a graphite miniature cup. Bunseki Kagaku 33: 301–305

332. Johnson D, Headridge JB, McLeod CW, Jackson KW, Roberts JA (1986) Direct determination of chromium in gallium arsenide by electrothermal atomisation atomic absorption spectrometry with Smith-Hieftje background correction. Anal. Proc. 23: 8–9

333. Guenais B, Poudoulec A, Minier M (1982) Dosage du chrome dans l'arséniure de gallium par spectrométrie d'absorption atomique en four graphite. Analusis 10: 78–82

334. Freedman YE, Ronen D, Long GL (1996) Determination of Cu and Cd content of groundwater colloids by solid sampling graphite furnace atomic absorption spectrometry. Environ. Sci. Technol. 30: 2270–2277

335. Desmorieux H (1983) Bestimmung von Schwermetallen mittels der Zeeman-Atomabsorptionsspektroscopie in der Zementindustrie. Diplomarbeit, Fachhochschule Münster
336. Docekal B, Krivan V, Franek M (1994) Separation of analyte and matrix for the direct analysis of high purity molybdenum-based materials by electrothermal atomic spectrometry methods – I. Radiotracer investigation of thermal extraction of impurities in a graphite cup. Spectrochim. Acta 49B: 577–582
337. Sneddon J (1990) Electrostatic precipitation atomic absorption spectrometry. Appl. Spectroscopy 44: 1562–1565
338. Sneddon J (1991) Direct and near–real-time determination of lead, manganese and mercury in laboratory air by electrostatic precipitation-atomic absoroption spectrometry. Anal. Chim. Acta 245: 203–206
339. Schmiedel G, Mainka E, Ache HJ (1989) Determination of molybdenium, ruthenium, rhodium, and palladium in radioinactive waste of the nuclear fuel cycle by solid sampling graphite furnace atomic absorption spectrometry (GFAAS). Fresenius Z. Anal. Chem. 335: 195–199
340. Scholl W (1985) Erfahrungen mit dem Zeeman-Atom-Absorptions-Spektrometer 5000 der Firma Perkin-Elmer bei der Feststoffanalyse. Fresenius Z. Anal. Chem. 322: 681–684
341. Krebs B (1987) Direct determination of nickel in margarine. Fresenius Z. Anal. Chem. 328: 388–389
342. Lücker E, Hornung E, Rosopulo A, Küllmer G, Busche C (1987) Feststoff-AAS mit Deuterium-Untergrund-Kompensation II. Laborpraxis 3: 176–183
343. Atsuya I, Itoh K, Ariu K (1991) Preconcentration by coprecipitation of lead and selenium with Ni/pyrrolidine dithiocarbamate complex and their simultaneous determination by internal standard atomic absorption spectrometry with the solid sampling technique. Pure & Appl. Chem. 63: 1221–1226
344. José Moura MJMP, Vasconcelos MTSD, Machado AASC (1987) Determination of lead in atmospheric aerosols by electrothermal atomisation atomic absorption spectrometry with direct introduction of filters into the graphite furnace. J. Anal. At. Spectrom. Vol.2: 451–454
345. Kowalska A, Kedziora M, Kedziora A (1980) Determination of lead in graphite by flameless atomic absorption and solid sampling technique. At. Spectrosc. 1: 33–34
346. Grobenski Z, Weber D, Welz B, Wolff J (1983) Determination of caesium and rubidium by flame and furnace atomic-absorption spectrometry. Analyst 108: 925–932
347. Oilunkaniemi R, Perämäki P, Lajunen LHJ (1994) Direct determination of selenium in solid biological materials by GFAAS using the Cup-in-Tube technique. Atomic Spectrosc. 15: 126–130
348. O'Haver TC (1979) Derivative and wavelength modulation spectrometry. Anal. Chem. 51: 91A–96A
349. Reed TB (1961) Induction-coupled plasma torch. J. Appl. Phys. 32: 821–824
350. Wendt RH, Fassel VA (1965) Induction-coupled plasma spectrometric excitation source. Anal. Chem. 37: 920–922
351. Montaser A, Golightly DW (1987) Inductively coupled plasma in analytical atomic spectrometry. VCH, New York, pp 268–274
352. Houk RS, Fassel VA, Flesch GD, Svec HJ, Gray AL, Taylor CE (1980) Inductively coupled argon plasma as an ion source for mass spectrometric determination of trace elements. Anal. Chem. 52: 2283–2289
353. Browner RF, Boorn AW (1984) Sample introduction: the achilles'heel of atomic spectrometry. Anal. Chem. 56: 787–798
354. Browner RF, Boorn AW, Smith DD (1982) Aerosol transport model for atomic spectrometry. Anal. Chem. 54: 1411–1419
355. Barth P, Krivan V (1994) Electrothermal vaporization inductively coupled plasma atomic emission spectrometric technique using a tungsten coil furnace and slurry sampling. J. Anal. At. Spectrom. 9: 773–777
356. Ali AH, Winefordner JD (1992) Microsample introduction by tungsten filament electrode into capacitively coupled microwave plasma for atomic emission spectroscopy: analytical figures of merit. Anal. Chim. Acta 264: 327
357. Ali AH, Wineforder JD (1992) Microsample introduction by tungsten filament electrode into capacitively coupled microwave plasma for atomic emission spectroscopy: diagnostics. Anal. Chim. Acta 264: 319
358. Buckley TB, Boss CB (1990) A tungsten filament vaporizer for sample introduction into a direct-current plasma. Applied Spectr. 44: 505

359. Dittrich K, Berndt H, Broeckaert JAC, Schaldach G, Tölg G (1988) Comparative study of injection into a pneumatic nebuliser and tungsten coil electrothermal vaporization for the determination of rare earth elements by ICP-OES. J. Anal. At. Spectrom. 3: 1105

360. Erwen M, Zucheng J, Zhenhuang L (1992) Determinations of trace amounts of rare earth and other elements in rice samples by ICP-AES with introduction by tungsten-coil electrothermal vaporization. Fresenius J. Anal. Chem. 344: 54

361. Evans EH, Caruso JA, Satzger RD (1991) Evalation of a tantalum-tip electrothermal vaporization sample introduction device for microwave induced plasma mass spectrometry and atomic emission spectrometry. Appl. Spectrosc. 45: 1478

362. Tsukahara R, Kubota M (1990) Some characteristics of inductively coupled plasma-mass spectrometry with sample introduction by tungsten furnace electrothermal vaporization. Spectrochim. Acta 45B: 779–787

363. Okamoto Y, Murata H, Yamamoto M, Kumamaru T (1990) Determination of vanadium and titanium in steel by inductively coupled plasma atomic emission spectrometry with modified use of a tungsten boat furnace atomizer for atomic absorption spectrometry. Anal. Chim. Acta 239: 139–143

364. Nixon DE, Fassel VA, Kniseley RN (1974) Inductively coupled plasma – optical emission analytical spectroscopy: tantalum filament vaporization of microliter samples. Anal. Chem. 46: 210–213

365. Shibata N, Fudagawa N, Kubota M (1992) Effects of hydrogen mixed with argon carrier gas in ETV-ICP-MS. Spectrochim. Acta 47B: 505–516

366. Park CJ, Van Loon JC, Arrowsmith P, French JB (1987) Sample analysis using plasma source mass spectrometry with electrothermal sample introduction. Anal. Chem. 59: 2191–2196

367. Fricke FL, Rose O, Caruso JA (1975) Simultaneous multielement determination of trace metals by microwave induced plasma coupled to vidicon detector: carbon cup sample introduction. Anal. Chem. 47: 2018–2020

368. Richts U, Broekaert JAC, Tschöpel P, Tölg G (1991) Comparative study of a beenakker cavity and a surfatron in combination with electrothermal evaporation from a tungsten coil for microwave plasma optical emission spectrometry (MIP-AES). Talanta 38: 863–869

369. Marawi I, Olson LK, Wang J, Caruso JA (1995) Utilization of metallic platforms in electrothermical vaporization inductively coupled plasma mass spectrometry. J. Anal. At. Spectrom. 10: 7–14

370. Shen W-L, Caruso JA, Fricke FL, Satzger RD (1990) Electrothermal vaporisation interface for sample introduction in inductively coupled plasma mass spectrometry. J. Anal. At. Spectrom. 5: 451–455

371. Carey JM, Caruso JA (1992) Electrothermal vaporization for sample introduction in plasma source spectrometry. Crit. Rev. Anal. Chem. 23: 397–439

372. Park CJ, Hall GEM (1987) Analysis of geological materials by inductively coupled plasma mass spectrometry with sample introduction by electrothermal vaporisation. Part 1. Determination of Molybdenum and Tungsten. J. Anal. At. Spectrom. 2: 473–480

373. Hall GEM, Pelchat J, Boomer DW, Powell M (1988) Relative merits of two methods of sample introduction in inductively coupled plasma mass spectrometry: electrothermal vaporization and direct sample insertion. J. Anal. At. Spectrom. 3: 791–797

374. Park CJ, Hall GEM (1988) Analysis of geological materials by inductively coupled plasma mass spectrometry with sample introduction by electrothermal vaporization. Part 2: Determination of thallium. J. Anal. At. Spectrom. 3: 355–361

375. Darke SA, Pickford CJ, Tyson JF (1989) Study of electrothermal vaporisation sample introduction for plasma spectrometry. Anal. Proc. 26: 379–381

376. Alvarado J, Cavalli P, Omenetto N, Rossi G (1987) Peak-area measurements in electrothermal atomisation inductively coupled plasma atomic emission spectrometry. J. Anal. At. Spectrom. 2: 357–363

377. Schmertmann SM, Long SE, Browner F (1987) Sample introduction studies with a graphite rod electrothermal vaporiser for inductively coupled plasma atomic emission spectrometry. J. Anal. At. Spectrom. 2: 687–693

378. Sneddon J, Bet-Pera F (1986) Electrothermal vaporization-inductivel coupled plasma emission spectrometry. Trends Anal. Chem. 5: 110–114

379. Ng KC, Caruso JA (1985) Electrothermal vaporization for sample introduction in atomic emission Spectrometry. Appl. Spectrosc. 719–726

380. Ng KC, Caruso JA (1982) Microliter sample introduction into an inductively coupled plasma by electrothermal carbon cup vaporization. Anal. Chim. Acta 143: 209–222

381. Kirkbright GF, Snook RD (1979) Volatilization of refractory compound forming elements from a graphite electrothermal atomization device for sample introduction into an inductively coupled argon plasma. Anal. Chem. 51: 1938–1941

382. Wu M, Carnahan JW (1990) Trace determination of Cd, Cu, Br, and Cl with electrothermal vaporization into a helium microwave-induced plasma. Appl. Spectrosc. 44: 673–678

383. Grégoire DC, M.Lamoureux, Chakrabarti CL, Al-Maawali S, Byrne JP (1992) Electrothermal vaporisation for inductively coupled plasma mass spectrometry and atomic absoption spectrometry: Symbiotic analytical techniques. J. Anal. At. Spectrom. 7: 579–585

384. Abdillahi MM (1993) Analytical feasibility of graphite rod vaporization – helium microwave – induced plasma (GRV/He-MIP) for some nonmetal determinations. Appl. Spectrosc. 47: 366–374

385. Alvarado J, Wu M, Carnahan JW (1992) Electrothermal vaporization and ultrasonic nebulisation for the determination of aqueous sulfur using a kilowatt-plus helium microwave-induced plasma. J. Anal. At. Spectrom. 7: 1253–1256

386. Byrne JP, Chakrabarti CL, Grégoire DC, Lamoureux M, Ly T (1992) Mechanisms of choloride interferences in atomic absorption spectrometry using a graphite furnace atomizer investigated by electrothermal vaporization inductively coupled plasma mass spectrometry. Part I: modifier on Mg. J. Anal. At. Spectrom. 7: 371–381

387. Grégoire DC (1990) Sample introduction techniques for the determination of osmium isotope ratios by ICP-MS. Anal. Chem. 62: 141–146

388. Ng KH, Caruso JA (1983) Determination of trace metals in synthetic ocean water by inductively coupled plasma atomic emission spectrometry with electrothermal carbon cup vaporization. Anal. Chem. 55: 1513–1516

389. Sanz-Mendel A, Roza RR, Alonso RG, Vallina AN, Cannata J (1987) Atomic spectrometric methods (atomic absorption and inductively coupled plasma atomic emission) for the determination of aluminium at the parts per billion level in biological fluids. J. Anal. At. Spectrom. 2: 177–184

390. Dean JR, Snook RD (1986) A study of the formation of atoms and dry aerosols above a graphite rod sample introduction device used for inductively coupled plasma atomic emission spectrometry. J. Anal. At. Spectrom. 1: 461–465

391. Gunn AM, Millard DL, Kirkbright GF (1978) Optical emission spectrometry with an inductively coupled radiofrequency Argon plasma source and sample introduction with a graphite rod electrothermal vaporisation device. Analyst 103: 1066–1073

392. van Berkel WW, Maessen FJMJ (1988) Use and evaluation of poly (dithiocarbamate) in electrothermal vaporization inductively coupled plasma atomic emission spectrometry for simultaneous trace analysis of seawater and related biological materials. Spectrochim. Acta 43B: 1337–1347

393. Aziz A, Broekaert JAC, Leis F (1982) A contribution to the analysis of microamounts of biological samples using a combination of graphite furnace and microwave induced plasma atomic emission spectrometry. Spectrochim. Acta 37B: 381–389

394. Aziz A, Broekaert JAC, F.Leis (1982) Analysis of microamounts of biological samples by evaporation in a graphite furnace and inductively coupled plasma atomic emission spectrometry. Spectrochim. Acta 37B: 369–379

395. van Berkel WW, Balke J, Maessen FJMJ (1990) Introduktion of analyte-loaded poly (dithiocarbamate) into inductively coupled argon plasma by electrothermal vaporization. Spatial emission characteristics of the resulting dry plasma. Spectrochim. Acta 45B: 1265–1274

396. Heltai G, Broekaert JAC, Burba P, Leis F, Tschöpfel P, Tölg G (1990) Study of a torodal argon MIP and a cylindrical helium MIP for atomic emission spectrometry-II. Combination with graphite furnace vaporization and use for biological samples. Spectrochim. Acta 45B: 857–866

397. Carey JM, Evans EH, Caruso JA, Shen W-L (1991) Evaluation of a modified commercial graphite furnace for reduction of isobaric interferences in argon inductively coupled plasma mass spectroscopy. Spectrochim. Acta 46B: 1711–1721

398. Kantor T, Zaray G (1992) Graphite furnace for alternative combination with d.c. arc or inductively coupled plasma. Introduction and analysis of solid samples. Fresenius J. Anal. Chem. 324: 927–935

399. Kantor T (1988) Interpreting some analytical characteristics of thermal dispersion methods used for sample introduction in atomic spectrometry. Spectrochim. Acta 43B: 1299–1320

400. Ulrich A, Meiners S, Völlkopu, Dannecker W (1991) ETV-ICP-MS Einzelelement- oder Multielementtechnik. In: Welz B (ed) 6. Coll. Atomspektrometrische Spurenanalytik. Perkin Elmer, Überlingen, pp 160–166

401. Voellkopf U, Guensel A, Paul M, Wiesmann H (1991) Applications of ICP-MS with sample introducton by electrothermal vaporization and flow injection techniques. In: The Royal Society of

Chemistry: Applications of plasma source mass spectrometry, 162–177

402. Kantor T (1992) Considerations in the gas flow design of a graphite furnace vaporization interface: effects of a halocarbon atmosphere and sample matrix. J. Anal. At. Spectrom. 7: 219–224

403. Ulrich A, Huchulski C, Dannecker W, Völlkopf U (1992) Use of electrothermal vaporization for inductively coupled plasma mass spectrometry for element determinations in complex matrices such as sandstone samples. Anal. Proc. 29: 282–284

404. Ulrich A, Dannecker W, Meiners S, Völlkopf U (1992) Use of electrothermal vaporization for inductively coupled plasma mass spectrometry for single – element and multi – element determinations. Anal. Proc. 29: 284–286

405. Casetta B, Di Pasquale G, Soffientini A (1985) Setting up an ICP coupled with HGA for characterization of sulfur in polymeric materials. Fresenius Z. Anal. Chem. 319: 62–64

406. Ediger RD, Beres SA (1992) The role of chemical modifiers in analyte transport loss interferences with electrothemal vaporisation ICP-mass spectrometry. Spectrochim. Acta 47B: 907–922

407. Blakemore WM, Casey PH, Collie WR (1984) Simultaneous determination of 10 elements in wastewater, plasma, and bovine liver by inductively coupled plasma emission spectrometry with electrothermal atomization. Anal. Chem. 56: 1376–1379

408. Swaidan HW, Christian GD (1984) Optimization of electrothermal atomization – inductively coupled plasma atomic emission spectrometry for simultaneous multielement determination. Anal. Chem. 56: 120–122

409. Matusiewicz H, Brovko IA, Sturgeon RE, Luong T (1990) Aerosol transport interface for electrothermal vaporization-microwave-induced plasma emission spectrometry. Appl. Spectrosc. 44: 736–739

410. Bin H, Zucheng J, Yun'e Z (1991) Slurry sample introduction with fluorinating electrothermal vaporization for the direct ICP-AES determination of boron in plant leaves. Fresenius J. Anal. Chem. 340: 435–438

411. Kumamaru T, Okamoto Y, Matsuo H (1987) A versatile sample introduction system with graphite tube furnace for inductively coupled plasma-atomic emission spectrometry. Appl. Spectrosc. 41: 918–920

412. Bulska E, Tschöpel P, Broekaert JAC, Tölg G (1993) Different sample introduction systems for the determination of As, Sb and Se by microwave – induced plasma atomic emission spectrometry. Anal. Chim. Acta 271: 171–181

413. Hull DR, Horlick G (1984) Electrothermal vaporization sample introduction system for the inductively coupled plasma. Spectrochim. Acta 39B: 843–850

414. Matousek JP, Orr BJ, Selby M (1986) Spectrometric analysis of non-metals introduced from a graphite furnace into a microwave induced plasma. Talanta 33: 875–882

415. Reisch M, Nickel H, Mazurkiewicz M (1989) Volatilization studies for Fe and Cd in their direct determination by inductively coupled plasma-atomic emission spectrometry using direct sample insertion or external electrothermal evaporation of powders. Spectrochim. Acta 44B: 307–315

416. Nickel H, Reisch M, Mazurkiewicz M (1989) Investigation of alumina-based ceramic materials using ICP-OES with external electrothermal vaporization. Fresenius Z. Anal. Chem. 335: 631–636

417. Ida I, Yoshikawa H, Ishibashi Y, Gunji N (1989) Trace element analysis by inductively coupled plasma atomic emission spectrometry with electrothermal vaporization method. Anal. Sci. 5: 615–618

418. Atsuya I, Itoh T, Kurotaki T (1991) Inductively coupled plasma atomic emission spectrometry of powdered samples using the miniature cup technique. Spectrochim. Acta 46: 103–108

419. Mitchell PG, Sneddon J (1987) Direct determination of metals in milligram masses and microlitre volumes by direct-current argon-plasma emission spectrometry with sample introduction by electrothermal vaporization. Talanta 34: 849–855

420. Ohls KD, Hütsch B (1986) Electrothermal vaporization analysis using solid/liquid sample insertion simultaneous ICP emission spectrometry. ICP Inform. Newsl. 12: 170–176

421. Ohls KD (1989) Sample introduction into ICP-OES for metallic samples. Mikrochim. Acta (Wien) III: 337–346

422. Elliot WG, Matusiewicz H, Barnes RM (1986) Electrothermal vaporization for sample introduction into a three – electrode direct current argon plasma. Anal. Chem. 58: 1264–1265

423. Verrept P, Galbács G, Moens L, Dams R, Kurfürst U (1993) Solid sampling electrothermal vaporisation inductively plasma atomic emission spectrometry (ETV-ICP-AES): influence of some ICP operating parameters. Spectrochim. Acta 48B: 671–680

424. Zaray G, Kantor T, Wolff G, Zadgorska Z, Nickel H (1992) ICP-AES detection of silicon carbide impurities volatilized in a graphite furnace with the use of carbon tetrachloride vapour. Mikrochim. Acta 107: 345–358
425. Hulmston P, Hutton RC (1991) Analytical capabilities of electrothermal vaporization-inductively coupled plasma-mass spectrometry. Spectrosc. Intern. 3: 35–38
426. Haßler J, Perzl PR (1996) Direkte Feststoff-Spektralanalyse am Beispiel eines Kombinationsspektrometer. GIT Fachz. Lab. 10: 989–995
427. Ren JM, Salin ED (1994) Direct solid sample analysis using furnace vaporization with freon modification and inductively coupled plasma atomic emission spectrometry – I. Vaporization of oxides and carbides. Spectrochim. Acta B 49: 555–566
428. Vanhaecke F, Boonen S, Moens L, Dams R (1995) Solid sampling electrothermal vaporization inductively coupled plasma mass spectrometry for the determination of arsenic in standard reference materials of plant origin. J. Anal. At. Spectrom 10: 81–87
429. Boonen S, Vanhaecke F, Moens L, Dams R (1996) Direct determination of Se and As in solid certified reference materials using electrothermal vaporization inductively coupled plasma mass spectrometry. Spectrochim. Acta B 51: 271–278
430. Grégoire DC, Miller-Ihli NJ, Sturgeon RE (1994) Direct analysis of solids by ultrasonic slurry electrothermal vaporization inductively coupled plasma mass spectrometry. J. Anal. At. Spectrom. 9: 605–610
431. Grégoire DC, Goltz DM, Lamoureux MM, Chakrabarti CL (1994) Vaporization of acids and their effect on analyte signal in electrothermal vaporization inductively coupled plasma mass spectrometry. J. Anal. At. Spectrom. 9: 919–926
432. H. Naka, Grégoire DC (1996) Determination of trace amounts of sulfur in steel by electrothermal vaporization-inductively coupled plasma mass spectrometry. J. Anal. At. Spectrom. 11: 359–363
433. Barnes RM, Fodor P (1983) Analysis of urine using inductively-coupled plasma emission spectrometry with graphite rod electrothermal vaporization. Spectrochim. Acta 38B: 1191–1202
434. Karanassios V, Ren JM, Salin ED (1991) Electrothermal vaporization sample indroduction system for the analysis of pelletized solids by inductively coupled plasma atomic emission spectrometry. J. Anal. At. Spectrom. 6: 527–533
435. Nickel H, Zadgorska Z, Wolff G (1993) Optimisation of electrothermal vaporisation of impurity elements in ceramic powders using inductively coupled plasma atomic emission spectroscopy. Spectrochim. Acta 48B: 25–38
436. Matusiewicz H, Fish J, Malinski T (1987) Electrothermal preconcentration of metals using mercury film electrodes followed by electrothermal vaporization into an inductively coupled plasma and determination by atomic emission spectrometry. Anal. Chem. 59: 2264–2269
437. Lamoureux MM, Grégoire DC, Chakrabarti CL, Goltz DM (1994) Modification of a commercial electrothermal vaporizer for sample introduction into an inductively coupled plasma mass spectrometer. 2. Performance evaluation. Anal. Chem. 66: 3217–3222
438. Nickel H, Zadgorska Z (1993) A new interface device for electrothermal vaporization – inductively coupled plasma atomic emission spectrometry (ETV-ICP-AES). ICP Inform. Newsl. 19: 71–73
439. Aziz A, Broekaert JAC, Laqua K, Leis F (1984) A study of direct analysis of solid samples using spark ablation combined with excitation in an inductively coupled plasma. Spectrochim. Acta 39B: 1091–1103
440. Daniels RS, Wigfield DC (1991) Gas-phase adsorptional losses of elemental mercury in cold vapor atomic absorption spectrometry. Anal. Chim. Acta 248: 575–577
441. Lamoureux MM, Grégoire DC, Chakrabarti CL, Goltz DM (1994) Modification of a commercial electrothermal vaporizer for sample introduction into an inductively coupled plasma mass spectrometer. 1. Characterization. Anal. Chem. 66: 3208–3216
442. Boumans PWJW (1987) Inductively coupled plasma emission spectrometry. Part 1: Methodology, Instrumentation and Performance. J. Wiley & Sons, New York
443. Dawson JB, Snook RD, Price WJ (1993) Background and background correction in analytical atomic spectrometry. Part I: Emission spectrometry, a tutorial review. J. Anal. At. Spectrom. 8: 517–534
444. Snellman W, Rains TC, Yee KW, Cook HD, Menis O (1970) Flame emission spectrometry with repetitive optical scanning in the derivative mode. Anal. Chem. 42: 394–398
445. Verrept P, Courtijin E, Vandecasteele C, Windels G, Dams R (1992) Modification of a commercial inductively coupled plasma atomic emission spectrometer for fast wavelength scanning by using a quarz refractor plate to measure background-corrected transient signals. Anal. Chim. Acta 257: 223–228

446. Verrept P, Vandecasteele C, Windels G, Dams R (1991) Use of a refractor plate for automatic background correction in electrothermal vaporisation inductively coupled plasma atomic emission spectrometry. Spectrochim. Acta 46B: 99–102

447. Benzur L, Marshall J, Ottaway JM (1984) A square-wave wavelenght modulation system for automatic background correction in carbon furnace atomic emission spectrometry. Spectrochim. Acta 39B: 787–805

448. Gökmen A, Delzendeh M, Volkan M, Ataman OY, Ottaway JM (1986) Background correction by wavelenght modulation using a microcomputer in atomic emission spectrometry. Mikrochim. Acta (Wien) II: 357–367

449. Ishii H, Satoh K (1982) Development of a high – resolution inductively coupled argon plasma apparatus for derivative spectrometry and its application to the determination of hafnium in high – purity zirconium oxide. Talanta 29: 243–248

450. Koirtyohann SR, Glass ED, Yates DA, Hinderberger EJ, Lichte FE (1977) Effect of modulation wave form on the utility of emission background corrections obtained with an oscillating refractor plate. Anal. Chem. 49: 1121–1126

451. Marshall J, Littlejohn D, Ottaway JM, Thorburn-Burns D (1986) Continuum source AAS. Chem. in Britain 2: 6–8

452. Michel RG, Sneddon J, Hunter JK, Ottaway JM, Fell GS (1981) A novel method of wavelength modulation for atomic spectrometry – some preliminary experiments. Analyst 106: 288–298

453. O'Haver TC, Kindervater JM (1988) Closed-loop feedback controlled data acquisition system for ensemble averaging repetitively scanned spectra with wavelength modulation. Appl. Spectrosc. 42: 183–186

454. Ottaway JM, Benzur L, Marshall J (1980) An echelle monochromator system for the measurement of sensitive carbon furnace atomic – emission signals. Analyst 105: 1130–1136

455. Skogerboe RK, Lambothe PJ, Bastiaans GJ, Freeland SJ, Coleman GN (1976) A dynamic background correction system for direct reading spectrometry. Appl. Spectrosc. 30: 495–500

456. Wohlers CC (1976) Reduction of stray light in inductively coupled plasma – atomic emission spectrometry. ICP Inform. Newsl. 2: 196–197

457. Wohlers CC, Schleicher RG, Sainz MA, Nygaard DD (1986) Reduction of interferences in inductively coupled plasma – electrothermal vaporization analysis (ICP-EVA). ICP Inform. Newsl. 12: 190

458. Wohlers CC, Schleicher RG, Nygaard DD, Smith SB (1986) Overcoming interferences in inductively coupled plasma – electrothermal vaporization analysis (ICP-EVA). ICP Inform. Newsl. 11: 803–804

459. Ren JM, Salin ED (1994) Direct solid sample analysis using furnace vaporization with freon modification and inductively coupled plasma atomic emission spectrometry – II. Analysis of real samples. Spectrochim. Acta B 49: 567–575

460. Driscoll WG, Vaughan WV (1978) Handbook of optics. Mcgraw – Hill Book Company, New York,

461. Busch KW, Busch MA (1990) Multielement detection systems for spectrochemical analysis. Vol. 107. J. Wiley & Sons, New York, pp 423–489

462. Denoyer ER (1994) Optimization of transient signal measurments. At. Spectrosc. 15: 7–16

463. Hinds WC (1982) Aerosol technology; properties, behavior and measurment of airborne particles. J. Wiley & Sons, New York

464. Towomey S (1977) Atmospheric aerosols. Elsevier, Amsterdam

465. Ren JJ, Salin ED (1993) Evaluation of a modified electrothermal vaporization sample introduction system for the analysis of liquids by ICP-AES. J. Anal. At. Spectrom. 8: 59–63

466. Sparks CM, Holcombe J, Pinkston TL (1993) Particle size distribution of sample transport from an electrothermal vaporizer to an inductively coupled plasma mass spectrometer. Spectrochim. Acta 48B: 1607–1615

467. Dittrich K, Mroczek A (1994) Untersuchungen zur Anwendbarkeit der elektrothermischen Verdampfung als Probeneinführungstechnik für die ICP-AES. In: Dittrich K, Welz B (eds) CARNAS'93. Universität Leipzig, Leipzig, pp 809–816

468. Hu B, Jiang ZC, Zeng Y (1994) Direct determination of aluminium in biological materials by electrothermal vaporization inductively coupled plasma atomic emission spectrometry with polytetrafluoroethylene as chemical modifier. Anal. Chim. Acta 269: 213–218

469. Qin YC, Jiang ZC, Zeng YE, Hu B (1995) Effect of particle size in the analysis of botanical samples by slurry sampling and fluorination-electrothermal vaporization inductively coupled plasma atomic emission spectrometry. J. Anal. At. Spectrom. 10: 455–457

470. Hinds MW, Grégoire DC, Ozaki EA (1996) Direct determination of volatile elements in nickel

alloys by electrothermal vaporization inductively coupled plasma mass spectrometry. J. Anal. At. Spectrom. 11: in press

471. Nickel H, Zadgorska Z, Hassler J, Hemel V (1994) Direkte Analyse metallischer Verunreinigungen in Siliziumnitrid und Borcarbid über ETV-ICP-OES. In: Dittrich K, Welz B (eds) CANAS'93. Universität Leipzig, Leipzig, pp 781–786

472. Hinds MW, Kogan VV (1995) Assay of precious metals by solid sampling atomic spectrometry. Spectroscopy 10: 14–18

473. Millard DL, Shan HC, Kirkbright GF (1980) Optical emission spectrometry with an inductively coupled radiofrequency argon plasma source and sample introduction with a graphite rod electrothermal vaporisation device. Analyst 105: 502–509

474. Argentine MD, Barnes RM (1994) Electrothermal vaporization-inductively coupled plasma mass spectrometry for the analysis of semiconductor-grade organometallic materials and process chemicals. J. Anal. At. Spectrom. 9: 1371–1378

475. Plantikow-Voßgätter F, Denkhaus E (1996) Application of an ETV-ICP system for the determination of elements in human hair. Spectrochim. Acta B 51: 261–270

476. Wolff G (1994) Feststoffprobenaufgabensysteme für die ICP-OES und ICP-MS. In: Dittrich K, Welz B (eds) CANAS'93. Universität Leipzig, Leipzig, pp 281–290

477. Huang M, Hanselman DS, Yang P, Hieftje GH (1992) Isocontour maps of electron temperature, electron number density and gas kinetic temperature in the Ar ICP obtained by laser-light Thomson and Rayleigh scattering. Spectrochim. Acta 47B: 765–785

478. Matousek JP, Mermet JM (1993) The effect of added hydrogen in electrothermal vaporization inductively coupled plasma atomic emission spectrometry. Spectrochim. Acta 48B: 835–850

479. Matousek JP, Mermet JM (1992) Signal enhancement with added hydrogen in electrothermal vaporization – inductively coupled plasma atomic emission spectrometry. ICP Inform. Newsl. 17: 802

480. Fonseca RW, Miller-Ihli NJ (1996) Influence of sample matrix components of the selection of calibration strategies in electrothermal vaporization inductively coupled plasma mass spectrometry. Spectrochim. Acta B 51: 1591–1599

481. Marshall J, Franks J (1990) Mutielement analysis and reduction of spectral interferences using electrothermal vaporization inductively coupled plasma-mass spectrometry. J. Anal. At. Spectrom. 11: 177

482. Wang J, Carey JM, Caruso JA (1994) Direct analysis of solid samples by electrothermal vaporization inductively coupled plasma mass spectrometry. Spectrochim. Acta 49B: 193–203

483. Moens L, Verrept P, Dams R, Greb U, Jung G, Laser B (1994) New high resolution ICP-MS technology applied for the determination of V, Fe, Cu, Zn and Ag in human serum. J. Anal. At. Spectrom.

484. Dittrich K, Walther A (1994) Untersuchungen zur direkten Feststoffanalyse durch ETV-ICP-AES. In: Dittrich K, Welz B (eds) CANAS'93. Universität Leipzig, Leipzig, pp 793–800

485. Moens LJ, Verrept P, Boonen S, Vanhaecke F, Dams RFJ (1995) Solid sampling electrothermal vaporization for sample introduction in inductively coupled plasma atomic emission spectrometry and inductively coupled plasma mass spectrometry. Spectrochim. Acta 50B: 463–475

486. Boonen S, Verrept P, Moens L, Dams R (1993) Advantages and limitations of different standardization methods in solid sampling electrothermal vaporization (SS-ETV) inductively coupled atomic emission spectrometry (ICP-AES). J. Anal. At. Spectrom. 8: 711

487. Galbács G, Vanhaecke F, Moens L, Dams R (1997) Determination of cadmium in solid certified reference materials using solid sampling electrothermal vaporisation inductively coupled plasma mass spectrometry complemented with thermogravimetric experiments. Microchem. J. accepted for publ.

488. Byrne JP, Hughes DM, Chakrabarti CL, Grégoire DC (1994) Mechanism of volatilization of tungsten in the graphite furnace investigated by electrothermal vaporization inductively coupled plasma mass spectrometry. J. Anal. At. Spectrom. 9: 913–917

489. Okamoto Y, Sugawa K, Kumamaru T (1994) Chemical modification for inductively coupled plasma atomic emission spectrometric determination of boron with tungsten boat furnace vaporizer. J. Anal. At. Spectrom. 9: 89–92

490. Manninen PKG (1994) Determination of extractable organic chlorine by electrothermal vaporization inductively coupled plasma mass spectrometry. J. Anal. At. Spectrom. 9: 209–211

491. Argentine MD, Krushevska A, Barnes RM (1994) Determination of trace impurities in organometallic semiconductor-grade reagents and process chemicals with electrothermal

vaporization-inductively coupled plasma atomic emission spectrometry. J. Anal. At. Spectrom. 9: 1121–1128

492. Darke S. A. and Tyson J. F. (1994) A Review on solid sampling plasma spectrometry and a comparison of results for laser ablation, electrothermal vaporisation and slurry nebulization. Microchem. J 50: 310–336

493. Moens L, Verrept P, Boonen S, Vanhaecke F, Dams R (1995) Solid sampling electrothermal vaporization for sample introduction in inductively coupled plasma atomic emission spectrometry and in inductively coupled plasma mass spectrometry. Spectrochim. Acta 50: 463–475

494. Voellkopf U, Paul M, Denoyer ER (1992) Analysis of solid samples by ICP-mass spectrometry. Fresenius Z. Anal. Chem. 324: 917–923

495. Grégoire DC, Lee J (1994) Determination of cadmium and zinc isotope ratios in sheep's blood and organ tissue by electrothermal vaporization inductively coupled plasma mass spectrometry. J. Anal. At. Spectrom. 9: 393

496. Ohls K, Dewies J, Loepp H (1986) Direct sample vapor introduction into an Argon ICP. ICP Inform. Newsl. 12: 177–179

497. Bendicho C, De Loos Vollebregt TC (1990a) The influence of pyrolysis and matrix modifieres for analysis of glass materials by GFAAS using sample introduction. Spectrochim. Acta 45B: 679–693

498. De Benzo ZA, Fernandez MR, Carrión N, Eljuri E (1988) Determination of Cu, Zn, Fe and Mn in slurries of ashed plant tissue by atomic absorption sprectrometry. At. Spectrosc. 9: 87–92

499. Sturgeon RE (1989) Graphite furnace atomic absorption analysis of marine samples for trace metals. Spectrochim. Acta 44B: 1209–1220

500. Miller-Ihli NJ (1992a) Solids analysis by GFAAS. Anal. Chem. 64: 964A–968A

501. Miller-Ihli NJ (1993) Advances in ultrasonic slurry graphite furnace atomic absorption spectrometry. Fresenius J. Anal. Chem. 345: 482–489

502. Miller-Ihli NJ (1994) Influence of slurry preparation on the accuracy of ultrasonic slurry electrothermal atomic absorption spectrometry. J. Anal. At. Spectrosc. 9: 1129–1134

503. Dannecker W, Ulrich A (1993) Fortschritte bei der Anwendung der Elementaranalytik auf Proben aus dem Umweltbereich. In: Dittrich K, Welz B (eds) CANAS'93. Universität Leipzig, Leipzig, pp 13–36

504. Carrión N, De Benzo ZA, Eljuri EJ, Ippoliti F, Flores DJ (1987) Determination of manganese, calcium, magnesium and potassium in pine (Pinus Caribaea) needle samples by flame atomic absorption spectrometry with slurry sample introduction. J. Anal. At. Spectrom. 2: 813-817

505. De Andrade JC, Strong FC, Martin NJ (1990) Rapid determination of zinc and iron in foods by flow-injection analysis with flame atomic-absorption spectrometry and slurry nebulization. Talanta 37: 711–718

506. López-García I, Sobejano FO, Hernández-Córdoba M (1991) Use of flow injection flame atomic absorption spectrometry for slurry atomization. Determination of copper, manganese, chromium and zinc in iron oxide pigments. Analyst 116: 517-520

507. Viñas P, Campillo N, López-García I, Hernández-Córdoba M (1993) Flow-injection flame atomic absorption spectrometry for slurry atomization. Determination of calcium, magnesium, zinc and manganese in vegetables. Anal. Chim. Acta 283: 393-400

508. Ebdon L, Foulkes ME, Hill SJ (1990) Direct atomic spectrometric analysis by slurry atomisation. Part 9. Fundamental studies of refractory samples. J. Anal. At. Spectrom. 5: 67–73

509. Goodall P, Foulkes ME, Ebdon L (1993) Slurry nebulization inductively coupled plasma spectrometry – the fundamental parameters discussed. Spectrochim. Acta 48B: 1563–1577

510. Madrid Y, Bonilla M, Cámara C (1989) Determination of lead in foodstuffs and biological samples by hydride generation atomic absorption spectrometry using an aqueous slurry technique. J. Anal. At. Spectrom. 4: 167–169

511. Calle Guntinas M, Dala B, Madrid Y, Cámara C (1991) Determination of total available antimony in marine sediments by slurry formation-hydride generation atomic absorption spectrometry. Applicability to the selective determination of antimony(III) and antimony(V). Analyst 116: 1029–1032

512. López-García I, Contéz JA, Hernández-Córdoba M (1993) Generation of vapors from slurried samples for the fast determination of As and Hg in coal fly ash and diatomaceous earth. Atomic Spectrosch. 14: 144–147

513. Hinds MW, Jackson KW (1987) Lead atomisation from soil by slurry introduction electrothermal atomisation atomic absorption spectrometry. Part 1. Effects of matrix components on the absorbance versus time profile. J. Anal. At. Spectrom. 2: 441–445

514. Karwowska R, Jackson KW (1987) Atomisation characteristics of lead determination in alumina matrices by slurry-electrothermal atomisation atomic absorption spectrometry. J. Anal. At. Spectrom. 2: 125–129
515. Ebdon L, Parry HGM (1988) Direct atomic spectrometic analysis by slurry atomisation. Part 4. Determination of selenium of coal by electrothermal atomisation atomic absorption spectrometry. J. Anal. At. Spectrom. 3: 131–134
516. Dobrowolski R, Mierzwa J (1993) Determination of trace elements in plant materials by slurry sampling graphite furnace AAS – some analytical problems. Fresenius Z. Anal. Chem. 346: 1058–1061
517. Mierzwa J, Dobrowolski R (1994) Silica gel analysis by slurry-sampling graphite-furnace atomic absorption spectrometry. Fresenius J. Anal. Chem. 348: 422–425
518. Hauptkorn S, Krivan V (1994) Determination of silicon in boron nitride by slurry sampling electrothermal atomic absorption spectrometry. Spectrochim. Acta 49B: 221–228
519. Fuller CW, Thompson I (1977) Novel sampling system for the direct analysis of powders by atomic-absorption spectrometry. Analyst 102: 141–143
520. Ebdon L, Pearce WC (1982) Direct determination of arsenic in coal by atomic absorption spectrometry using solid sampling and electrothermal atomisation. Analyst 107: 942–950
521. Littlejohn D, Stephen SC, Ottaway JM (1985) Slurry sample introduction procedures for the analysis of foodstuffs by electrothermal atomisation atomic absorption spectrometry. Anal. Proc. 22: 376–378
522. Olayinka KO, Haswell SJ, Grzeskoiak R (1986) Development of a slurry technique for the determination of cadmium in dried foods by electrothermal atomization atomic absorption spectrometry. J. Anal. At. Spectrom. 1: 297–300
523. Hinds MW, Mohan K, Jackson KW (1988) The effectiveness of palladium plus magnesium as a matrix modifier for the determination of lead in solutions and soil slurries by electrothermal atomic absorption spectrometry. J. Anal. At. Spectrom. 3: 83–87
524. Lynch S, Littlejohn D (1989) Palladium as a chemical modifier for the determination of lead in food slurries by electrothermal atomisation atomic absorption spectrometry. J. Anal. Atom. Spectrom. 4: 157–161
525. López-García I, Córtez JA, Hernández-Córdoba M (1993a) Slurry-electrothermal atomic absorption spectrometry of samples with large amounts of silica. Determination of cadmium, zinc and mangananese using fast temperature programmes. Anal. Chim. Acta 283: 167–174
526. Schneider G, Krivan V (1995) Slurry and liquid sampling using electrothermal atomic absorption spectrometry for the analysis of zirconium dioxide based materials. Spectrochim. Acta B 50: 1557–1571
527. Miller-Ihli NJ (1988) Slurry sample preperation for simultaneous multielement graphite furnace atomic absorption spectrometry. J. Anal. At. Spectrom. 3: 73–81
528. López-García I, Hernández-Córdoba M (1990) Determination of arsenic in commercial iron (III) oxide pigments by electrothermal atomic absorption spectrometry with slurry sample introduction. J. Anal. At. Spectrom. 5: 647–650
529. Sperling M (1989) Automatische Aufschlämmung von pulverförmigen Proben und Dosierung von Suspensionen in der Graphitrohrofen-AAS. In: Welz B (ed) 5. Colloquium Atomspektrometrische Spurenanalytik. Bodenseewerk Perkin-Elmer GmbH D-7770 Überlingen, pp 531–541
530. Jordan P, Ives JM, Carnrick GR, Slavin W (1989) Manganese determination in biological materials using fast automated slurry sampling and GFAAS. At. Spectrosc. 10: 165–169
531. Shengjun M, Holcombe JA (1990) Preconcentration of copper absorption by slurry graphite furnace atomic absorption spectrometry. Anal. Chem. 62: 1994–1997
532. Shengjun M, Holcombe JA (1991) Preconcentration of nickel and cobalt on algae and determination by slurry graphite-furnace atomic absorption spectrometry. Talanta 38: 503–510
533. Hinds MW, Jackson KW (1991) Determination of lead in soil by vortex mixing slurry-graphite furnace atomic absorption spectrometry. At. Spectrosc. 17: 109–110
534. Ohta K, Aoki W, Mizuno T (1990) Direct determination of cadmium in biological material using electrothermal atomisation atomic absorption spectrometry with a metal tube atomizer and a matrix modifier. Mikrochim. Acta [Wien] I: 81–86
535. Bendicho C, De Loos Vollebregt TC (1990b) Metal extraction by hydrofluoric acid from slurries of glass materials in graphite furnace atomic absorption spectrometry. Spectrochim. Acta 45B: 695–710

536. Yu Z, Vandecasteele C, Desment B, Dams R (1990) Determination of lead in environmental reference materials (plant materials) by slurry-ET-AAS. Mikrochim. Acta [Wien] I: 41–48

537. Fernández CA, Fernández R, Carrión N, Loreto D, Benzo Z, Fraile R (1991) Metals determination in atmospheric particulates by atomic absorption spectrometry with slurry sample introduction. At. Spectrosc. 12: 111–117

538. Hinds MW, Latimer KE, Jackson KW (1991) Determination of lead in soil by slurry-electrothermal atomic absorption spectrometry with a fast temperature programme. J. Anal. At. Spectrom. 6: 473–476

539. Van den Akker AH, Van den Heuvel H (1992) Using an ultrasonic slurry sampler with graphite furnace AAS for the determination of heavy metals in sediments. At. Spectrosc. 13: 72–73

540. Bendicho C, Sancho A (1993) Determination of selenium in wheat flour by GFAAS using automated ultrasonic slurry sampling. At. Spectrosc. 14: 187–190

541. Docecal B, Krivan V (1993) Determination of trace elements in high-purity molybdenum trioxide by slurry sampling electrothermal atomic absorption spectrometry. J. Anal. At. Spectrom. 8: 637–641

542. Docekal B, Krivan V (1993) An improved electrothermal atomic absorption spectrometry method for the determination of lithium in molybdenum oxide using slurry sampling and a tungsten atomizer. Spectrochimica Acta 48B: 1645–1649

543. Hauptkorn S, Schneider G, Krivan V (1994) Determination of silicon in titanium dioxide and zirconium dioxide by electrothermal atomic absorption spectrometry using the slurry sampling technique. J. Anal. At. Spectrom. 9: 463–468

544. Stephen CC, Littlejohn D, Ottaway JM (1985) Evaluation of a slurry technique for the determination of lead in spinach by electrothermal atomic-absorption spectrometry. Analyst 110: 1147–1151

545. Brady DV, Montalvo JGJr, Jung J, Curran. RA (1974) Direct determination of lead in plant leaves via graphite furnace atomic absorption. At. Absorpt. Newsletter 13: 118–119

546. Ebdon L, Fisher AS, Hill SJ (1993) Use of hydrogen in electrothermal atomic absorption spectrometry to decrease the background signal arising from environmental slurry samples. Anal. Chim. Acta 282: 433–436

547. Langland JK, Harrison SH, Kratochvil B, Zeisler R (1983) Cryiogenetic homogenisation of biological tissues. In: Zeisler R, Harrison SH, Wise SA (eds) The pilot national environmental specimen bank – analyses of human liver specimens. NBS Special Publ. 656, US Dept. of Commerce, pp 21–34

548. Schladot JD, Backhaus F (1988) Preparation of sample material for environmental specimen bank purposes – milling and homogenization at cryoggenic temperatures. In: Wise SA, Zeisler R, Goldstein GM (eds) Progress of speciment banking. NBS Special Publ. 740, US Dept. of Commerce, pp 184–193

549. Schladot JD, Backhaus F, Burow M, Froning M, Mohl C, Ostapczuk P, Roßbach M (1993) Collection, preparation and characterization of fresh, marine candidate reference materials of the German environmental specimen bank. Fresenius J. Anal. Chem. 345: 137–139

550. Kramer GN, Pauwels J, Belliardo JJ (1993) Preparation of biological and environmental reference materials. J. Anal. Chem. 145: 133–136

551. Barnett NW, Ebdon L, Evans EH, Ollivier P (1988) Determination of boron in plants by graphite furnace atomic absorption spectrometry with slurry atomization using matrix modifikation and totally pyrolytic graphite tube. Anal. Proc. 25: 233–235

552. Carrión N, De Benzo ZA, Moreno B, Fernandez EJ, Flores D (1988) Determination of copper, chromium, iron and lead in pine needles by electrothermal atomisation spectrometry with slurry sample introduction. J. Anal. At. Spectrom. 3: 479–483

553. Ebdon L, Fisher A, Parry HGM, Brown AA (1990) Direct atomic spectrometric analysis by slurry atomisation. Part 10. Use of an air-ashing stage in electrothermal atomic absorption spectrometry. J. Anal. At. Spectrom. 5: 321–324

554. Hernández-Córdoba M, López-García I (1991) A fast method for the determination of lead in paprika by electrothermal atomic absorption spectrometry with slurry sample introduction. Talanta 38: 1247–1251

555. Majidi V, J.A, Holcombe (1990) Error analysis for sampling of slurries: sedimentation errors. Spectrochim. Acta 45 B: 753–761

556. López-García I, Viñas P, Hernández-Córdoba M, (1992) Slurry procedure for the determination of titanium in plant material using electrothermal atomic absorption spectrometry. J. Anal. At. Spectrom. 7: 529–532

557. Hinds MW, Jackson KW (1990) Comparison of palladium nitrate and chloride as a chemical

modifier for the determination of lead in solutions and soil slurries by electrothermal atomic absorption spectrometry. J. Anal. At. Spectrom. 5: 199–202

558. Karwowska R, Jackson KW (1986) Lead atomtization in the presence of aluminium matrices by electrothermal atomic absorption spectrometry. A comparative study of slurry vs solution sample introduction. Spectrochim. Acta 41B: 947–957

559. Haraldsen L, Pougnet MAB (1989) Direct determination of beryllium in coal slurrys using graphite furnace atomic absorption spectrometry with automatic injection. Analyst 114: 1331–1333

560. Lester JN, Harrison RM, Perry R (1977) Rapid flameless atomic absorption analysis of the metallic content of sewage sludges I. Lead, cadmium and copper. Sci. Total Environ. 8: 153–158

561. Stoveland S, Astruc M, Perry R, Lester JN (1978) Rapid flameless atomic absorption analysis of the metallic content of sewage sludge. II. Chromium, nickel, zinc. Sci. Total Environ. 9: 263–269

562. Hoenig M, Regnier P, Chou L (1991) Determination of the high aluminium content in suspended matter samples collected in natural waters by slurry sampling-electrothermal atomic absorption spectrometry. J. Anal. At. Spectrom. 6: 273–275

563. Epstein MS, Carnick GR, Slavin W, Miller-Ihli NJ (1989) Automated slurry sample introduction for analysis of a river sediment by graphite furnace atomic absorption spectrometry. Anal. Chem. 61: 1414–1419

564. Bradshaw D, Slavin W (1989) Rapid slurry analysis of solid coal and fly ash samples. Spectrochim. Acta 44B: 1245–1256

565. Richter U, Dannecker W (1994) Feststoffanalytik mit der Slurry-GF-AAS in anorganischen Proben. In: Dittrich K, Welz B (eds) CANAS'93. Universität Leipzig, Leipzig, pp 639–644

566. Kelemen J, Szakácz O and Lásztity A (1990) Determination of barium by atomic absorption spectrometry with electrothermal atomisation. Part 1. Study of the volatilisation and atomisation of solid barium phosphate, silicate, molybdate and molybdovanadate compounds. J. Anal. At. Spectrom. 5: 377–384

567. Bargouth I (1993) Vergleichende Untersuchungen zur Homogenität von biologischen Materialien mit den Atomabsorptionsspectrometern SM 20 (Feststoffgerät) und Z 5100 PC (Slurry). Diplomarbeit FH Aachen

568. López-García I, Córtez Ja, Hernández-Córdoba M, (1993) Rapid furnace programmes for the slurry-electrothermal atomic absorption spectrometrc determination of chromium, lead and copper in diatomaceous earth. J. Anal. At. Spectrom. 8: 103–108

569. López-García I, Navarro E, Viñas P, Hernández-Córdoba M, (1997) Rapid determination of lead, cadmium and thallium in cements using electrothermal atomic absorption spectrometry with slurry sample introduction. Fresenius J. Anal. Chem. 357: 642–646

570. Hartley JHK, Hill SJ, Ebdon L (1993) Analysis of slurries by inductively coupled plasma mass spectrometry using desolvation to improve transport efficiency and atomization efficiency. Spectrochim. Acta 48B: 1421–1433

571. Schäffer U, Krivan V (1996) Slurry sampling electrothermal atomic absorption spectrometry for the analysis of graphite powders. Spectrochim. Acta B 51: 1211–1222

572. Holcombe JA, Majidi V (1989) Error analysis for sampling of slurries: volumetric errors. J. Anal. Spec. 4: 423–425

573. van Dalen G, de Galan L (1994) Direct determination of particulate elements in edible oils and fats using an ultrasonic slurry sampler with graphite furnace atomic absorption spectrometry. Spectrochim. Acta 49B: 1689–1693

574. Huang MD, Krivan V, Welz B, Schlemmer G (1997) Determination of trace impurities in titanium dioxide by slurry sampling electrothermal atomic absorption spectrometry. Spectrochim. Acta, submitted

575. Hoenig M, Regnier P, Wollast R (1989) Automated trace metal analyses of slurried solid samples by graphite furnace atomic absorption spectrometry with application to sediments and suspended matter collected in natural waters. J. Anal. At. Spectrom. 4: 631–634

576. Hoenig W, Van Hoeyweghen P (1986) Alternative to solid sampling for trace metal determination by platform electrothermal atomic absorption spectrometry: direct dispensing of powdered samples suspended in liquid medium. Anal. Chem. 58: 2614–2617

577. Lynch S, Littlejohn D (1990) Development of a slurry atomization method for the determination of cadmium in food samples by electrothermal atomization atomic-absorption spectrometry. Talanta 37: 825–830

578. Ebdon L, Parry HGM (1987) Direct atomic spectrometric analysis by slurry atomisation. Part

2. Elimination of interferences in the determination of arsenic in whole coal by electrothermal atomisation atomic absorption spectrometry. J. Anal. At. Spectrom. 2: 131–133

579. Hinds MW, Jackson KW (1988) Lead atomisation from soil by slurry introduction electrothermal atomisation atomic absorption spectrometry Part 2. Atomisation characteristics with various matrix modifiers. J. Anal. At. Spectrom. 3: 997–1003

580. Papaspyrou M (1993) Untersuchung über den Einbau und Transport von Bor in Melanomzellen mit Hilfe von borierten Stoffwechselvorläufen für die Neutroneneinfangtherapie. Dissertation, Heinrich-Heine-Universität, Düsseldorf, Germany

581. Docekal B (1993) Simple stirring device for a slurry sampling technique in electrothermal atomic absorption spectrometry. J. Anal. At. Spectrom. 8: 763–765

582. Stoeppler M, Kampel M, Welz B (1976) Studies on automated flameless atomic absorption spectrometry. Fresenius Z. Anal. Chem. 282: 369–378

583. Hoenig M, Cilissen A (1993) Robotized sampling device for graphite furnace atomic absorption spectrometry slurry analysis with Varian SpectrAA instruments. Spectrochim. Acta 48B: 1303–1306

584. Hoenig M, Cilissen A (1993) Electrothermal atomic absorption spectrometry: fast or conventional programs?. Spectrochim. Acta 48B: 1003–1012

585. Carnick GR, Daley G, Fortinopoulos A (1989) Design and use of a new automated ultrasonic slurry sample for graphite furnace atomic absorption. At. Spectrosc. 10: 170–174

586. US patent (June 5, 1990) No 4930898 awarded

587. Miller-Ihli NJ (1989) Automated ultrasonic mixing accessory for slurry sampling into a graphite furnace atomic absorption spectrometer. J. Anal. At. Spectrom. 4: 295–297

588. Slavin W, Miller-Ihli NJ, Carnick GR (1990) Fast furnace analyses and slurry sampling. Fresenius Z. Anal. Chem.

589. Hauptkorn S, Krivan V (1996) Solution and slurry sampling electrothermal atomic absorption spectrometry for the analysis of high purity quartz. Spectrochim. Acta B 51: 1197–1210

590. Bermejo-Barrera P, Lorenzo-Alonso MJ, Aboal-Somoza M, Bermejo-Barrera A (1994) Determination of arsenic in mussels by slurry sampling and electrothermal atomic absorption spectrometry. Mikrochim. Acta 117: 49–64

591. Beaty M, Barnett W (1980) Techniques for analyzing difficult samples with the HGA graphite furnace. Atomic Spetrosc. 1: 72–77

592. Steiner JW, Kramer HL (1983) In situ gaseous pretreatment of liver extracts in a modified carbon rod atomiser during the determination of cadmium and lead. Analyst 108: 1051–1059

593. Eaton DK, Holcombe JA (1983) Oxygen ashing attachment for a furnace atomizer power supply. Anal. Chem. 55: 1821–1823

594. Narres, H.D., Mohl, C., Stoeppler, M. (1985) Metal analyses in difficult materials with platform furnace Zeeman-atomic-absorbtion spectroscopy. 2. Direct determination of cadmium and lead in milk. Z. Lebensm. Unters. Forsch. 181: 111–116

595. Narres HD, Mohl C, Stoeppler M (1986) Direktbestimmung von Cadmium in Rohöl und Ölprodukten mit Zeeman-Atom-Absorptionsspectromerie in der Graphitküvette mit L'vov-Plattform. Erdöl und Kohle- Erdgas- Petrochemie mit Brennstoffchemie 39: 193–194

596. Müller C (1987) Cadmium content of human milk. Trace elements in medicine 4: 4–7

597. Larsen EH, Rasmussen L (1991) Chromium, lead and cadmium in Danish milk products and cheese determined by Zeeman graphite furnace atomic absorption spectrometry after direct injection or pressurized ashing. Z. Lebensm. Unters. Forsch. 192: 136–141

598. Novak L, Stoeppler M (1986) Use of hydrogen for the elimination of matrix interferences in the determination of lead by graphite furnace atomic absorption spectrometry. Fresenius Z. Anal. Chem. 323: 737–741

599. Eaton DK, Holcombe JA (1983) Oxygen ashing and matrix modifiers in graphite furnace atomic absorption spectrometric determination of lead in whole blood. Anal. Chem. 55: 946–950

600. Salmon SG, Holcombe JA (1982) Alteration of metal release mechanismus in graphite furnace atomizers by chemisorbed oxygen. Anal. Chem. 54: 630–634

601. Slavin W (1984) Graphite furnace AAS – a source book. Perkin-Elmer Corp. Norwalk, USA,

602. Slavin W (1994) Atomic absorption spectrometry. In: Stoeppler M, Herber RFM (eds) Trance element analysis in biological specimens. Elsevier, Amsterdam, pp 53–90

603. Robinson JW (1990) Atomic spectroscopy. Marcel Dekker, New York

604. Stoeppler M (1991) Analytical chemistry of metals and metal compounds. In: Merian E (ed)

Metals and their compounds in the environment – occurence, analysis and biological relevance. Verlag Chemie, Weinheim, Germany, pp 105–206

605. Stoeppler M (1992) Analytical methods and instrumentation – a summarizing overview. In: Stoeppler M (ed) Hazardous metals in the environment. Elsevier, Amsterdam, pp 97–132

606. Viñas P, Campillo N, López-García I, Hernández-Córdoba M (1995) Slurry electrothermal atomic absorption spectrometric determination of aluminium and chromium in vegetables using hydrogen peroxide as a matrix midifier. Talanta 42: 527-533

607. Docekal B, Krivan V (1992) Direct determination of impurities in powdered silicon carbide by electrothermal atomic absorption spectrometry using the slurry sampling technique. J. Anal. At. Spectrom. 7: 521–528

608. Bermejo-Barrera P, Barciela-Alonso C, Bermejo-Barrera A (1996) Determination of cadmium in slurries of marine sediment samples by electrothermal atomic absorption spectrometry using palladium and phosphate as chemical modifiers. Mikrochim. Acta 124: 251–261

609. Klemm W, Bombach G (1995) Trace element determination in contaminated sedimenrs and soils by ultrasonic slurry sampling and Zeeman graphite furnace atomic absorption spectrometry. Fresenius J. Anal. Chem. 353: 12–15

610. Bermejo-Barrera P, Barciela-Alonso MC, Ferrón-Novais M, Bermejo-Barrera A (1995) Speciation of arsenic by the determination of total arsenic and arsenic (III) in marine sediment samples by electrothermal atomic absorption spectrometry. J. Anal. At. Spectrom. 10: 247-252

611. Bermejo-Barrera P, Barciela-Alonso MC, Moreda-Piñeiro J, González-Sixto C, Bermejo-Barrera A (1996) Determination of trace metals (As, Cd, Hg, Pb and Sn) in marine sediment slurry samples by electrothermal atomic absorption spectrometry using palladium as a chemical modifier. Spectrochim. Acta 51B: 1235-1244

612. Robles LC, García-Olalla C, Aller AJ (1993) Determination of gold by slurry electrothermal atomic absorption spectrometry after preconcentration by escherichia coli and pseudomonas putida. J. Anal. At. Spectrom. 8: 1015–1022

613. Papaspyrou M, Feinendegen LE, Mohl C, Schwuger MJ (1994) Determination of boron in cell suspensions using electrothermal atomic absorption spectrometry. J. Anal. At. Spectrom. 9: 791–795

614. Robles LC, Aller AJ (1994) Preconcentration of beryllium on the outer membrane of escherichia coli and pseudomonas putida prior to determination by electrothermal atomic absorption spectrometry. J. Anal. At. Spectrom. 9: 871–879

615. Slovák Z, Docekal B (1981) Determination of trace metals in aluminium oxide by electrothermal atomic absorption spectrometry with direct injection of aqueous suspensions. Anal. Chim. Acta 129: 263–267

616. Akatsuka K, Nobuyama N, Atsuya I (1988) Atomic absorption spectrometry of nanogram amounts of cadmium, lead and zinc after precipitation with 8-quinolinol. Anal. Sci. 4: 281–285

617. Viñas P, Campillo N, López-García I, Hernández-Córdoba M (1994) Slurry procedures for the determination of cadmium and lead in cereal-based products using electrothermal atomic absorption spectrometry. Fresenius J. Anal. Chem. 349: 306-310

618. Viñas P, Campillo N, López-García I, Hernández-Córdoba M (1994) Slurry atomization of vegetables for the electrothermal atomic absorption spectrometric analysis of lead and cadmium. Food Chemistry 50: 317-321

619. Yanxi T, Marshall WD, Blais J-S (1996) Slurry preparation by high-pressure homogenization for cadmium, copper and lead determination in cervine liver and kidney by electrothermal atomic absorption spectrometry. Analyst 121: 483–488

620. Viñas P, Campillo N, López-García I, Hernández-Córdoba M (1994) Slurry-electrothermal atomic absorption spectrometric methods for the determination of copper, lead, zinc, iron and chromium in sweets and chewing gum after partial dry ashing. Analyst 119: 1119-1123

621. Viñas P, Campillo N, López-García I, Hernández-Córdoba M (1995) Rapid procedures for cobalt and nickel determination in slurried food samples by electrothermal atomic absorption spectrometry. Atomic Spectrosc. 16: 86–89

622. Viñas P, Campillo N, López-García I, Hernández-Córdoba M (1993) Analysis of copper in biscuits and bread using a fast-program slurry electrothermal atomic absorption procedure. J. Agric. Food Chem. 8: 2024-2027

623. Bermejo-Barrera P, Moreda-Piñeiro J, Moreda-Piñeiro A, Bermejo-Barrera A (1994) Palladium as a chemical modifier for the determination of mercury in marine sediment slurries by electrothermal atomization atomic absorption spectrometry. Anal. Chim. Acta 296: 181-193

624. Lamberty A, De Bièvre P, Götz A (1993) The international measurement evaluation programme IMEP-2: Cd in polyethylene. Fresenius J. Anal. Chem. 345: 310–313
625. Fuller CW (1974) A kinetic theory of atomisation for non-flame atomic-absorption spectrometry with a graphite furnace. The kinetics and mechanism of atomisation for copper. Analyst 99: 739–744
626. López-García I, Hernández-Córdoba M (1989) Fast determination of lead in commercial iron oxide pigments by graphite furnaceatomic absorption spectrometry using a slurry technique. J. Anal. At. Spectrom. 4: 701-704
627. Chiricosta S, Cum G, Gallo A, Spadaro A, Vitarelli P (1982) Polymer-supported catalysts in organic synthesis: direct determination of palladium by atomic absorption spectroscopy. Atomic Spectrosc. 3: 185-187
628. López-García I, Viñas P, Campillo N, Hernández-Córdoba M (1996) Determination of selenium in seafoods using electrothermal atomic absorption spectrometry with slurry sample introduction. J. Agric. Food Chem. 44: 836-841
629. Langmyhr FJ (1985) The solid sampling technique of atomic absorption spectrometry – what can the method do? Fresenius Z. Anal. Chem. 322: 654–656
630. Pesch H-J, Bloß S, Schubert J, Seibold H (1992) The mercury, cadmium and lead content of cigarette tobacco: Comparative analytical-statistical studies in 1987 and 1991 employing Zeeman-AAS. Fresenius J. Anal. Chem. 343: 152–153
631. Horner E, Kurfürst U (1987) Cadmium in wheat from biological cultivation. Fresenius Z. Anal. Chem. 328: 386–387
632. Kurfürst U, Beck A (1995) Cadmiumgehalte in ökologisch angebautem Weizen. Lebendige Erde 6: 477–479
633. Steubing L, Grobecker KH (1985) Blei- und Cadmiumbelastung in verschiedenen Ökosystemen. Fresenius Z. Anal. Chem. 322: 692–696
634. Gellert G, Wittassek R (1985) Kupferverteilung in einem Flußsediment und seine biologische Verfügbarkeit. Fresenius Z. Anal. Chem. 322: 700–703
635. Ellenberg H, Dietrich J, Stoeppler M, Nürnberg HW (1986) Habicht-Mauserfedern als hochintegrierende Biomonitoren für die Schadstoffbelastung von Landschaftsausschnitten. Allg. Forstz. 41: 23–25
636. Hahn H (1991) Schwermetallgehalte in Vogelfedern – ihre Ursache und der Einsatz von Federn standorttreuer Vogelfedern im Rahmen von Bioindikationsverfahren. KFA Jülich, Bericht 2493
637. Hahn E, Hahn K, Mohl C, Stoeppler M (1990) Zeeman SS-GFAAS – An ideal method for the evalution of lead and cadmium profiles in bird's feathers. Fresenius J. Anal. Chem. 337: 306–309
638. Hahn E, Hahn K, Stoeppler M (1993) Bird feathers as bioindicators in areas of the german environmental specimen bank – bioaccumulation of mercury in food chains and exogenous deposition af atmospheric pollution with lead and cadmium. Sci. Total Environ. 139/140: 259–270
639. Klüßendorf B, Rosopulo A, Kreuzer W (1985) Untersuchungen zur Verteilung und Schnellbestimmung von Blei, Cadmium und Zink in Lebern von Schlachtschweinen mittels Feststoff-Zeeman-Atomabsorptionsspektrometrie. Fresenius Z. Anal. Chem. 322: 721–727
640. Lücker E, Meithen J, Kreuzer W (1993) Distribution of Pb and Cd in equine liver – direct determination by means of solid sampling ZAAS. Fresenius J. Anal. Chem. 346: 1068–1071
641. Lücker E, Rosopulo A, Kreuzer W (1987) Analyses of the distribution of lead and cadmium in fresh renal tissue by means of solid sampling Zeeman-AAS. Part I. Fresenius Z. Anal. Chem. 328: 370–377
642. Lücker E, Gerbig C, Kreuzer W (1993) Distribution of Pb and Cd in the liver of the mallard – direct determination by means of solid sampling ZAAS. Fresenius J. Anal. Chem. 346: 1062–1067
643. Lücker E, Thorius-Ehrler S (1993) Solid sampling ZAAS determintion of endogenous Pb contamination in muscle tissue caused by calcification. Fresenius J. Anal. Chem. 346: 1072–1076
644. Fleckenstein J, Graff O (1982) Schwermetallaufnahme aus Müllkompost durch den Regenwurm Eisenia foetida. Landbauforschung Völkenrode 32: 198–202
645. Kjer I, Meier-Ploeger A, Voigmann H (1994) Hopfen und Gerste aus ökologischem Anbau – Qualität und Einfluß auf die Roh-, Zwischen- und Endprodukte der Bierherstellung. Brauwelt Nr. 12: 462–469
646. Fuchshofen W, Schüler C, Vogtmann H (1993) Zur Schwermetallbelastung von Boden und Pflanze bei Düngung mit Bioabfallkompost, Rindermistkompost und mineralischer Düngung in zwei Versuchen mit unterschiedlicher Fruchtfolge. VDLUFA, Darmstadt Kongreßband: pp 209–212

647. Esser P (1987) Solid sampling AAS methods in industrial product control. Fresenius Z. Anal. Chem. 328: 410–412

648. Esser P (1986) Direkte Feststoffanalyse von anorganischen Materialien im Labor eines Zementwerkes. In: Welz B (ed) Fortschritte in der atomspektroskopischen Spurenanalyse. Vol. 2, Verlag Chemie, Weinheim Germany, pp 307–316

649. Rühl WJ (1985) Cadmium in polymers – product control in the automobile industry. Fresenius Z. Anal. Chem. 322: 710–712

650. Shan XQ, Wang W, Wen B (1992) Determination of gallium in coal and coal fly ash by electrothermal atomic absorption spectrometry using slurry sampling and nickel chemical modification. J. Anal. At. Spectrom. 7: 761–764

651. Van Loenen DC, Weers CA (1986) The determination of arsenic, cadmium and lead in large series of samples from pulverized coal fly ash (PFA) by Zeeman graphite furnace AAS and microliter injection of ultrasonic agitated PFA suspensions. In: Welz B (ed) Fortschritte in der atomspektroskopischen Spurenanalyse, vol. 2. Verlag Chemie, Weinheim, Germany, pp 635–647

652. Lücker E, Schuierer O (1996) Sources of error in direct solid sampling Zeeman atomic absorption spectrometry analyses of biological samples with high water content. Spectrochim. Acta B51: 201–210

653. Bortoli A, Gerotto M, Marchiori M, Muntau H, Rehnert A (1995) Critical comparison of methods for mercury determination in fish. Microchim. Acta 119: 305–310

654. Cervera ML, Navarro A, Montoro R, La Guardia Mde, Salvador A (1991) Platform in furnace Zeeman-effect atomic absorption spectrometric determination of arsenic in beer by atomization of slurries of sample ash. J. Anal. At. Spectrom. 6: 477–481

655. Viñas P, Campillo N, López-García I, Hernández-Córdoba M (1995) Determiantion of aluminium in chewing gum samples using electrothermal atomic-absorption spectrometry and slurry sample introduction. Fresenius J. Anal. Chem. 351: 695-696

656. Herber RFM (1993) Solid sampling analysis. In: Seiler HG (ed) Handbook on metals in clinical chemistry. Vol. 7, Marcel Dekker, New York, 1993

657. Belysev YI, Koveshnikowa TA, Kostin BI (1973) Anal. Khim. 28: 2111–2117

658. Sneddon J (1985) Use of an impaction-electrothermal atomization atomic absorption spectrometric system for the direct determination of cadmium, copper, and manganese in the laboratory atmosphere. Anal. Lett. 18: 1261–1280

659. Sotera JJ, Cristiano LC, Conley MK, Kahn HL (1983) Reduction of matrix interferences in furnace atomic absorption spectrometry. Anal. Chem. 55: 204–208

660. Kurfürst U, Rehnert A, Muntau H (1996) Uncertainty in the analytical results from solid materials with electrothermal atomic absorption spectrometry: a comparison of methods. Spectrochim. Acta B 51: 229–244

661. Pauwels J, Hofmann C, Vandecasteele C (1994) On the usefulness of SS-ZAAS for the microhomogeneity control of CRM's. Fresenius J. Anal. Chem. 348: 418–421

662. Bagschik U, Quack D, Stoeppler M (1990) Homogeneity studies in a variety of reference materials using solid sampling Zeeman graphite furnace AAS. Fresenius J. Anal. Chem. 338: 386–389

663. Ihnat M, Stoeppler M (1990) Preliminary assessment of homogeneity of new candidate agriculture/food reference materials. Fresenius J. Anal. Chem. 338: 455–460

664. Rossbach M, Ostapczuk P, Schladot JD, Emons H (1995) Ein Beitrag zur Qualitätssicherung in der Umweltanalytik. Homogenität von Referenzmaterialien. UWSF – Z. Umweltchem. Ökotox. 7: 365–370

665. Natrella MG (1966) Experimental statistics. NBS Handbook. 91. NIST, Washington DC

666. Kurfürst U (1993) Potential relevance of imprecise data from solid sample analysis with graphite furnace AAS. Fresenius J. Anal. Chem. 346: 556–559

667. Baumgardt B (1985) Spurenanalytische Untersuchung zur Schwermetallbelastung von Lungengeweben – ein Beitrag zur Objektivierung beruflicher Gefährdung. Ph D Thesis, Ruhr-Universität Bochum

668. Herber RFM (1995) Use of solid sampling analysis for the determination of trace elements in tissues. Microchem. J. 518: 46–52

669. Spruit D, Bongaarts PJM (1977) Nickel content of plasma, urine and hair in contact dermatitis. dermatologica 154: 291–300

670. Alder JF, Batoreu MCC (1983) Determination of lead and nickel in epithelial tissue by electrothermal atomic absorption spectrometry. Anal. Chim. Acta 155: 199–207

671. Strübel G, Rzepka-Glinder V, Grobecker KH, Jarrar K (1990) Heavy metals in human urinary

calculi. Determination of chromium, lead, cadmium, nickel and mercury in human urinary calculi by direct solid sampling atomic absorption spectrometry using Zeeman backround correction. Fresenius J. Anal. Chem. 337: 316–319

672. Strübel G, Rzepka-Glinder V, Grobecker KH (1987) Heavy metals in human salivary calculi. Determination of cadmium, lead and zinc in human salivary calculi by direct solid sampling atomic absorption spectrometry using Zeeman effect background correction. Fresenius Z. Anal. Chem. 328: 382–385

673. Herber RFM, Roelofsen AM, Roelfzema WH, Peereboom-Stegeman JHJC (1985) Direct determination of cadmium in placenta. Comparison with a destruction atomic absorption spectrometric method. Fresenius Z. Anal. Chem. 322: 743–746

674. Aadland E, Aaseth J, Radziuk B, Saeed K, Thomassen Y (1987) Direct electrothermal atomic absorption spectrometric analysis of biological samples and its application to the determination of selenium in human liver biopsy specimens. Fresenius Z. Anal. Chem. 328: 362–366

675. Aaseth J, Thomassen Y, Aadland E, Fausa O, Schrumpf E (1995) Hepatic retention of copper and selenium in primary sclerosing cholangitis. Scand. J. Gastroenterol 30: 1200–1203

676. Pesch H-J, Kraus T (1991) Schwermetalle in Leber und Umwelt. Postmortal–analytische Untersuchungen mittels Zeeman-AAS. Eine Übersicht. In: Welz B (ed) 6. Colloquium Atomspektrometrische Spurenanalytik. Bodenseewerk Perkin-Elmer, Überlingen, pp 867–878

677. Pesch H-J, Palesch T, Seibold H (1989) Zum Cadmium-Gehalt menschlicher Organe. Eine analytisch-statistische Untersuchung Verstorbener aus dem Raum Franken 1985 In: Riewenherm S, Lieth H (eds) XIX/III. Gesellschaft für Ökologie, Osnabrück, pp 427–436

678. Pesch H-J, Kraus T, Palesch T, Seibold H (1991) Cadmium determination in human organs by Zeeman AAS. A postmortem inverstigation. In: Graumann W, Drukker J (eds) Progress in Histo- and Cytochemistry, vol. 23. Fischer, Stuttgart, pp 365–371

679. Pesch H-J, Kraus T, Biermann K, Knapp B (1992) Renal post-mortem cadmium concentration in Erlangen und Leipzig. Fresenius J. Anal. Chem. 343: 150–151

680. Pesch H-J, Kraus T, Schweinzer M (1992) Concentration of copper in the placenta and liver of foetuses, infants and young children a comparative analytical study with Zeeman-AAS. In: Pauwels J et al. (eds) Book of abstracts. 5. Intern. Solid Sampling Coll., Geel, 30–31

681. Sansoni B, Brunner W (1981) Modellversuche zur instrumentellen Festkörper-Atomspektrometrie im Graphitrohr unter Verwendung standardisierter Ionenaustauscher. In: Welz B (ed) Fortschritte in der atomspekrometrischen Spurenanalytik. Verlag Chemie, Weinheim Germany, pp 187–208

682. De Kersabiec AM, Blanc G, Pinta M (1985) Water analysis by Zeeman atomic absorption spectrometry. Fresenius Z. Anal. Chem. 322: 731–735

683. Takada T, Koide T (1987) Atomic absorption spectrometric determination of trace copper in water by sorption on an ion-excange resin and direct atomization of the resin. Anal. Chim. Acta 198: 303–308

684. Slovák Z(1979) Anal. Chim. Acta 110: 301

685. Slovák Z, Docekalová H(1980) Anal. Chim. Acta 115: 111

686. Slovák Z, Docekal B (1980) Sorption of arsenic, antimony and bismuth on glycolmethacrylate gels with bound thiol groups for direct sampling in electrothermal atomic absorption spectrometry. Anal. Chem. Acta 117: 293–300

687. Atsuya I, Itoh K (1982) Direct determination of lead in the NBS bovine liver by Zeeman atomic absorption spectrometry using the graphite miniature cup. Bunseki Kagaku 31: 708–712

688. Akatsuka K, Atsuya I (1987) Preconcentration by coprecipitation of submicrogram amounts of copper and manganese with 8-quinolinol and direct electrothermal atomic absorption spectrometry of the precipitates. Anal. Chim. Acta 202: 223–230

689. Akatsuka K, Nobuyama N, Atsuya I (1988) Atomic absorption spectrometry of nanogram amounts of cadmium, lead and zinc after precipitation with 8-Quinolinol. Anal. Sci. 4: 281–285

690. Akatsuka K, Nobuyama N, Atsuya I (1989) Preconcentration of submicrogram cobalt with 8-Quinolinol from natural waters for electrothermal atomic absorption spectrometry. Anal. Sci. 5: 475–479

691. Sturgeon RE, Berman SS, Willie SN, Desaulniers JAH (1981) Preconcentration of trace elements from seawater with silica-immobilized 8–hydroxyquinoline. Anal. Chem. 53: 2337–2340

692. Atsuya I, Itoh K (1988) Fundamental studies on the coprecipitation of nanogram quantities of some metals with the dimethylglyoxime/Ni/1-(2-pyridylazo)-2-naphthol complex and their direct determination by atomic absorption spectrometry using an inner miniature cup

for the solid sampling technique; application to natural waters. Fresenius Z. Anal. Chem. 329: 750–755

693. Atsuya I, Itoh K, Ariu K (1991) Preconcentration by coprecipitation of lead and selenium with Ni/pyrrolidine dithiocarbamate complex and their simultaneous determination by internal standard atomic absorption spectrometry with the solid sampling technique. Pure & Appl. Chem. 63: 1221–1226

694. Nakamura T, Oka H, Ishii M, Sato J (1994) Direct atomization atomic absorption spectrometric determination of Be, Cr, Fe, Co, Ni, Cu, Cd, and Pb in water with zirconium hydroxide coprecipitation. Analyst 119: 1397–1401

695. Shengjun M, Holcombe JA (1991) Preconcentration of nickel and cobalt on algae and determination by slurry graphite-furnace atomic-absorption spectrometry. Talanta 38: 503–510

696. Dobrowolski R, Mierzwa J (1992) Application of activated carbon for the enrichment of some heavy metals and their determination by atomic spectrometry. Vestn. Slow. Kem. Drus. 39: 55–64

697. Wenz HW, Lichtenberg WJ, Katterwe H (1991) Surface analysis and surface measuring techniques in firearm offences. Fresenius Z. Anal. Chem. 341: 155–165

698. Lichtenberg W (1985) Schußspurenuntersuchung in Verbindung mit Munitionsarten mit "bleifreiem" Pulverschmauch. Beitr. gerichtl. Medizin XLIII: 293–300

699. Leszczynski C (1959) Bestimmung der Schußentfernung. Kriminalistik 9: 377–382

700. Lichtenberg W (1994) Concerning automation in detection, identification and determination of the distribution of "GSR" on the hand of a shooter. In: Direzione centr. d. polizia criminale servicio Rome: European Network of Forensic Science Institutes, pp 75–82

701. Lichtenberg W (1990) Zur Problematik der Schußspurenuntersuchung. NStZ 4: 159–164

702. Lichtenberg W (1994) Concerning automation in detection, identification and determination of the distribution of "GSR" on the hand of a schooter. In: Direzione centr. d. polizia criminale servicio Rome: European Network of Forensic Science Institutes, 75–82

703. Lichtenberg W (1991) Schußspurenuntersuchung mittels Direkter Zeeman-AAS (DZ-AAS). Labor 2000, 124–131

704. Castledine CG, Robbins CJ (1983) Practical field-portable atomic absorption analyser. J. Geochem. Explor. 19: 1–16

705. Gottlieb J, Huck K, Maurer A (1997) Field Screening – neue Methoden und Strategien bei der Umweltanalytik vor Ort. GIT Lab. Fachz. 48–50

706. Vanhaecke F, Galbács G, Boonen S, Moens L, Dams R (1995) Use of the Ar^{2+2} signals as a diagnostic tool in solid sampling electrothermal vaporization inductively coupled plasma mass spectrometry. J. Anal. At. Spectrom. 10: 1047

707. Grégoire DC, Lee J (1994) Determination of cadmium and zinc isotpe ratios in sheeps blood and organ tissue by electrothermal vaporization inductively coupled plasma mass spectrometry. J. Anal. At. Spectrom. 9: 393

708. Grobecker K-H (1990) Schwermetallbelastung durch Blei und Quecksilber in zwei terrestrischen und einem aquatischen Ökosystem. Ph D Thesis, Universität Gießen

709. Kessel R (1997) Evaluation of uncertainty in measurements. Version 1.0, D-31028 Gronau

710. Erb R (1985) Rohstoff- und Produktkontrolle mit Zeeman-Feststoff-AAS in einem chemischen Betrieb. Direkte Bestimmung von Kupfer, Mangan, Eisen, Cadmium und Blei in Klebebändern und anderen Materialien. Fresenius Z. Anal. Chem. 322: 719–720

711. Lichtenberg W (1987) Determination of gunshot residues (GSR) in biological samples by means of Zeeman atomic absorption spectrometry Fresenius Z. Anal. Chem. 328: 367–369

712 Schäfer U, Krivan V (1996) Automated large volume slurry preparation in slurry sampling electrothermal atomic absorption spectrometry. J. Anal. At. Spectrom. 11:1119–1120

713 López-García I, Sánchez-Merlos M, Hernández-Córdoba M (1996) Rapid determination of selenium in soils and sediments using slurry sampling-electrothermal atomic absorption spectrometry. J. Anal. At. Spectrom. 11:1003–1006

714. Ming-Jyh L, Shiuh-Jen J (1996) Determination of copper, cadmium and lead in sediment samples by slurry sampling electrothermal vaporization inductively coupled plasma mass spectrometry. J. Anal. At. Spectrom: 1: 555–560

715. Bermejo-Barrera P, Moreda-Piñeiro A, Moreda-Piñeiro J, Bermejo-Barrera A (1996) Determination of nickel in human scalp hair by slurry sampling-electrothermal atomic absorption spectrometry. Anal. chim. Acta 349: 319–325

716 Slaveykova VI, Hoenig M (1997) Electrothermal atomic absorption spectrometric determination of lead and tin in slurries; optimisation study. Analyst 122: 337–343

Subject Index

M. Stoeppler (Ed.)

Sampling and Sample Preparation

Practical Guide for Analytical Chemists

1997. XIV, 202 pages. 75 figures, 43 tables.
Hardcover DM 148
ISBN 3-540-61975-5

This book describes in great detail sampling and sample preparation in routine and research. All chapters are written for practitioners by experts in the field. The book provides well-documented and illustrated procedures for sampling and sample preparation prior to trace metal analysis. A careful selection of appropriate references to pitfalls, methods and strategies furnishes the information for a proper performance of the most crucial steps of any analytical procedure.

Springer

Springer-Verlag, P. O. Box 31 13 40, D-10643 Berlin, Germany

H. Günzler (Ed.)

Accreditation and Quality Assurance in Analytical Chemistry

1996. XIV, 266 pages, 53 figures, 11 tables.
Hardcover DM 148
ISBN 3-540-60103-1

Quality assurance and accreditation in analytical chemistry laboratories is an important issue on the national and international scale. The book presents currently used methods to assure the quality of analytical results and it describes accreditation procedures for the mutual recognition of these results. The book describes in detail the accreditation systems in 13 European countries and the present situation in the United States of America. The editor also places high value on accreditation and certification practice and on the relevant legislation in Europe. The appendix lists invaluable information on important European accreditation organizations.

Please order from
Springer-Verlag Berlin
Fax: + 49 / 30 / 8 27 87- 301
e-mail: orders@springer.de
or through your bookseller

Price subject to change without notice.
In EU countries the local VAT is effec-

Springer

Springer-Verlag, P. O. Box 31 13 40, D-10643 Berlin, Germany

G. W. H. Höhne, W. Hemminger, H.-J. Flammersheim

Differential Scanning Calorimetry

An Introduction for Practitioners

1996. XII, 222 pages, 112 figures, 13 tables.
Hardcover DM 178
ISBN 3-540-59012-9

The urgent need for a clear and practice-oriented guide to differential scanning calorimetry (DSC) is resolved by this book. The authors provide the newcomer and the experienced practitioner with a comprehensive insight into all important DSC methods, including a sound presentation of the theoretical basis of DSC calorimeters. Emphasis is layed on instrumentation, the underlying measurement principles, metrologically correct calibrations, factors influencing the measurement process, and on the exact interpretation of the results.

The information given enables the research scientist, the analyst and experienced laboratory staff to apply DSC methods successfully and to measure correctly thermodynamic values.

Please order from
Springer-Verlag Berlin
Fax: + 49 / 30 / 8 27 87- 301
e-mail: orders@springer.de
or through your bookseller

Price subject to change without notice.
In EU countries the local VAT is effec-

Springer

Springer-Verlag, P. O. Box 31 13 40, D-10643 Berlin, Germany

Printing: Saladruck, Berlin
Binding: Buchbinderei Lüderitz & Bauer, Berlin